Advances in Neuroethics

Advances in neuroscience research are bringing to the forefront major benefits and ethical challenges for medicine and society. The ethical concerns related to patients with mental health and neurological conditions, as well as emerging social and philosophical problems created by advances in neuroscience, neurology and neurotechnology are addressed by a specialized and interdisciplinary field called neuroethics.

As neuroscience rapidly evolves, there is a need to define how society ought to move forward with respect to an ever growing range of issues. The ethical, legal and social ramifications of neuroscience, neurotechnology and neurology for research, patient care, and public health are diverse and far-reaching — and are only beginning to be understood.

In this context, the book series "Advances in Neuroethics" addresses how advances in brain sciences can be attended to for the benefit of patients and society at large.

More information about this series at http://www.springer.com/series/14360

Fabrice Jotterand • Marcello Ienca
Editors

Artificial Intelligence in Brain and Mental Health: Philosophical, Ethical & Policy Issues

Springer

Editors
Fabrice Jotterand
Medical College of Wisconsin
Center for Bioethics and Medical Humanities
Milwaukee, WI
USA

Institute for Biomedical Ethics
University of Basel
Basel
Switzerland

Marcello Ienca
Department of Health Sciences and Technology
ETH Zurich
Zürich, Switzerland

ISSN 2522-5677 ISSN 2522-5685 (electronic)
Advances in Neuroethics
ISBN 978-3-030-74187-7 ISBN 978-3-030-74188-4 (eBook)
https://doi.org/10.1007/978-3-030-74188-4

This Springer imprint is published by the registered company Springer Nature Switzerland AG
The registered company address is: Gewerbestrasse 11, 6330 Cham, Switzerland

Acknowledgments

The editors would like to thank the many people and institutions that have contributed to the realization of this edited volume. First of all, we are indebted to Prof. Dr. Mingyu Liang, Professor and Kohler Co. Chair in Physiology and Director of the Center of Systems Molecular Medicine (CoSMM) at the Medical College of Wisconsin, for his constant support in the conceptualization and preparation of this volume. We would also like to thank the CoSMM and the Swiss Academy of Medical Sciences for supporting the editorial work of this book. Further thanks go to the Center for Bioethics and Medical Humanities at the Medical College of Wisconsin and the Health Ethics and Policy Lab, Department of Medical Science and Technology, ETH Zurich, for providing us with an optimal academic environment during this editorial journey. A special mention goes to Sylvana Freyberg, and Deepika Devan and her team at Springer Nature for their outstanding editorial help. Last but not least, we would like to thank all authors for their excellent chapters, and for the constructive patience they exercised to complete their contributions during the worst pandemic in recent history.

Contents

About the Authors

Mark V. Albert is the director of the Biomedical AI Lab at the University of North Texas and holds a dual appointment in the Department of Computer Science and Engineering and the Department of Biomedical Engineering. He leverages machine learning to automate the collection and inference of clinically useful health information to improve clinical research. His projects in wearable sensor analytics have improved the measurement of health outcomes for individuals with Parkinson's disease, stroke, transfemoral amputations, and cerebral palsy. Current projects include video-based activity tracking and mobile robotic platforms, all to improve measures of clinical outcomes to support therapeutic interventions.

Julia Amann is a postdoctoral researcher at the Health Ethics and Policy Lab at the Swiss Federal Institute of Technology (ETH Zurich). She holds a PhD in Health Sciences and Health Policy from the University of Lucerne, Switzerland, and is currently co-leading the Technologies for Public Health special interest group of Public Health Schweiz. Her research focuses on the impact of digital technologies on the doctor-patient relationship and healthcare, more generally. As part of her work on the Horizon2020 project PRECISE4Q, Julia is investigating the opportunities and challenges of machine learning in stroke medicine with a particular focus on the ethical implications for research and clinical practice. Julia previously held a postdoctoral appointment at Swiss Paraplegic Research. She is also a member of the research committee of the International Association for Communication in Healthcare (EACH).

Emily E. Anderson is associate professor of bioethics and medical education at Loyola University Chicago's Stritch School of Medicine. She teaches courses in research ethics and responsible conduct of research to graduate and medical students. Her areas of interest and expertise include researcher and physician professionalism and misconduct, ethics and community engagement, research with vulnerable populations, informed consent, and institutional review board (IRB) policy. Dr. Anderson serves as associate editor for *Narrative Inquiry in Bioethics* and *Progress in Community Health Partnerships* and is coauthor of *100 Questions and Answers About Research Ethics* (2018, SAGE) with Amy Corneli, PhD.

Timothy Brown is currently a postdoctoral scholar working primarily on a National Institutes of Health–funded project on the effect of neurotechnologies on user agency. Further, he is a long-time contributor to the Center for Neurotechnology's (CNT) Neuroethics Thrust—where he supports efforts to teach neuroethics to young investigators, catalyze ethics investigations through interdisciplinary collaborations, and promote the field of neuroethics through public outreach. More generally, Tim's work lies at the intersection of biomedical ethics, philosophy of technology, (black/latinx/queer) feminism, and aesthetics.

Ishan Dasgupta is a postdoctoral scholar in the Department of Philosophy and the Center of Neurotechnology at the University of Washington. He works at the intersection of law, ethics, and public health policy as it relates to emerging technology. His past work has focused on ethical issues surrounding induced pluripotent stem cells, inclusion of pregnant women in biomedical research, and the use of tissue samples in genetics research. At UW, Ishan focuses on issues of agency in relation to neurotechnology.

Sara Goering is Professor of Philosophy and the Program on Ethics and has affiliations with the Department of Bioethics and Humanities, and the Disability Studies Program. In addition, she currently leads the ethics thrust at the UW Center for Neurotechnology. She teaches courses in bioethics, ethics, philosophy of disability, feminist philosophy, and philosophy of medicine. She also spends time discussing philosophy with children in the Seattle public schools, through her role as the Program Director for the UW Center for Philosophy of Children.

Sarah Graham is a postdoctoral fellow in geriatric mental health in the Department of Psychiatry and the Stein Institute for Research on Aging at UC San Diego. She conducts research using biosensors and artificial intelligence to better understand aging from a multimodal perspective involving mental, cognitive, and physical health and well-being with the ultimate goal of enabling older adults to live independently longer with a high quality of life.

Pim Haselager is Associate Professor at the Donders Institute for Brain, Cognition and Behaviour and the Department of Artificial Intelligence at the Radboud University, Nijmegen. He focuses on the ethical and societal implications of cognitive neuroscience and AI. He publishes in journals such as *Nature: Biotechnology, Science and Engineering Ethics*, *American Journal of Bioethics*, *Neuroethics*, *Journal of Cognitive Neuroscience*, and *Journal of Social Robotics*.

José Hernández-Orallo is Professor at the Universitat Politècnica de València, Spain, and Senior Research Fellow at the Leverhulme Centre for the Future of Intelligence, University of Cambridge, UK. He received a BSc and a MSc in Computer Science from UPV, partly completed at the École Nationale Supérieure de l'Électronique et de ses Applications (France), and a PhD in Logic with a doctoral extraordinary prize from the University of Valencia. His academic and research activities have spanned several areas of artificial intelligence, machine learning, data science, and intelligence measurement. He has published five books and more than two hundred journal articles and conference papers on these topics. His research in the area of machine intelligence evaluation has been covered by several popular outlets, such as *The Economist*, *New Scientist*, or *Nature*. His most recent book addressed an integrated view of the evaluation of natural and artificial intelligence (Cambridge University Press, 2017, PROSE Award 2018).

Hudlicka is a member of the International Society for Research on Emotion, the Society for Affective Science, the Association for the Advancement of Artificial Intelligence, and the Coalition for Technology in Behavioral Science. She was a member of the National Research Council committee on "Behavioral Modeling and Simulation" and is an Associate Editor of the *International Journal of Synthetic Emotions* and a member of the Editorial Board of the *Journal of Cognitive Systems Research* and the *Oxford Series on Cognitive Models and Architectures*. She has authored over 50 journal and conference papers and numerous book chapters.

Dr. Hudlicka was born in Prague, Czech Republic, and received her BS in Biochemistry from Virginia Tech, MS from The Ohio State University in Computer Science, PhD in Computer Science from the University of Massachusetts-Amherst, and MSW from the Simmons School of Social Work.

Eva Hudlicka has a dual career combining research in affective computing and clinical work in psychotherapy. She is a Principal Scientist at Psychometrix Associates, which she founded in 1998 to pursue research in computational models of emotion, and since 2014 she is also a psychotherapist in private practice in Amherst, MA. During the 2017/2018 academic year she was a Fulbright Canada-Palix Foundation Distinguished Research Chair at the University of Alberta, Edmonton, and University of Lethbridge, Lethbridge, exploring the applications of affective computing in behavioral health technologies. Between 2013 and 2018 she was a Visiting Lecturer at the College of Information and Computer Sciences at the University of Massachusetts-Amherst, where she taught courses in affective computing and human-computer interaction. Prior to founding Psychometrix, she was a Senior Scientist at BBN in Cambridge, MA.

Marcello Ienca is a Senior Researcher in the Department of Health Sciences and Technology at ETH Zurich, Switzerland. His research focuses on the ethical, legal, and social implications of neurotechnology and artificial intelligence, with particular focus on big data trends in neuroscience and biomedicine, human-machine interaction, social robotics, digital health, and cognitive assistance for people with intellectual disabilities. He is interested in comparative approaches to the study of human and artificial cognition. Ienca is the Principal Investigator of multidisciplinary federal research projects and has received several awards for social responsibility in science and technology such as the *Prize Pato de Carvalho* (Portugal), the *Vontobel Award for Ageing Research* (Switzerland), and the *Paul Schotsmans Prize* from the *European Association of Centres of Medical Ethics* (EACME). He has authored one monograph, one edited volume (*Intelligent Assistive Technologies for Dementia, Oxford University Press*, 2019), +50 scientific articles in peer-reviewed journals, several book chapters, and is a frequent contributor to *Scientific American*. His research was featured in academic journals such as *Nature Medicine*, *Nature Biotechnology*, *Neuron*, *the Lancet Digital Health*, and the *American Journal of Bioethics* and media outlets such as *Nature, The New Yorker, The Guardian, The Times, Die Welt, The Independent, the Financial Times*, and others. Ienca is serving as appointed member or expert advisor in a number of national and international governance bodies including the Steering Group on *Neurotechnology and Society* of the Organisation for Economic Co-operation and Development (OECD) and the Council of Europe's Ad Hoc Committee on Artificial Intelligence. He is also a former Board Member of the International Neuroethics Society.

Fabrice Jotterand is Professor of Bioethics and Medical Humanities and Director of the Graduate Program in Bioethics at the Center for Bioethics and Medical Humanities, Medical College of Wisconsin. He holds a second appointment as Senior Researcher at the Institute for Biomedical Ethics at the University of Basel, Switzerland. His scholarship and research interests focus on issues including neuroethics including the ethics of AI in medicine, ethical issues in psychiatry and mental health, the use of neurotechnologies in psychiatry, medical professionalism, neurotechnologies and human identity, and moral and political philosophy. He has published more than 65 articles and book chapters as well as reviews and edited five books. His present research focuses on an examination of the ethical, regulatory, and social issues arising from the use of emerging neurotechnologies in psychiatry and neurology.

Eran Klein is a neurologist specializing in dementia at Oregon Health and Sciences University (OHSU) and the Portland VA Medical Center. He is part of the Neuroethics thrust at the NSF Center for Sensorimotor Neural Engineering (CSNE) at the University of Washington. He works at the intersection of neurology, neuroscience, and philosophy.

Karola Kreitmair is an assistant professor in bioethics at the University of Wisconsin—Madison. She completed a PhD in philosophy and a clinical ethics fellowship at Stanford University. Her research includes neuroethics, in particular issues surrounding consciousness, digital behavioral technology, and citizen science.

David D. Luxton is a nationally recognized expert and trainer in suicide prevention, telehealth, and innovative technologies in behavioral healthcare. He is Director of Research and Data Analytics, Washington State Department of Corrections, and Associate Professor in the Department of Psychiatry and Behavioral Sciences at the University of Washington School of Medicine in Seattle. Dr. Luxton previously served as a Research Health Scientist at the Naval Health Research Center in San Diego, CA, and Research Psychologist and Program Manager at the National Center for Telehealth and Technology (Defense Health Agency), Joint Base Lewis-McChord. A seasoned researcher, he has authored more than 100 scientific articles and book chapters and published three books: *Artificial Intelligence in Behavioral and Mental Health Care* (2015), *A Practitioner's Guide to Telemental Health* (2016), and *Behind the Machine* (2020). He has also helped to develop national guidelines for telemental health and clinical best practices in the use of technology in behavioral healthcare. In 2015 he was awarded the American Psychological Association Division 19 (Military Psychology) Arthur W. Melton Award for Early Career Achievement. He is a licensed clinical psychologist and served in the U.S. Air Force.

Gary Marchant teaches, researches, and speaks about the governance of a variety of emerging technologies, including biotechnology, genomics, neuroscience, nanotechnology, artificial intelligence, and blockchain. Prior to joining Arizona State University in 1999, he was a partner at the Washington, D.C., office of Kirkland & Ellis, where his practice focused on environmental and administrative law. He received his JD from Harvard Law School, where he was awarded the Fay Diploma (awarded to top graduating student at Harvard Law School). He also has a PhD in genetics from the University of British Columbia and a Master of Public Policy from the Kennedy School of Government. Professor Marchant frequently lectures about the intersection of law and science at national and international conferences, including speaking at over 75 judicial conferences. He has authored more than 200 articles and book chapters on various issues relating to emerging technologies. Among other activities, he has served on six National Academy of Sciences committees, has been the principal investigator on several major grants, and has organized over 50 academic conferences and workshops on law and science issues. He is an elected lifetime member of the American Law Institute and Fellow of the Association for the Advancement of Science.

Nicole Martinez-Martin is an assistant professor at Stanford University's Center for Biomedical Ethics, with a secondary appointment in the Department of Psychiatry. She is conducting research for an NIMH grant-funded study of the ethics of digital mental health technologies, and her areas of scholarship include neuroethics, digital health, and artificial intelligence in healthcare.

Giulio Mecacci is Assistant Professor at the Donders Institute for Brain, Cognition and Behaviour, and the Department of Artificial Intelligence at the Radboud University, Nijmegen. He investigates the impact of intelligent technologies and neurotechnologies on human values such as responsibility and privacy.

Camille Nebeker is Associate Professor of Behavioral Medicine in the Department of Family Medicine and Public Health, School of Medicine, UC San Diego. Her research and teaching focus on two intersecting areas: (1) research capacity building (e.g., participant-led, community-engaged research) and (2) digital health research ethics (e.g., consent, risk/benefit, data management). She directs the Research

Center for Optimal Digital Ethics (ReCODE.Health) and is affiliated faculty with the UCSD Design Lab, the Stein Institute for Research on Aging, and the Center for Wireless and Population Health Systems. Dr. Nebeker has received continuous support for her research from government, foundation, and industry sources since 2002.

Yair Neuman is a Full Professor at the Department of Cognitive and Brain Sciences and a member of the Zlotowski Center for Neuroscience, Ben-Gurion University of the Negev. He received his BA in Psychology (Major) and Philosophy (Minor) and his PhD in Cognition (Hebrew Univ. 1999), and his expertise is in interdisciplinary research where he draws on diverse disciplines to address problems from an unusual perspective. More specifically, his expertise is in studying complex textual-symbolic, social, psychological, and cognitive systems, with a specific emphasis on the development of novel research methodologies and computational models. Prof. Neuman has published numerous papers and six academic books and was a visiting scholar/Prof. at M.I.T, University of Toronto, University of Oxford, and Weizmann Institute of Science. Beyond his purely academic work, he developed state-of-the-art algorithms for social and cognitive computing such as those he developed for the IARPA metaphor project (ADAMA group) and for his innovative work in computational personality analysis.

Emma M. Parrish is a graduate student in the San Diego State University/University of California San Diego Joint Doctoral Program in Clinical Psychology, and a T32 Predoctoral Fellow. Emma's research interests lie in real-time interventions and assessments through technology for people with serious mental illness, with a focus on suicide prevention, functioning, and cognition, as well as the ethics of digital mental health interventions.

Andreas Schönau is a postdoctoral scholar in the Department of Philosophy and Center for Neurotechnology at the University of Washington. His past research focused on the clarification of conceptual theories and empirical methods in philosophical and neuroscientific research, the interdisciplinary combination of their respective insights, and the generation of conclusions towards understanding the phenomenon of free will from an action-theoretical perspective. At UW, he continues working on agency-related issues in the intersection of Neuroscience and Philosophy.

Naveen Shamsudhin was born in Kerala, India. He received his BTech in Instrumentation and Control Engineering from the National Institute of Technology—Tiruchirappalli and his MSc in Micro and Nanosystems from ETH Zurich. He is a recipient of the NITT Alumni Award (2009) and the Swiss Government Scholarship (ESKAS 2009–2011) for excellence in undergraduate and graduate studies. He gained experience in the design, development, and quality control of MEMS technology through internships at the Indian Institute of Science (Bangalore) and at the Automotive Electronics division of Robert Bosch GmbH (Reutlingen). Working in close association with IBM Research Zurich and the

Institute of Plant Biology at the University of Zurich, he developed micro- and nanorobotic tools for single cell mechanics, completing his doctoral degree at the Multi-Scale Robotics Lab in January 2017. He was awarded the Prix OMEGA scientifique 2018 for his doctoral thesis. He joined the Multi-Scale Robotics Lab as a postdoctoral researcher in November 2017 and currently is a lecturer for two courses, Introduction to Robotics and Mechatronics (Spring Semester) and Microrobotics (Fall Semester). He is also the co-founder of The Origin AG. His current academic interests are the history and philosophy of technology in particular robotics and artificial intelligence (e.g., www.roboethics.ch), engineering for the developing world, and revitalizing inter-(trans-) disciplinarity in academic education and teaching (e.g., ETH Cortona Week, Roboethics Workshop, Kaleido).

Ryan Spellecy is the Ursula von der Ruhr Professor of Bioethics in the Center for Bioethics and Medical Humanities at the Medical College of Wisconsin, where he chairs one of the IRBs. His work focuses on research ethics, informed consent, ethical issues in psychiatry, and community involvement in research. He advised the Patient-Centered Outcomes Research Institute regarding engaging stakeholders in the peer review process, the Association of American Medical Colleges on IRBs, and community-based research. Recently, he was the co-PI for an NIH-funded national study evaluating a novel, easier to read consent form for blood and marrow transplant trials and currently leads a project to create a community-engaged research ethics training program as well as a study to evaluate the strengths and barriers regarding cancer clinical trial participation in African American churches.

Mónika Sziron is a Hungarian-American technology and humanities PhD candidate at Illinois Institute of Technology in Chicago, Illinois. Generally, her research focuses on the influence of various technologies on our daily lives today and throughout history. More recently, her PhD research focuses on how AI influences our daily lives, specifically considering moral status, ethics, and human rights issues and considerations in AI and robotics.

Lucille Nalbach Tournas researches and lectures in the intersection of law and emerging technologies, with an emphasis on artificial intelligence, big data, and neurotechnologies. She received her JD from the Sandra Day O'Connor College of Law at Arizona State University, where she was awarded the Strouse Prize for excellence in law, science, and technology. She is currently a PhD student in the School of Life Sciences at Arizona State University, where she is working on the global governance of neurotechnology and the data it provides.

Karina Vold is Assistant Professor at the Institute for the History and Philosophy of Science and Technology and a Faculty Affiliate at the Centre for Ethics, University of Toronto. She was previously a Research Fellow at the Leverhulme Centre for the Future of Intelligence, University of Cambridge, UK, and a Digital Charter Fellow at the Alan Turing Institute, the UK's national center for data science and artificial intelligence. Vold has been a visiting scholar at the University of Oslo, Ruhr

University, Duke University, and the Australian National University. She received an Honors BA in Philosophy and Political Science from the University of Toronto and a PhD in Philosophy from McGill University. Her current research spans across topics in philosophy of cognitive science, the ethics of artificial Intelligence, and the limits of machine learning.

Ting Xiao is a research assistant professor in the Department of Computer Science and Engineering at the University of North Texas. She was formally a postdoctoral researcher in experimental particle physics at Northwestern University applying statistical and computational tools to extract meaningful signals from large (~10 TB) data sets. In recent years she leveraged those skills to a variety of projects in computer science applications, generally applying signal processing and machine learning to video, audio, and wearable sensor data. She has published numerous papers with over 1800 citations. Her most cited individual effort involves the first observation of a new particle—Zc0 (3900).

Jie Yin is Associate Professor at the School of Philosophy and Center for Biomedical Ethics, Fudan University in China. She works on a wide range of topics in bioethics, philosophy of medicine, and Kant. She received training from medical school (B.M., Fudan University) as well as philosophy department (MPhil, Fudan University; PhD, SUNY Albany). Dr. Yin teaches undergraduate and graduate courses on Kant's *Critique of Practical Reason*, political philosophy, just health, bioethics, neuroethics, and nursing philosophy. Recent publications in Chinese include several articles on neuroethics and a textbook on philosophy of medicine.

Introduction

1

Fabrice Jotterand and Marcello Ienca

Artificial intelligence (AI) has the potential to transform the delivery and management of health care and improve biomedical research. Brain and mental health could significantly benefit from this technological transformation. Some of the most promising applications of AI in brain and mental health include the use of deep learning algorithms for early detection and diagnosis, as well as automated learning and the infusion of AI capabilities in everyday technologies such as smartphones, assistive social robots, and intelligent assistive technologies for continuous health monitoring and screening (e.g., Alzheimer's disease and schizophrenia) or for the assistance of psychogeriatric and neurorehabilitation patients. In addition, machine learning (ML) can also be used to improve existing neuropsychiatric therapies and allow new indications for existing drugs and tailor them to the individual patient through precision medicine approaches. For example, Watson, an AI-driven question-answering computing system developed by IBM, has proven to make similar treatment recommendations as human experts in 99% of the cases, and in 30% of the cases, Watson found treatment options missed by human physicians [1]. In addition, Watson can perform tasks such as data integration and aggregation, assessment of patients' risk to develop a particular disease or to require high cost treatment [2].

Further, big data analytics can be helpful to improve the epistemic power of neuropsychological explanations and unlock the etiology of brain and mental disorders by revealing relevant patterns across big and heterogeneous data volumes. In particular, multidimensional models integrating multiple biomarker data—for

F. Jotterand (✉)
Medical College of Wisconsin, Milwaukee, WI, USA

Institute for Biomedical Ethics, University of Basel, Basel, Switzerland
e-mail: fjotterand@mcw.edu

M. Ienca
Department of Health Sciences and Technology, ETH Zurich, Zürich, Switzerland
e-mail: marcello.ienca@hest.ethz.ch

F. Jotterand, M. Ienca (eds.), *Artificial Intelligence in Brain and Mental Health: Philosophical, Ethical & Policy Issues*, Advances in Neuroethics,
https://doi.org/10.1007/978-3-030-74188-4_1

example, neuroimaging biomarkers and digital phenotyping data—could help scientists overcome current reductionist approaches based on single explanatory neurobiological hypotheses. The automation of healthcare management processes via intelligent software to optimize healthcare delivery and reduce administrative cost is another promising implementation of AI technology.

The transformative potential of AI in brain and mental health does not limit to transforming the mode of generating scientific knowledge or assisting medical decision-making. In addition to that, it also portends to transform social and professional practices. For example, AI could redefine the therapeutic relationship. A study performed by researchers from the Dartmouth-Hitchcock health system, the American Medical Association (AMA), Sharp End Advisory, and the Australian Institute of Health Innovation revealed that physicians spend on overage 27% of their total time on direct clinical face time and 49.2% of their time on administrative work and Electronic Health Records (EHRs) [3]. The incorporation of AI in medical practice could help clinicians spend more time with patients and make health care more personal, albeit using more technology [4].

Such promissory outlook, however, has not materialized yet, at least, not entirely. The deployment of AI in neurology, psychiatry, neuropsychology, and brain research is still limited to sparse domains of application, often with suboptimal outcomes. Whether AI will re-humanize or de-humanize health care remains an open question as it is too early to understand the real impact long term of AI on clinical practice [5]. It is therefore paramount to cast light on emerging AI approaches in brain and mental health and provide an anticipatory impact assessment, with special focus on the assessment of emerging technical, scientific, ethical, and regulatory challenges. Such assessment is needed not only to chart the route ahead for scientific innovation in this domain but also to appraise such innovative dynamics within its broader socio-cultural and regulatory context. A broad spectrum of philosophical, ethical, regulatory, and social implications is rapidly emerging at the cross-section of AI and brain and mental health. Many of these implications have not been assessed in a comprehensive and systemic way. To this end, this unique volume provides an interdisciplinary collection of essays from leaders in various fields to address the current and future challenges arising from the implementation of AI in brain and mental health.

The volume is structured according to three main sections, each of them focusing on different types of AI technologies. Part I, *Big Data and Automated Learning: Scientific and Ethical Considerations*, specifically addresses issues arising from the use of AI software, especially machine learning, in the clinical context or for therapeutic applications. In Chap. 2 ("Big Data in Medical AI: How Larger Datasets Lead to Robust, Automated Learning for Medicine"), Ting Xiao and Mark V. Albert review the implications of the use of vast data sets in the context of medical research and clinical practice. They show how machine learning strategies can assist clinicians in various ways such as helping in the process of automatizing data selection for better diagnosis, improving the predictive power of statistical models tailored to specific hospitals or patient groups, or establishing the factor(s) that explains symptoms. However, Xiao and Albert point out that the collection of massive data sets is not without challenges such as data security, the interpretation and validation of data, and the accuracy of automated decision-making. In Chap. 3 ("Automatic Diagnosis and Screening of Personality Dimensions and Mental Health Problems"), Yair Neuman likewise addresses issues related to automatic diagnosis and screening but in the context of personality research. Computational Personality Analysis, as Neuman puts it, refers to the use of machine learning algorithms to measure variables in personality dimensions and disorders. As one can expect, such approach for the diagnosis of mental disorders or antisocial behaviors must be scientifically valid, ethically safe, and pragmatically relevant. So while "the promise of computational personality analysis is huge," Neuman concludes that the implementations of such technologies must be sensitive and critical to some of its challenges such as a good understanding of the complexity of human personality in light of the fact that automatic analysis of personality relies on "low-level features" in its categorization of personality. The other challenge is the fact that personality is a cluster of dynamic phenomena difficult to capture without a clear sense of the trajectory of the mental state captured. In Chap. 4 ("Intelligent Virtual Agents in Behavioral and Mental Healthcare: Ethics and Application Considerations"), David Luxton and Eva Hudlicka provide an overview of embodied Intelligent Virtual Agents (IVAs) and non-embodied conversational agents and examine the implications of their use in the context of behavior and mental health care. In particular, their analysis focuses on concerns about risks associated with the breach of privacy, the safety of individuals interacting with IVAs, and the ethical issues arising from artificial relationships. In Chap. 5 ("Machine Learning in Stroke Medicine: Opportunities and Challenges for Risk Prediction and Prevention"), Julia Amman examines issues related to the use of risk prediction and prevention tools such as novel machine learning-driven methods to reduce the global burden of stroke (incidence and mortality rates). There are many advantages for physicians and researchers to use such approaches as the increased accuracy of their predictions allow them to suggest interventions tailored to the specific needs of patients predisposed to strokes. But the implementation of such technology is not without challenges and limitations. These include issues of data sourcing, application development, and implementation in clinical setting, which, in Amman's estimation, should be fully recognized and addressed in order to benefit maximally from ML approaches to stroke predication and prevention. In the final chapter of the first section (Chap. 6, "Respect for Persons and Artificial Intelligence in the Age of Big Data"), Ryan Spellecy and Emily E. Anderson explore the extent to which traditional ways to honor respect for persons (in particular, informed consent) are challenged by AI and big data. In particular, they point out that in big data models where consent is not practicable due to the high data volume and velocity, waiving consent can be tempting for researchers for practical reasons but is ethically inadequate. They therefore argue that alternative approaches should be explored to hold the ethical standard of respect for persons. According to Spellecy and Anderson, "in discussions of ethics of AI and big data health research," there should be "less focus on the technical aspects of informed consent and more imagination regarding ways to demonstrate respect for persons" (p. 10 manuscript).

Part II, *AI for Digital Mental Health and Assistive Robotics: Philosophical and Regulatory Challenges*, examines philosophical, ethical, and regulatory issues arising from the use of an array of technologies beyond the clinical context. In Chap. 7 ("Social Robots and Dark Patterns: Where Does Persuasion End and Deception Begin?"), Naveen Shamsudhin and Fabrice Jotterand look at some of the challenges associated with the deployment of social robots for applications in areas such as entertainment, companionship, mental health, and well-being. The anthropomorphic design of these robots takes advantage of insights gained through human and social psychology, communication, and behavior which makes human beings vulnerable to manipulation and deception. Using digital media and web technologies, *dark patterns* are developed to deceive people to behave certain ways leading to addictive demeanor, hence undermining the autonomy of the users. The authors conclude that advances in robotics (i.e., social robots) should move forward but without the use of dark patterns. Nicole Martinez-Martin in Chap. 8 ("Minding the AI: Ethical Challenges and Practice for AI Mental Health Tools") directs her attention to fundamental questions of privacy, bias, and the potential impact of AI in the therapeutic relationship within the context of mental health. She contends that biases (i.e., systematic errors in a computer system that can cause unfair outcomes) may occur in the process of gathering data and health information and/or may depend on how algorithms are configured. These biases can cause inequities in the delivery of or access to mental health services. However, she also points out that the use of AI can be designed to address injustices. Martinez-Martin also examines how the implications of AI tools might affect the clinical encounter and provide recommendations for best practices. The use of digital behavioral technology (DBT) in combination with deep learning is the focus of Chap. 9 ("Digital Behavioral Technology, Deep Learning, and Self-Optimization"), authored by Karola Kreitmair. In her analysis, she considers technologies such as wearables, mobile health technologies, various smartphone apps, and noninvasive neurodevices that collect a large amount of data about individuals including brain activity, bodily functions, and behavioral patterns. Her analysis shows how the preferred way to process the data and make it relevant and useful for self-optimization (for instance, change of behavior through neurostimulation) is through an approach to AI known as deep learning. However, such technology presents many ethical challenges that are evaluated carefully by Kreitmair. In the next contribution, (Chap. 10, "Mental Health Chatbots, Moral Bio-enhancement and the Paradox of Weak Moral AI"), Jie Yin provides a philosophical exploration of the implications of the potential use of chatbots to enhance behavior in mental health. Hypothetically, her idea would be to use "a weak moral artificial intelligence" to enhance cognitive capacities, in particular moral deliberation. In principle, if such technology would be available, be safe, and respect human agency, it could be used for therapeutic purposes, although Yin argues, such approach would undermine essential elements of morality (such as motivation). However, she notes that mere philosophical argumentation is not sufficient for a final assessment of a weak moral artificial intelligence. Only once empirical evidence is available, we will be able to determine whether this type of technology ought to be implemented. In Chap. 11 ("The AI-Powered Digital Health

Sector: Ethical and Regulatory Considerations When Developing Digital Mental Health Tools for the Older Adult Demographic"), Camille Nebeker, Emma Parrish, and Sarah Graham examine the social benefits but also the potential ethical and regulatory pitfalls and risks associated with a widespread implementation of AI in day-to-day living, including "airline reservation systems, loan eligibility programs, college admissions, transportations systems, judicial decisions, and healthcare." In their analysis, they specifically focus on questions associated with the development of tools to help elderly people suffering from dementia which raise ethical questions regarding informed consent and agency. As more AI tools find their way into the marketplace and more data is collected, Nebeker et al. argue that new approaches to the governance of these technologies are needed in order to optimize their responsible implementation in the social context. Extra layers of protection should be put in place, particularly when dealing with vulnerable population such as elderly people with dementia. In Chap. 12 ("AI Extenders and the Ethics of Mental Health"), Karina Vold and José Hernandez-Orallo consider the extended mind thesis in the context of mental health and in light of AI technology. They examine the use of what they call "AI extenders" which is, in their view, different from previous cognitive extension based on simple technologies like a notebook or a smartphone. As they note, the "increased use of machine learning, and other functionalities brought by artificial intelligence, is importantly different from the kinds of cognitive extension that preceded it in many ways: these system can perceive, navigate, make complex decisions, understand and produce language, plan, understand emotions, etc., all in complex and changing situation". When applied to mental health to better diagnose and treat mental disorders, these technologies offer many opportunities to improve care but also raise many ethical challenges carefully outlined by Vold and Hernandez-Orallo.

In the final section of the volume, Part III entitled *AI in Neuroscience and Neurotechnology: Ethical, Social and Policy Issues*, contributions examine some of the implications of AI in neuroscience and neurotechnology and the regulatory gaps or ambiguities that could potentially hamper the responsible development and implementation of AI solutions in brain and mental health. The first contribution of this section by Pim Haslager and Giulio Mecacci (Chap. 13, "The Importance of Expiry Dates: Evaluating the Societal Impact of AI-Based Neuroimaging") analyzes the ethical and societal implications emerging from AI-powered neuroimaging. Such technology increases our ability to make predictive inferences about mental information and to recognize behavioral dispositions based on brain activity. However, Haselager and Mecacci argue that as more advances in AI-powered neuroimaging occur, further analysis must take place concerning the future implications of technologies for brain reading and the evaluative framework used in computational processing regarding neuroimaging. To this end, their contribution offers some fundamental recommendations for the regulation of the technology with a specific caveat: expiry dates for informed consent, data storage, and data analysis. In the next contribution, (Chap. 14, "Does Closed-Loop Deep Brain Stimulation for Treatment of Psychiatric Disorders Raise Salient Authenticity Concerns?"), Ishan Dasgupta, Andreas Schoenau, Tim Brown, Eran Klein, and Sara

Goering investigate issues associated with the new generation of deep brain stimulation (DBS) technology for the treatment of psychiatric disorders that employs artificial intelligence technologies as a means to "facilitate closed-loop implants that are adaptive and continuously modified by neural feedback". One major issue they examine is the impact of closed-loop DBS on authenticity. This chapter provides a salient empirical and philosophical analysis of the phenomenological implications of closed-loop neurostmulation for neuropsychiatric patients. Next, in Chap. 15 ("Matter Over Mind: Liability Considerations Surrounding Artificial Intelligence in Neuroscience"), Lucy Tournas and Gary Marchant address issues of liability. They recognize the benefits of the implementation of AI in the clinical setting for diagnostic and therapeutic purposes, but they also point out that there are risks and potential harms associated with the collection of neurological health data and an eagerness to deploy the technology without a careful consideration of liability concerns. They suggest building a "liability framework" that reconsiders informed consent in light of AI technology, increased education of physicians about AI, and an update of FDA regulations to include AI technology. In the last contribution of the volume (Chap. 16, "A Common Ground for Human Rights, AI and Brain and Mental Health"), Monika Sziron explores international regulations of AI in the context of health care and how human rights may be integrated in regulatory frameworks. The integration of human rights in international guidelines, however, is confronted to an important challenge: There are no agreed-upon international standards that regulate health care and AI. As she points out, "as philosophical and ethical environments vary across nations, subsequent policies reflect varying conceptions and fulfillments of human rights". She argues that despite this challenge, the development of ethical guidelines that encompass human rights may be possible at an international level if variations in their application and understanding are carefully acknowledged, which provide the common ground necessary to adapt policies and regulations. Finally, *the epilogue* ("Brains, Minds, and Machines: Brain and Mental Health in the Era of Artificial Intelligence") by Marcello Ienca concludes the volume by taking stock retrospectively of the work contained in this book and outlining the open challenges for future research in this field.

In light of its comprehensiveness and multidisciplinary character, this book marks an important milestone in the public understanding of the ethics of AI in brain and mental health and provides a useful resource for any future investigation in this crucial and rapidly evolving area of AI application.

References

1. Lohr S. IBM is counting on its Bet on Watson, and paying big money for it. The New York Times, October 17, 2016.
2. IBM Watson Health. Available online: https://www.ibm.com/watson/health/value-based-care/.
3. Sinsky C, Colligan L, Li L, et al. Allocation of physician time in ambulatory practice: a time and motion study in 4 specialties. Ann Intern Med. 2016;165:753–60. https://doi.org/10.7326/M16-0961.

4. Vize R. Technology could redefine the doctor-patient relationship. The Guardian. March 11, 2017.
5. Jotterand F, Bosco C. Keeping the 'human in the loop' in the age of artificial intelligence: accompanying commentary for "correcting the brain?" by Rainey and Erden, Sci Eng Ethics. 2020. https://doi.org/10.1007/s11948-020-00241-1.

Part I

Big Data and Automated Learning: Scientific and Ethical Considerations

Big Data in Medical AI: How Larger Data Sets Lead to Robust, Automated Learning for Medicine

2

Ting Xiao and Mark V. Albert

2.1 Why the Big Data Revolution?

Machine learning is having a dramatic impact on the way we leverage information to make decisions [1, 2]. The success has been obvious in commercial business settings where data from advertising [3], supply logistics [4], and even social media [5, 6] is collected and processed in real time, enabling decisions at speeds and scales that would be impossible for hired employees. Medical applications present unique challenges due to risks but also provide satisfying targets due to the potential for improving health outcomes [7–10].

Many steps of the medical decision-making process can benefit from the tools of machine learning (Table 2.1). For example, we can consider a common sequence of choices made during the course of a medical treatment.

1. The clinician is tasked with collecting the relevant information.
2. A judgement about the cause is made based on the information available.
3. A treatment is proposed when possible.
4. Response to treatment is periodically evaluated and altered when needed.

T. Xiao
Department of Computer Science and Engineering, University of North Texas, Denton, TX, USA

Department of Information Science, University of North Texas, Denton, TX, USA
e-mail: ting.xiao@unt.edu

M. V. Albert (✉)
Department of Computer Science and Engineering, University of North Texas, Denton, TX, USA

Department of Biomedical Engineering, University of North Texas, Denton, TX, USA

Department of Physical Medicine and Rehabilitation, Northwestern University Feinberg School of Medicine, Evanston, IL, USA
e-mail: mark.albert@unt.edu

F. Jotterand, M. Ienca (eds.), *Artificial Intelligence in Brain and Mental Health: Philosophical, Ethical & Policy Issues*, Advances in Neuroethics,
https://doi.org/10.1007/978-3-030-74188-4_2

Table 2.1 Definitions

Artificial intelligence (AI)	The development of computer systems performing tasks commonly associated with intelligent beings, either through explicit programming or by learning from data
Machine learning	A large subset of AI which makes data-driven inferences. Notably, this is the area in which the vast majority of AI advances are made
Big data	A term to describe the tools and techniques of inference that are particular to large data sets, which enable more robust, automated learning
Deep learning	Machine learning using multilayer ("deep") neural networks. Currently the state of the art in solving challenging inference problems with large data sets by learning intermediate features directly from raw data
TensorFlow, PyTorch	The two dominant deep learning frameworks
GPU	Graphics Processing Unit. A processor designed to handle graphics operations that can be used to dramatically speed up neural network training due to the similarly simple, distributed processing needs

Medical professionals are trained to perform each of these steps taking into account what they observe directly or measure, and they then relate that information to their own personal experience and the medical research. However, it is worth noting that each of these steps can loosely be associated with a related approach used in machine learning techniques which are particularly valuable for large data sets and suggest recommendations for complex decision-making problems. For example, here we can list four machine learning strategies that can be directly mapped to the four steps above to assist the clinician in certain cases:

1. *Feature selection*: With enough data, the process of determining which information is more or less important can be automated. If the data is difficult or invasive to collect, a ranking of the importance can be provided to help the clinician choose the best measures to collect for a diagnosis [11].
2. *Factor analysis*: Notwithstanding the philosophical arguments of truly establishing cause and effect relationships, much of approach to understand a collection of symptoms is finding the underlying factor or factors explaining the symptoms presented. This goes well beyond disease diagnosis. Underlying factors may be more fine-grained than disease states, or emerge from comorbid diseases—a factor analysis would be able to identify groups of common concern in an automated way to allow patients with similar conditions to be grouped and treated more effectively [12].
3. *Predictive modeling*: The choice of treatment relies on the belief of which option is expected to lead to the greatest improvement, while weighing appropriate risks. Clinical researchers use statistical models to evaluate the superiority of one treatment over another, and in ambiguous cases, medical practitioners also use internal estimates of future improvement through their years of medical experience. However, with larger data sets, such predictions can be explicit and even tailored to the particular hospital, patient group, clinician, surgical technique using available data on past outcomes to provide an additional point of reference to help make a treatment recommendation [13].

4. *Automated outcome data collection and synthesis*: For long-term treatments, follow-up is necessary to judge compliance, efficacy, and make adjustments as needed. However, visits to the clinic are costly in terms of clinician time and associated financial costs. Questions regarding symptoms in a clinical visit can be subjective or incomplete, and physical measures may differ based on a variety of factors. Sensor technologies exist now which enable convenient, continuous, and objective measures of a variety of symptoms, with associated analytics to distill the measures to clinically relevant information [14].

In short, machine learning, and the associated use of large data sets to improve the process of learning, can augment the process of clinical decision-making. Such analytics provide a unique perspective for each decision. Notably, such tools perform a similar function to a secondary consult or collective review among clinicians, without the associated time, costs, or overhead—enabling rapid, often automated assistance to inform medical care.

2.1.1 More Samples, More Features

One of the reasons for the explosion of machine learning is the availability of data for training decision-making systems. The amount of data varies along two dimensions that are particularly relevant to learning systems—additional samples and additional features. Samples generally represent more examples or cases. Features, on the other hand, are new types of information that can be collected for each sample. Modern technology has made it possible to dramatically increase both dimensions of data to build learning models. More data enable systems to be more capable of automated decision-making.

To understand why this is the case, let us begin with a common rule of thumb for collected data to train many standard machine learning prediction models.

$$n_{\text{samples}} \gg \left(n_{\text{features}}\right)^2$$

That is, the number of samples collected should be substantially greater than the square of the number of features. Double the number of features, and so the number of samples has to be quadruped, etc. Note this is only a rough “rule of thumb” with many exceptions. This is not as critical for some simpler prediction algorithms (such as Naive Bayes), but it is reasonably accurate for a number of common machine learning models which are sufficiently flexible and powerful to learn for a wider variety of prediction problems. Why is this true? That is beyond the scope of this chapter, but some motivation is provided in the footnote.[1]

[1] Succinctly, the goal of machine learning is roughly stated as the ability to group similar sample points together in a n_{features} dimensional space. Most ways of flexibly grouping points in a n-dimensional space require more than n^2 parameters (groups of planes, multidimensional ellipses, etc.), and a well-known fact of estimation is that you generally need more data points than you

The implication of a rule like this is that if there are not many samples, it is necessary to a priori select features for a learning model based on prior experience and intuition (which is often built based on prior experience); otherwise, there is not enough data for reliable learning. It is this limited hand-selection of features that generally leads to weaker performance when more data is available. If a massive amount of data can be collected prior to hand-selecting features, the process of selecting features becomes automated and in many cases more reliable than personal judgement.

In fact, the advantage of big data as a tool in medical decision-making primarily comes from this ability to automatically select, weigh, and combine features. Although many statistical tools have existed which can use features similarly, modern machine learning approaches built upon these techniques enable a slightly larger number of features relative to samples through a variety of strategies. Recall that including more relevant features universally improves predictive accuracy—including features that may seem irrelevant according to personal intuition. There are approaches which stray far from the approach of building a single model. We will discuss ensemble learning, which effectively polls disparate machine learning algorithms to arrive at an answer better than any algorithm alone [15]. Similarly, deep learning can combine the features present in raw data to create more complex features sets that can be leveraged to improve learning [16]. However, both ensemble learning and deep learning require a larger amount of data as they are hierarchical in nature needing well-trained component models to perform well.

Ultimately, however, even without these recent modeling advances, much of the improvement relies on the availability of larger data sets and computing systems capable of processing these vast quantities of data efficiently.

2.1.2 Hardware Improvements

The collection and processing of massive data sets has required new paradigms of data management. In industry, this has led to the creation of positions for data engineers whose primary role is to collect and manage the acquired data for later machine learning researchers and data scientists [17]. In addition to collection, the processing tasks are challenging with substantially more data requiring parallel architectures for processing, ranging from distribution between machine, cores, or even across GPUs for very low-level distributed processing. We will address each of these in turn.

Data collection has been made simpler and more standardized through enterprise data systems with shared resources. In commercial systems, this has occurred through large cloud-based data management architectures with scalable data repositories and shared processing repositories. To enable efficient use of centralized data management systems, virtualization has made it possible to access such systems

have parameters to estimate, suggesting most learning algorithms require significantly more than $(n_{\text{features}})^2$ samples.

with a variety of tools. Different operating systems, with different analysis tools and programming languages, can work in tandem on a shared repository of information, significantly speeding up adoption of centralized data management systems. Additionally, the ability to allocate a variable number of processors to tasks allows not only efficient data storage, but also processing resources which can scale larger for data mining and model learning tasks, and scale smaller for daily incremental development and deployment tasks.

High-performance computing has also enabled real-time processing tasks over terabytes of data. Architectures based on the concept of map-reduce (e.g., Hadoop, Spark) speed up queries through separating the computation on different processors and/or sections of the data set (map) and combining the results (reduce) [18, 19]. With a sufficiently resourced architecture, queries that traditionally would take hours or days are shortened to seconds. Tasks that involve preparing data for analysis, including data visualization and cleaning, are significantly sped up using such strategies.

One of the most recent advances in machine learning is the supremacy of deep learning for complex learning problems such as visual object recognition, speech recognition, and natural language processing. Deep learning neural network models are capable of learning directly from raw acquired data by building layers of features derived from earlier layers. This permits successively more complex feature extraction. However, due to the large number of learned parameters, deep learning neural networks are particularly resource intensive to train. Luckily, because the computations of individual neural elements are relatively simple, it is possible to distribute them among simpler processing units. As graphics processing traditionally relied on a large series of fast, simple linear computations, graphics processing units (GPUs) were built to efficiently distribute such low-level processing tasks. Deep learning neural network frameworks, such as TensorFlow and PyTorch, are able to shift processing from CPUs to GPUs with the speed of processing increasing by orders of magnitude.

These hardware advances enable large, centralized data repositories with shareable and configurable processor allocation. Standard data analysis tasks can be distributed among processors across terabytes of data in seconds speeding data preparation and standard analysis approaches. And finally, readily available GPU processing has enabled high-speed training of deep learning neural network models for progressively more challenging modeling tasks.

2.2 Precision Medicine

The traditional approach to medicine has been to treat patients in a similar manner to how all patients with similar causes or measured symptoms would be treated. However, this one-size-fits-all approach to medicine has been gradually replaced by precision medicine approaches which attempt to tailor prevention, diagnosis, treatment, and evaluation to the particular patient under consideration [20, 21].

On one extreme for preventative care, advances in genomics provided an idealized case of data-driven medical intervention. For example, by identifying particular genetic anomalies, cancer risks can be assessed, and patients can make informed decisions to minimize the chance of developing a cancer before any symptoms have been identified. However, lifestyle analytics provide an alternate extreme where the data may be noisy and causal inferences unclear, but the resolution of advice about changing lifestyle may permit improved health outcomes on a massive scale—for example, precisely determining what quantity and manner of exercise is ideal for each person given their health needs, career requirements, and compliance concerns.

Precision diagnosis and treatment can rely on more than simply demographic information to increase or decrease the probability of a particular diagnosis. Clinicians are aware that individuals respond to different treatments depending on aspects of their physiology, mental health, compliance, and lifestyle. Instead of relying on clinician judgement, given these factors to influence how treatment choices are selected and presented to the patient, predictive models can be provided additional information on the state of the patient to systematically rank potential treatments.

Precision evaluation and follow-up depend upon reliable, readily acquired information on a subject's well-being. Reliability can be obtained by more frequent, more objective measures. Such information can often be acquired by passively worn wearable devices that can measure movements [22]; these devices can range from simple consumer step counters and calorie estimators [23–25] to research-grade wearables [26, 27] or even smartphones [28–31]. The interpretation of such movements to clinically relevant activities can often require analytics tailored to the individual population movements [32, 33] and potentially even the context such as at-home versus in the clinic [27, 34]. The convenience of wearable devices and other passively recorded health outcome measures has a definite impact on the future of medicine.

2.2.1 Challenges and Opportunities Unique to Mental Health and Wellness

Mental health has particular challenges in precision medicine. The first distinction from physical health is the lack of straightforward physical metrics to measure. For example, someone who suffers from depression can readily list a variety of symptoms which can be used to make a clinical diagnosis; however, affixing a sensor to them to collect data to make as certain of a diagnosis would be challenging.

One way of addressing limited mental health measures is to provide ways of allowing people undergoing treatment to periodically self-assess. However, this suffers from a number of drawbacks. First, questions require self-reflection or a recall of past experiences, both of which are known to be error, subjective and prone to error, or misrepresentation [35]. Additionally, compliance can be challenging, particularly in the case of individuals whose desire to respond regularly is affected by the very mental state that is being assessed.

Wearable device analytics paired with predictive methods on summary metrics prove a potential avenue for real-time assessment. For example, subjects suffering from depressive episodes may move less, speak less, and engage in more passive forms of entertainment. Any individual measure alone may not provide sufficient information for an estimate that a depressive episode occurred, however, taken in aggregate in the form of a prediction algorithm they would provide a reliable way of probabilistically estimating a depressive episode occurred. This could provide real-time scoring and large-scale data collection to assess the efficacy of group therapies and construct interventions particular to individual populations.

2.3 Tools for Big Data in Medicine

2.3.1 Standardization Tools

One of the greatest impairments in large-scale machine learning projects is the capture of data in formats readily available for analysis. This is particularly problematic in medical contexts where clinicians in different hospitals may record information in ways which are not compatible—for example, using different metrics or different units to quantify a given symptom. Even with the same type of data, the electronic systems may use different data formats, query languages, or permissions. This heterogeneity limits the application of big data for solving more challenging problems in medicine.

There are different approaches to standardizing data sets. One is to use a shared vocabulary for acquired data. For example, the International Classification of Diseases (ICD) and the Systematized Nomenclature of Medicine and Clinical Terms (SNOMED CT) are available standards [36, 37]; however, these are limited to fairly narrow data sets and terminology. More expansive common data models are being developed, including the Observational Medical Outcomes Partnership (OMOP) [38]. Notably, the standardization of data fields is not only helpful for big data analytics approaches. Standardization also enhances the ability for models to be trained, validated, and tested in different settings, increasing the level of scientific rigor possible prior to standardization. In addition to shared standards, shared access to computational tools for data storage, queries, and analytics support the use of centralized access of the data collected across institutions. This standardization of data and analytics tools enables rapid development and testing of predictive models on acquired health data.

2.3.2 Analytics to Leverage Big Data

A variety of machine learning techniques are available to exploit collected data for improved understanding; however, a subset of them have been particularly valuable in making inferences on large, often noisy data sets as is typical in medical contexts.

2.3.2.1 Unsupervised and Semi-Supervised Learning Methods

Measuring the extent of diseases or disease progression, particularly with mental health, can be challenging. Many more types of information can be collected relative to the number of individuals participating in the study—this is particularly the case with continuously collected observational data (e.g., from wearable devices). As discussed previously, in order to make valid inferences, the number of features must be reduced when fewer samples are available.

One approach to shrink the number of features in a learning model is to synthesize a large number of features into fewer, more reliable aggregate factors. Principal components analysis (PCA) and related factor analysis techniques provide one means of achieving this; however, a variety of methods now exist to parameterize a larger data set with a large number of features into a smaller, more meaningful set of features. Some are based on analytical assumptions about the distribution of underlying causes, such as with PCA or ICA (independent components analysis), while other reductions are made possible through a nonparametric combination of features using the hidden layers of neural network models to effectively compress the data.

Additionally, one of the challenges in building machine learning models in medicine is properly vetting the quality of decisions the model is using for learning. For example, some scoring methods may have low inter-rater reliability and require group consensus for high reliability. When high-quality labeled data is scarce for training, but unlabeled collected data is plentiful, there are two particular machine learning strategies that work well to exploit the large amount of unlabeled data—anomaly detection and semi-supervised learning. If the positive diagnoses are exceptionally rare, and are the main source of "unusual" observations, anomaly detection techniques use the deviation from typical to better identify potential future cases. However, if there are multiple classes and there is sufficient labeled data for all classes being considered, semi-supervised learning techniques provide a means of estimating labels for all samples. In label spreading, a standard semi-supervised learning technique, the unlabeled samples that are easiest to classify are labeled first until all samples have been classified. In general, this allows clusters around labeled values in the feature space to be grouped according to similarity, giving standard machine learning models access to larger, labeled data sets.

2.3.2.2 High-Throughput Model Selection and Testing

A standard approach to statistical modeling involves selecting the features and statistical model prior to analysis—this greatly simplified the problems inherent in iteratively trying many different models when making statistical inferences. However, now there are a wide variety of models with many modeling parameters and options, which can all be trivially tried and adapted to each data set by altering, in many cases, only a single line of code. For example, the following variations can be easily attempted to fit the best model:

1. Changing the input features through feature selection or feature engineering (e.g., dimensionality reduction or clustering).

2. Setting hyperparameters of a model (e.g., the degree of a polynomial, the regularization strength of lasso regression).
3. Selecting different models (e.g., support vector machines vs. random forest).
4. Combining models through ensemble learning (more in the next subsection).

If we systematically test all these variants on a single set of data, the best fitting model would likely overfit. Generally, the most complex models can readily fit a given data set for training, but without proper tuning to prevent overfitting, they would perform poorly on newly acquired samples.

A regiment of separating the data sets into training, validation, and testing sets is standard practice in machine learning, and critical for proper selection among all these alternate models and parameterizations. This will be addressed further in Sect. 2.4.1.1.

Fortunately, each combination of modeling options can be tested independently. The space of options can be explored rigorously through a grid search or can be optimized adaptively using a variety of available optimization strategies based on model performance. Different model combinations can readily be tasked to different CPU cores, greatly increasing the opportunity to tune a variety of models for a given predictive problem.

2.3.2.3 Ensemble Methods

In predictive analytics competitions, it is not uncommon for competing groups to "team up" to improve their predictive accuracy beyond either classifier alone using ensemble methods. This is a common enough strategy that many competitions ban the practice; however, if the goal is to increase prediction quality, this is precisely a strategy worth exploring. Models which may bear no resemblance to each other analytically can combine their results to form a prediction which is generally at least as good as the best individual model. In fact, this ensemble approach is the basis for a very effective predictive model, random forest, which is aptly built from the systematic combination of predictions from individual decisions trees.

2.3.2.4 Deep Learning

Generally, machine learning algorithms require designers to extract features from the raw signals using their intuition or previous experience to identify features where a fair amount of the extracted features are at least weakly capable of assisting classification. For many problems, this procedure works well—it is the standard means of generating predictive models for hundreds of years. However, this requirement for human implementation and selection of features is a major impediment in many machine learning tasks. Many problems require a complex series of intermediate feature combinations which are not readily built by human designers. For example, object identification occurs in the brain through a series of processing stages moving from simple features to complex feature combinations. Similarly, the best performing visual object identification systems also process visual information through a series of layers of processing of artificial neurons. The intermediate features are often impossible to describe plainly, but such data-driven low-level features are essential for the algorithm to make distinctions in rich, complex, raw data sets.

Currently, the field of machine learning has developed many tools related to deep learning analytics to speed programming and processing. TensorFlow and PyTorch are the de facto standards of frameworks for coding and training deep learning neural network models. By creating models in TensorFlow or PyTorch, designers also benefit from an ability to readily power their models to run on available GPUs for significant speed improvements.

Ultimately, these data management and analytic tools provide a great deal of power to model designers, particularly for those with access to large data sets.

2.4 Big Data Concerns

Medicine presents a unique set of challenges in the development of predictive models. There are high stakes in assuring that model predictions not only are accurate in the original context but also function robustly in other contexts. In some cases, lives literally depend on high accuracy or robust behavior. Beyond concerns of predictive ability, the medical data necessary to make such predictions is particularly sensitive and personal, requiring extra care to keep it secure. And finally, even with accurate models and adequate precautions for individual data security, the results of predictive models can be as biased as the human medical decisions on which they are based—care must be taken to avoid systematic bias and identify when particular populations may not be adequately represented which affects accuracy as well as the potential for bias.

2.4.1 The Need for Proper Validation

Accuracy is paramount in medical decision-making; however, there are a variety of factors in machine learning which can affect the accuracy of real-world applicability of machine learning models. We address each major concern in turn.

2.4.1.1 Test Models with Separating Training, Validation, and Test Sets

When a large number of models are tested, it becomes more critical to have separate training and test sets to avoid overfitting which causes both inflated expected performance and the selection of models which perform poorer on new data sets. However, when an exceptionally large number of models are iteratively tested using a validation set (a test set used for model selection), as is the case with modern automated machine learning model fitting and selection procedures, it is critical to properly evaluate any selected model on an independent test set. This is easily automated with sufficient data availability, though many cases of applied machine learning models do not evaluate their models with proper test sets after model selection. We recommend anyone using machine learning approaches to not only consider traditional cross-validation in model selection with a hold-out data set, but also consider nested cross-validation techniques to avoid overfitting and reporting accurate quality metrics for the selected, trained models.

2.4.1.2 Assure Samples Adequately Represent the Reported Populations and Contexts

Many researchers are aware of the need to have an appropriate representation of sexes and races for their results to generalize across the population; however, there are a number of other factors to consider that can be equally important for proper statistical representation. For example, it may be necessary to design not only a model for a particular disease population but also a subpopulation of that group with particular characteristics or comorbidities. Similarly, if assessing behavior, it may be necessary to acquire data in particular contexts, for example, not only in the clinic but also at home.

2.4.1.3 Avoid Contamination Between Training and Test Sets

For proper model validation, the test set should be independent of the training and validation sets to avoid inflating reported model accuracy. However, this understood rule can often be broken without realizing it. One common example of this is having the same subject's data in both the training set and test set. Even if the data samples are independent, the individual may introduce a dependency between samples that promote overfitting. This may not be a concern with a model that will be trained with an individual's own data for personal use, but most models in medical practice are trained, fixed, and deployed without adaptation to a single individual. For this reason, training and test sets should involve separate individuals—commonly performed using subject-wise cross-validation. This often leads to lower predictive accuracy which makes it less palatable of a result to emphasize in publication, but provides a much more realistic assessment of models in a deployed context.

2.4.1.4 Select the Right Performance Metrics

The need for proper metrics to evaluate models is particularly strong in medical context and, without proper consideration metrics, can be selected by marketers of prediction systems that can border on fraud. For a classic example, a system can report a 99% accuracy for tumor detection in radiological scans, but fail to indicate the model just naively reports for every scan that no tumor is present, which is accurate for 99% of the population.

Various metrics exist to tease apart the behavior of a system for a clearer estimate of efficacy. The classic metrics for binary classification are sensitivity and positive predictive value. However, given the increase of multiclass classification problems, an introduction and emphasis of recall and precision, their multiclass equivalents, may be more appropriate. In cases where it is unclear whether recall or precision is favored, F1 score is the harmonic mean of precision and recall. By averaging across all classes, average F1 score provides a metric more in line with a sense of practical efficacy.

In regression, many models are designed to optimize mean squared error by default. However, there are a variety of other metrics which may be more appropriate. Mean absolute error is more forgiving of errors in outliers to better minimize more moderate errors. Additionally, there are log error metrics which are best to minimize the relative or percent error rather than the absolute scale of errors. For example, a model predicting medical care costs across a wide range of scales would likely benefit from using log error rather than mean squared error. A $1000 error in an expensive surgery should be penalized less than a $1000 error in low-cost routine care, whereas a 1% error for the surgery may be comparable to a 1% error in routine care depending on the use case of the cost of care model. The key point is that it may be important to select an appropriate metric rather than rely on defaults, as all metrics make assumptions either explicitly or implicitly about which errors are more important to improve the model.

2.4.2 Security

Models in medical decision-making often rely on sensitive information that is protected by various regulations including HIPAA, the Health Insurance Portability and Accountability Act in the United States [39], and GDPR, the General Data Protection Regulations in the European Union [40], which restrict the ways in which health information can be shared. Primary among the protocols for data sharing are the information is inaccessible outside the healthcare provider or intended original use, and the assumption that analytics and results derived from this data do not adversely affect an individual.

For data collected in a hospital setting and saved on centralized hospital servers, standard data security concerns are applicable with protocols vigilantly adhered to in order to avoid data security breaches. However, as we enter an era of wearable devices, phone-acquired data, smart home health, etc., the constellation of "Internet of Things" data security can be more challenging. This is particularly relevant as users may not be aware of all the inferences that can be made from data they consider innocuous. For example, on a mobile phone alone, the accelerometer can be used to not only estimate step counts and calories but also predict personal activities such as trips to the bathroom or bedroom sleeping habits. For this reason, it is important that all information used for medical decision making be collected securely and encrypted not only when stored locally but also en route to a centralized repository.

2.4.3 Ethical Challenges

Hardware and analytics advances have led to the potential for powerful and accurate automated decision-making; however, there are problems that more powerful prediction algorithms may not only avoid addressing but also make far worse. It is important to take the broader context of predictive algorithms into account. For example, imagine one group of patients is diagnosed improperly in a systematic way (e.g., based on race, sex), which can often occur when decision-making is done by humans with their own implicit biases. Due to the nature of machine learning algorithms, the training data in this case will also lead to systematic errors in this group of patients. One might naively assume that if we simply remove the label of that group from the features a machine learning system can use, that might solve the potential for discrimination, but that is not true. Even without use of the group label during training or application of a model, or even if it is not collected during the training phase, a developed system can become biased against that group by propagating the poor choices that were made in the original data set.

Although humans can report on aspects of their decision-making, and we have developed relatively sophisticated means of establishing if someone is biased through a combination of nonverbal cues, past decisions, and history, the same level of insight is more challenging for automatic decision systems. This is particularly true for deep learning neural networks which, though quite powerful for complex decision-making, are also more challenging to explain the steps of decision-making. Tools are available to observe bias in automated decision-making systems, but this requires additional data collection and discussion of fairness using statistical arguments that may not be standard practice in machine learning or clinical teams.

2.5 Conclusion

Medical decision-making has relied on collected evidence and experience to generate models to predict outcomes and influence the choice of treatment options. Until relatively recently in history, this process has been limited to human experience and decision-making; however, the advent of technologies to collect and curate massive data sets has led to new capabilities not previously possible. Consumers are observing cars that can drive themselves by identifying objects in real-time at high speeds, systems that can generate narrative prose and speech which sounds natural, and systems that regularly win at strategic games like chess or go. Medicine is currently embracing the same big data technologies that enable these feats that were previously thought unique to human capability and experience. By leveraging the larger data sets available, systems can learn to pay attention to relevant features, weigh them, and combine them systematically to arrive at more accurate decisions than human decision-making alone. This will not only aid clinical research to help provide human-interpretable insights to improve standard clinical care as it is practiced today, but it will increasingly be integrated in an automated way to clinical care and early prevention and screening, dramatically shaping the practice of medicine going forward.

References

1. Angra S, Ahuja S. Machine learning and its applications: a review. In: 2017 international conference on big data analytics and computational intelligence (ICBDAC). 2017. p. 57–60.
2. Portugal I, Alencar P, Cowan D. The use of machine learning algorithms in recommender systems: a systematic review. Expert Syst Appl. 2018;97:205–27.

3. Juan Y, Lefortier D, Chapelle O. Field-aware factorization machines in a real-world online advertising system. In: Proceedings of the 26th international conference on world wide web companion. International World Wide Web Conferences Steering Committee, Republic and Canton of Geneva, CHE. 2017. p. 680–8.
4. Wenzel H, Smit D, Sardesai S. A literature review on machine learning in supply chain management. In: Artificial intelligence and digital transformation in supply chain management: innovative approaches for supply chains. Proceedings of the Hamburg international conference of logistics (HICL), vol. 27. Berlin: epubli GmbH; 2019. p. 413–41
5. Romanov A, Semenov A, Mazhelis O, Veijalainen J. Detection of fake profiles in social media - literature review. In: Proceedings of the 13th international conference on web information systems and technologies. https://doi.org/10.5220/0006362103630369.
6. Singh J, Singh G, Singh R. Optimization of sentiment analysis using machine learning classifiers. HCIS. 2017;7:32.
7. Zhou H, Tang J, Zheng H. Machine learning for medical applications. ScientificWorldJournal. 2015;2015:825267.
8. Erickson BJ, Korfiatis P, Akkus Z, Kline TL. Machine learning for medical imaging. Radiographics. 2017;37:505–15.
9. Thrall JH, Li X, Li Q, Cruz C, Do S, Dreyer K, Brink J. Artificial intelligence and machine learning in radiology: opportunities, challenges, pitfalls, and criteria for success. J Am Coll Radiol. 2018;15:504–8.
10. Litjens G, Kooi T, Bejnordi BE, Setio AAA, Ciompi F, Ghafoorian M, van der Laak JAWM, van Ginneken B, Sánchez CI. A survey on deep learning in medical image analysis. Med Image Anal. 2017;42:60–88.
11. Gronsbell J, Minnier J, Yu S, Liao K, Cai T. Automated feature selection of predictors in electronic medical records data. Biometrics. 2019;75:268–77.
12. Caballero FF, Soulis G, Engchuan W, Sánchez-Niubó A, Arndt H, Ayuso-Mateos JL, Haro JM, Chatterji S, Panagiotakos DB. Advanced analytical methodologies for measuring healthy ageing and its determinants, using factor analysis and machine learning techniques: the ATHLOS project. Sci Rep. 2017;7:43955.
13. Tang F, Xiao C, Wang F, Zhou J. Predictive modeling in urgent care: a comparative study of machine learning approaches. JAMIA Open. 2018;1:87–98.
14. Dunn J, Runge R, Snyder M. Wearables and the medical revolution. Per Med. 2018;15:429–48.
15. Zhang C, Ma Y. Ensemble machine learning: methods and applications. Boston, MA: Springer; 2012.
16. LeCun Y, Bengio Y, Hinton G. Deep learning. Nature. 2015;521:436–44.
17. Provost F, Fawcett T. Data science and its relationship to big data and data-driven decision making. Big Data. 2013;1:51–9.
18. Zaharia M, Xin RS, Wendell P, et al. Apache spark: a unified engine for big data processing. Commun ACM. 2016;59:56–65.
19. Shvachko K, Kuang H, Radia S, Chansler R. The hadoop distributed file system. In: 2010 IEEE 26th symposium on mass storage systems and technologies (MSST). ieeexplore.ieee.org. 2010. p. 1–10.
20. Prosperi M, Min JS, Bian J, Modave F. Big data hurdles in precision medicine and precision public health. BMC Med Inform Decis Mak. 2018;18:139.
21. Collins FS, Varmus H. A new initiative on precision medicine. N Engl J Med. 2015;372:793–5.
22. Albert MV, Shparii I, Zhao X. The applicability of inertial motion sensors for locomotion and posture. Locomotion and posture in older adults. 2017.
23. Qiu S, Cai X, Chen X, Yang B, Sun Z. Step counter use in type 2 diabetes: a meta-analysis of randomized controlled trials. BMC Med. 2014;12:36.
24. Modave F, Guo Y, Bian J, Gurka MJ, Parish A, Smith MD, Lee AM, Buford TW. Mobile device accuracy for step counting across age groups. JMIR Mhealth Uhealth. 2017;5:e88.
25. Albert MV, Deeny S, McCarthy C, Valentin J, Jayaraman A. Monitoring daily function in persons with transfemoral amputations using a commercial activity monitor: a feasibility study. PM R. 2014;6:1120–7.

26. Albert MV, Sugianto A, Nickele K, Zavos P, Sindu P, Ali M, Kwon S. Hidden Markov model-based activity recognition for toddlers. Physiol Meas. 2020;41:025003.
27. Albert MV, Azeze Y, Courtois M, Jayaraman A. In-lab versus at-home activity recognition in ambulatory subjects with incomplete spinal cord injury. J Neuroeng Rehabil. 2017;14:10.
28. Shawen N, Lonini L, Mummidisetty CK, Shparii I, Albert MV, Kording K, Jayaraman A. Fall detection in individuals with lower limb amputations using mobile phones: machine learning enhances robustness for real-world applications. JMIR Mhealth Uhealth. 2017;5:e151.
29. Antos SA, Albert MV, Kording KP. Hand, belt, pocket or bag: practical activity tracking with mobile phones. J Neurosci Methods. 2014;231:22–30.
30. Albert MV, Kording K, Herrmann M, Jayaraman A. Fall classification by machine learning using mobile phones. PLoS One. 2012;7:e36556.
31. Albert MV, McCarthy C, Valentin J, Herrmann M, Kording K, Jayaraman A. Monitoring functional capability of individuals with lower limb amputations using mobile phones. PLoS One. 2013;8:e65340.
32. Albert MV, Toledo S, Shapiro M, Kording K. Using mobile phones for activity recognition in parkinson's patients. Front Neurol. 2012; https://doi.org/10.3389/fneur.2012.00158.
33. Sok P, Xiao T, Azeze Y, Jayaraman A. Activity recognition for incomplete spinal cord injury subjects using hidden markov models. IEEE Sensors J. 2018;
34. O'Brien MK, Shawen N, Mummidisetty CK, Kaur S, Bo X, Poellabauer C, Kording K, Jayaraman A. Activity recognition for persons with stroke using mobile phone technology: toward improved performance in a home setting. J Med Internet Res. 2017;19:e184.
35. Coughlin SS. Recall bias in epidemiologic studies. J Clin Epidemiol. 1990;43:87–91.
36. Bowman SE. Coordination of SNOMED-CT and ICD-10: getting the most out of electronic health record systems. Coordination of SNOMED-CT and ICD-10: getting the most out of electronic health record systems/AHIMA, American Health Information Management Association. 2005.
37. Nosek BA, Alter G, Banks GC, et al. Scientific standards. Promoting an open research culture. Science. 2015;348:1422–5.
38. Stang PE, Ryan PB, Racoosin JA, Overhage JM, Hartzema AG, Reich C, Welebob E, Scarnecchia T, Woodcock J. Advancing the science for active surveillance: rationale and design for the Observational Medical Outcomes Partnership. Ann Intern Med. 2010;153:600–6.
39. Annas GJ. HIPAA regulations - a new era of medical-record privacy? N Engl J Med. 2003;348:1486–90.
40. Tovino SA. The HIPAA privacy rule and the EU GDPR: illustrative comparisons. Seton Hall Law Rev. 2017;47:973–93.

Automatic Diagnosis and Screening of Personality Dimensions and Mental Health Problems

3

Yair Neuman

3.1 Introduction: Automatic Analysis of Personality

3.1.1 Personality and Diagnosis

The idea of personality [1–6] suggests that human beings, and other nonhuman organisms (e.g., [7]), exhibit unique patterns of thought, emotion, and behavior that may be used as a kind of a psychological "signature." These patterns are not unique, in the sense that a genetic signature is unique or a fingerprint is unique. For example, a person may be characterized as having an introvert type of personality—but being introvert does not uniquely characterize that particular individual, as there are many other introverts. Although each of us is unique in a very profound sense, the idea of personality suggests that we may be characterized by measuring a limited number of personality dimensions that are shared by other human beings. Measuring personality dimensions can be used for diagnosis or in screening, or in a clinical or a nonclinical context.

3.1.2 Diagnosis Versus Screening

The difference between *diagnosis* and *screening* should be clarified [8]. The aim of diagnosis is to confirm, or rule out, the hypothesis that a *specific individual* has a certain personality dimension. For example, a teenager is sent to a clinical psychologist, after being involved in too many accidents. Reasonably dismissing other explanations, the psychologist may hypothesize that the improbable cluster of

Y. Neuman (✉)
The Department of Cognitive and Brain Sciences and Zlotowski Center for Neuroscience, Ben-Gurion University of the Negev, Beer-Sheva, Israel
e-mail: yneuman@bgu.ac.il

F. Jotterand, M. Ienca (eds.), *Artificial Intelligence in Brain and Mental Health: Philosophical, Ethical & Policy Issues*, Advances in Neuroethics,
https://doi.org/10.1007/978-3-030-74188-4_3

accidents is indicative of a self-harming behavior. The psychologist may therefore use a variety of tools to diagnose the teenager as having some kind of *accidental personality*. By diagnosing this specific individual, the psychologist is trying to conclude whether a particular personality, or personality dimension, validly characterizes the teenager. Diagnosis is highly important, as it organizes a cluster of behaviors, emotions, and cognitions into an interpretable pattern that may be used for prognosis, treatment, and prediction.

Unlike diagnosis, which is focused on the individual, screening is broadly used to determine which member of *a large group of individuals* has the attribute in question [8]. In other words, the group—rather than the individual—is the focus of analysis. For instance, a computer program might automatically screen a large group of Facebook users for signs of depression and suicidal thoughts, rank them according to their depressivity and risk level, and send the top-ranked individuals a recommendation for an in-depth personal diagnosis. Such automatic screening for depression has indeed been found to be effective (e.g., [9]), and one may use it in cases where personal clinical diagnosis is not easily available, at least in the initial phase of a process.

3.1.3 The Clinical Versus the Nonclinical Context

Personality dimensions may have both clinical and nonclinical aspects. The clinical context of diagnosis focuses on emotional, mental, and behavioral *disorders*. In other words, the aim of a clinical diagnosis is to determine the presence of a certain personality disorder. In contrast, diagnosis in the nonclinical context is broadly used to measure the personality dimensions of an individual, for use outside the clinical context. For example, one of the dimensions measured by the SWAP-200 personality test [6] is the *narcissistic personality disorder* [10]. When imagining a narcissist personality, we usually think about an individual with an exaggerated sense of self-importance, accompanied by a nonadaptive behavior. However, narcissism is a spectrum, ranging from a healthy form of self-love, to a pathological conflict over self-value. The typical narcissist has an exaggerated sense of self-importance—but narcissism is not necessarily pathological, and identifying and measuring the "soft" levels of narcissism may be used for practical purposes. Here are two examples where we may be interested in measuring nonpathological versus pathological narcissism.

Example 1 An intelligent targeted advertising engine might analyze the texts written by its "target" in social media to conclude that she scored highly on both extroversion and narcissism. In this case, the engine might send her a personalized advertisement for a rock-n-roll concert, but when designing the ad text, it might place strong emphasis on themes that resonate with an extrovert narcissist personality—for example, by appealing to her sense of self-importance (e.g., "The Greatest Rock Concert, for the Greatest People"). By first identifying the target's personality dimensions, then appealing to those dimensions at the unconscious level, the engine appears to have better chances of achieving its major aim of "seducing" the individual to click on the ad.

Example 2 Let us imagine that we are asked to design an automated system for determining the risk factor of violent men who might pose a threat to their spouses, in order to take preventive steps and reduce the danger of homicide. A forensic psychologist might teach us that one of the dimensions worth examining is *pathological narcissism*—as a man who cannot see beyond his self-centered perspective is more dangerous than someone who cognitively, emotionally, and behaviorally understands that he is not the center of the universe. The engine that we may build should therefore run on data produced by violent men under inspection—such as the text messages they send to their (ex-)wives. By analyzing the texts, the engine should score the text for signs of pathological narcissism, and using machine learning (ML) algorithms, we may examine whether narcissism is a risk factor that distinguishes between dangerous husbands and those are merely "barking" with no real danger of "biting." In this context, of course, great emphasis should be placed on the selection and engineering of the appropriate features, as we are not simply seeking general signs of narcissism, but signs of pathological narcissism that may point to the risk of a potentially harmful husband. In this regard, the success of the automatic personality engine is measured by its ability to classify dangerous vs. non-dangerous husbands, based on their respective exhibited levels of pathological narcissism.

A similar idea may be applied to the context of depression and depressivity as a personality dimension. Depression, in its pathological form, is a risk factor for suicide. There are contexts in which we would like to screen for individuals who suffer from depression, as they may pose a threat to themselves and to their surroundings. To avoid a straw-man type of fallacious reasoning, I must emphasize that I am making no argument here about depression and dangerousness. The overwhelming majority of depressed individuals would not harm others, or themselves. However, in certain contexts, screening for individuals with pathological depressivity and suicidal intentions may save lives, and this is only one example in which computational personality analysis may contribute to the field of diagnosis [11, 12]. One manifest example is the one of the Germanwings Flight 9525. On March 2015, this Airbus plane crashed in the Alps, resulting in 150 casualties. This was not an accident: it was deliberately planned and executed by the co-pilot, Andreas Lubitz, who had been treated for suicidal tendencies. It was not only an act of suicide, but of murder (i.e., homicide-suicide), since by crashing the plane, Lubitz took the lives of many innocent people, who paid the price of the failure to screen out individuals such as himself, who suffered from suicidal tendencies, from serving as a pilot. As a rule, the aviation industry is highly sensitive to the safety of its passengers, as any mistake, improbable as it may be, may result in a humanitarian and economic catastrophe. The pilots are clearly one of the vulnerabilities of the system, as attested by the case of Germanwings Flight 9525. Had they been less zealous in protecting personal privacy, the German authorities could have used Lubitz's medical records—coupled with other sources of information—to prevent him from taking the lives of so many innocent people. In designing such an alarm system, ethical issues could have easily been resolved, and these should not be used to counter the necessity of verifying the mental health of pilots.

3.2 Computational Personality Analysis

In the previous sections, I explained the idea of personality, the difference between screening and diagnosis, and the expression of personality in clinical and non-clinical contexts. Traditionally, personality analysis is conducted by a human expert or by means of questionnaires that require the voluntary collaboration of the person being diagnosed, and the validity of her self-reported personality dimensions. However, when massive datasets are involved, the use of a human expert and manual data analysis is impractical. Moreover, in such contexts, it is usually extremely difficult, if not impossible, to gain the voluntary participation of the diagnosed subjects, or a valid measure of their personality dimensions. Here we have a clear answer to the questions "Why are automated approaches to personality analysis useful for mental health?" and "Why are they preferable to conventional (non-automated) approaches?" To clarify these points, let us examine an example.

It has been recently reported[1] that suicides among active-duty members of the U.S. Air Force surged to its highest level in over three decades. Given the ratio of mental health experts in the U.S. military to active-duty personnel, it is almost impossible to diagnose depression among the soldiers in a reasonable space of time. Moreover, using questionnaires may be the wrong strategy, as the soldiers may fail to answer them honestly, due to lack of self-awareness or fear of being "exposed" and dismissed from duty. In such contexts, automated approaches are the only solution, as they provide a valid means of diagnosing large numbers of people in a very short time, by using texts (written or spoken) that they *naturally* produce. Automated systems are therefore preferable whenever human experts cannot provide diagnosis, due to the constraints of number or time, or whenever the use of questionnaires or other conventional methods is less appropriate.

The same rationale and justification that apply to medical diagnosis also apply to the diagnosis of personality through automated tools. Google has recently demonstrated[2] how an automated system can identify skin diseases—a massive, quick, and valid diagnosis that can match the performance of human experts. Google justifies the use of such tools by the dearth of dermatologists, coupled with the relatively high number of individuals seeking diagnosis and treatment—the same justification as the one cited for using tools of computational personality analysis.

[1] https://time.com/5780447/air-force-suicide-surge/.

[2] https://ai.googleblog.com/2019/09/using-deep-learning-to-inform.html.

3.2.1 The Relevance of Computational Personality Analysis in the Current Culture

The idea of personality analysis, and automatic personality analysis in particular, is particularly relevant for the modern and post-modern societies, where the individual has become the focus of interest, and where digital traces of individuals are evident everywhere. In the past, the idea of "personalized medicine" was irrelevant, as medicine was at an embryonic stage and the individual was not at the center. However, today we expect medicine—as an advanced practice supported by a wealth of data and technological tools—to address our particular signatures, uniqueness, and needs, in a manner that maximizes the effectiveness of diagnosis and treatment. The same is true for other fields as well—particularly that of personality analysis.

When the famous Edward Bernays launched a sophisticated pro-smoking campaign for women at 1929, he marketed it under the slogan "Torches of Freedom." Instead of seeing cigarettes as deadly poison, women were encouraged to perceive smoking as a feminist and emancipatory act. The campaign was extremely successful, and in retrospect, one may wonder whether the benefits of emancipation were worth the price paid by women who have joined the "cancer club." In any case, Bernays's campaign was extremely clever in its targeting of women instead of the public at large—but today, a more individualized campaign would probably have been called for. Although women share certain distinct needs and characteristics as a group, the unique attributes of each individual woman are such that—as with personalized medicine—they should be taken into account for maximum effectiveness.

In sum, although the identification of personality dimensions for any practical purpose is a long-established practice, it has become particularly important in modern-day, technologically oriented societies, where it is easier to identify such patterns by analyzing the digital traces left by everyone almost everywhere. For this reason, we use tools of AI—or more specifically, machine-learning tools—to automatically analyze the data and perform screening or diagnosis. Computational personality analysis, as its name suggests, is the field where methodologies and tools are developed for automated analysis of personality dimensions. In the following sections, this general approach is presented in a nutshell through a specific example, then elaborated further.

3.2.2 Computational Personality Analysis in a Nutshell

A project in computational personality analysis usually starts with a clear idea of (1) *Why* (i.e., Why do we need automated personality analysis?), (2) *Which* (i.e., Which personality dimensions are relevant for the task?), and (3) *How* (i.e., How are we going to measure the personality dimensions?).

When constructing a system for automatically measuring personality dimensions, we usually use a *supervised* form of learning, where the ML algorithms are trained on a tagged dataset—namely a training set of examples and their diagnosis/tag. For example, if we would like to teach the computer to measure depression among teenagers, we might provide it with personal text passages (e.g., diary entries) that they have written. After each text is read and scored by several experts according to clear criteria, it is given a "depression score," on the assumption that the level of depression evident in the text reveals its author's depressivity. If possible, we can even diagnose the people who wrote the texts, and score them on a depressivity scale, to validate the performance of our algorithm.

The training set is therefore composed of a set of documents, each scored according to the depression level, as measured by the human experts. In a simpler case, we do not score the text on a spectrum, and the expert may be required only to tag the text as either clinically depressed, or not. In instances where we would like to predict a continuous score, we use an ML model fit for regression; in categorical instances, we use a ML algorithm designed for classification.

There are various ML algorithms for regression and classification—from Naïve Bayes, to SVM, XGBoost, and Deep Neural Networks (DNN). The decision as to which ML algorithm to use is governed by the particular details of our task and data, and usually several ML algorithms are trained and tested on the dataset.

To teach the computer to diagnose depression using texts, we should provide it with a set of features that characterize each document. These features, sometimes called *variables* or *attributes*, are supposed to reveal whether or not the text is indicative of depressivity. There are various features that we can measure in each text, and in this case, too, the decision which features to analyze is determined by the particular characteristics of the project. For the diagnosis of depression, for example, the most intuitive attributes that we may want to measure are content categories, such as those we can measure using *Empath*.[3] Let us assume the computer is provided with the following text, which an expert has tagged as "depressed":

> I' sad and lonely. No one loves me and I feel abandoned and neglected. Life is hopeless and there is no hope, just despair.

This almost caricature-like example clearly represents depressivity—one need not be a certified psychologist to see that. After running the text through automatic analysis, Empath provides a list of content categories, and the extent in which they are expressed in the text. This reveals the following content categories, and the extent (i.e., frequency) to which they are expressed in the text:

Content category	Score
Shame	2
Negative emotion	1
Body	1
Love	1
Violence	1
Sadness	1
Contentment	1
Pain	1
Emotional	1
Nervousness	1
Cold	1

[3] http://empath.stanford.edu/.

We can see that the content categories identified by the computer can be theoretically associated with depressivity. The computer than learns that a "depressed text" (i.e., a depressed person), at least according to the above example, has a "signature"—a particular combination of content categories and their "weight" in the text—which may be optimally used to classify a text as "depressed" or "nondepressed."

When fed with enough examples and with the appropriate features, the machine learning algorithm learns a model that optimally classifies a text as "depressed" or "nondepressed." To test how well the machine has learned to identify depressed texts, we present it with another set of texts, which serves as the test set. The machine learning algorithm then uses the model that it built in the previous learning phase to identify *nontagged* texts as "depressed" or "nondepressed." These are new texts that the algorithm has not seen before, and therefore its ability to successfully classify the new texts is an indication to the extent in which it can validly classify/diagnose a text as "depressed." At this point, we measure the performance of the model through various diagnostic measures—such as precision, and recall—and by validating the results.

One important way of validation is through the *k-fold cross-validation procedure*. This procedure aims to address the problem of over-fitting our model to the data. In each run of the cross-validation, we divide the dataset into a training set and a test set: we train the model on one set and test it on the other, and the performance of the model is tested by averaging the results over several runs. If the model performs well, we may apply it in practice, and use it as a kind of a "digital psychologist."

How good is the performance of such computational personality analysis tools? In many cases, they provide a highly successful and efficient diagnosis. For example, [9] have provided 84% accuracy in diagnosing depression, and current studies provide much better results. In a recent study [12], we designed a computer algorithm for identifying a psychopathic signature in texts. The test set included 2333 texts—only 4% of which were texts with a distinctive psychopathic signature. Identifying such a text by chance has a very low probability ($p = 0.04$), but when applying our automated methodology, we were able to identify them with 67% precision, which is an enormous improvement over the base-rate of "psychopathic" texts in the dataset.

Having presented the idea of computational personality analysis in a nutshell, I shall now detail and elaborate it in the next section.

3.3 Computational Personality Analysis Further Detailed

For automatically measuring personality dimensions and disorders, we need data, which may come in various forms and modalities.

It is generally assumed that the language we use is a window onto our personality. If someone says: "I'm depressed and lonely," then given the appropriate context of interpretation, we may hypothesize that he is trying to convey his despair, and when the incidence of words such as *depressed*, *lonely*, *helplessness*, etc. is measured automatically, we may score the depressivity level of the text as indicating the depressivity level of its author. However, the person may be joking, or being ironic, or simply citing something that he heard from someone else—which is why we must also take into account contextual knowledge to gain a valid conclusion about the depressivity level as expressed in the text, and whether it truly represents the depressivity level of the author. In any event, there is a wealth of evidence that the language that we use is an enormously rich mine of information for personality analysis. Other sources of information may also be identified and used if possible. For example, when analyzing depression among individuals, we may analyze their medical records and the visual images they upload to social media. In a past unpublished study, we analyzed the images uploaded to Instagram by young people—mostly women involved in self-harming behavior. It was clear from the images that these young people were depressed and self-harming: dark images, with signs of loneliness, blood, and cuts, were everywhere. An automatic image analysis algorithm could have easily classified them. Therefore, when using the term *text*, I may use it in the most generic sense to include visual images, facial expressions, body posture, and so on.

It is important to emphasize that when analyzing texts, we use a corpus of *personal* texts produced by the individuals. Why is it so important to use such personal texts? The reason is that a scientific report is probably not a good source for diagnosis, but personal texts—of the sort published in social media, diaries, stories, conversations in informal settings—are all better candidates, because in principle, at least, they reveal the individuals' *inner life*. Thus, if an accountant is preparing a financial statement for a company, we should not expect his inner life or personality dimensions to be expressed in the statement. However, if she keeps a journal, corresponds with others on Facebook, or writes a personal essay, then a personality signature should be evident. Gaining access to a personal text is a necessary step. In addition, and for the first phase of building a personality analysis system, each text is *labeled/tagged*. There are various ways in which we might do so, for example, by asking the subject who completed a personality questionnaire, by interviewing the subject, by scoring the text according to well-defined protocol and criteria.

At this point, and for each individual, we should have a personal text and personality tags and/or the specific score she has gained on each of the required personality dimensions. The text is then pre-processed, cleaned, and expanded upon using a variety of Natural Language Processing (NLP) tools, to prepare it for the main analysis. For example, we may be interested in analyzing only certain parts of speech of the text—such as nouns, verbs, or adjectives—in which case, we would use a Part-of-Speech Tagger. Next, various textual *features* are extracted from the text—such as the degree in which the person uses various words or word categories. For example, we may use LIWC [13], or Empath, [14] to measure the prevalence of positive vs. negative sentiment in the text—since a high level of negative emotion expressed in the text may be an important indicator of depressivity. Next, an ML algorithm is trained and tested to find the optimal model that can best "predict" (i.e., classify) the individuals' respective personality labels/scores. What do we mean by an optimal model? An ML algorithm is basically a sophisticated *optimization* engine. Given the tag of the text and the list of personality features and their score, it builds a classification model that assigns weights to the various features, so the classification performance is maximized. For example, the ML algorithm may show us that some features that we believed to be valuable in fact contribute nothing to the performance, and therefore can be ignored. The selection of attributes or features is therefore an important phase in constructing a successful model. Moreover, the algorithm can calculate the weights—or "importance"—that should be attributed to each feature. Different features may have a different predictive value, and the ML algorithm knows how to identify it. A computational personality project is ultimately judged by its success. The impetus for any given project is the specific of task that we would like to perform—such as choosing the best CEOs among many candidates, identifying depressed individuals, screening for lone-wolf perpetrators.

If the ML algorithm has produced good results according to some relevant standards, we can use the system. Deciding what performance is good enough must be clarified within a wider context of decision-making. For example, diagnosing PTSD through the use of human experts is costly. Let us assume that only 1% of people suffering from PTSD are diagnosed in time: if an automated system improves this diagnosis rate by 1%, should it be considered effective enough to be adopted? The answer depends on the wider context.

This, then, in a nutshell, is the essence of automatic personality analysis. In some contexts—the automatic profiling of shooters [15]; the measurement of disorders [16]; the screening of suicide ideation [17]; and the measurement of the "Big Five" personality dimensions (neuroticism, extraversion, consciousnesses, openness, agreeableness)—this general approach seems to work quite well.

In conclusion, here is an example, from a nonclinical context.

Targeted advertising is a type of online advertising that targets audiences with certain traits, based on the product or person the advertiser is promoting. By way of example, let us assume that we are developing a targeted advertisement engine that promotes music concerts. In a past study, in the context of computational personality analysis, we found a link between the lyrics of various music genres and certain personality types [18]. This finding can be used for automatically and optimally targeting advertisements, by analyzing texts written by individuals and deciding whether they are of the extrovert "Rock-n-Roll" type of person, or the introvert "Mellow" kind of personality. For a targeted advertising engine seeking to improve its performance, it may be highly informative to know whether a given individual is an extrovert or an introvert: if they are an extrovert, the engine might decide to send them an advertisement for a rock concert; if they are of the introvert type, they would get an advertisement for a mellow jazz show. In this case, determining the correct approach to the individual based on their particular personality type is justifiable.

3.4 A Critical Perspective

What are the problems in applying computational personality analysis? First, we should be careful when choosing a personality theory and personality dimensions. For example, the Big Five is a dogma with many theoretical and empirical problems

[19]. (See the paper by [20] for one possible criticism.) Therefore, although it is the main theory used in automatic personality analysis, one should critically decide whether and when to use it. Given the problem-oriented perspective that I have presented, the personality theory and personality dimensions that we choose should be carefully selected by their *clear relevance* to the challenge that we aim to address. The fact that the Big Five model is simple and easy to understand does not mean that it is relevant everywhere. For example, it is highly questionable whether it is of any relevance in the analysis of suicidal intentions. Clearly, one may find a statistical correlation between depressivity and neuroticism, since both involve negative emotion. However, the real challenge is not to identify statistical correlations of their p-values, but to construct methodologies that are meaningful in real-world challenges, by using the powerful tools of ML. From algorithmic finance to the automatic identification of lone-wolf perpetrators, one finds almost the same methodological criticisms and the same calls for a meaningful, relevant, and reality-based approach to the design of intelligent systems. In my experience of academic and non-academic/commercial projects to do with automatic personality analysis—including those in which we measured the Big Five—I must admit that, in hindsight, the Big Five model has no significant value for most real-world applications that I have encountered.

In the context of identifying suicidal intentions, for example, one may prefer the modern psychodynamic approach to personality [21], with its focus on the conflicts and defense mechanisms [22] that constitute the human personality. However, the psychodynamic approach is also fraught with difficulties, as it was designed for the clinical context. In addition, it is very difficult to translate the theory's ideas into measurable features. For example, *splitting*—seeing the world in binary terms of good and bad, black and white—is a primitive defense mechanism that some people use in order to cope with their anxieties. You can see it in action when you hear zealous ideologists—be they Islamic fundamentalists, zealous vegetarians, or fanatical BDS supporters—when presenting their worldview.

In some problem-oriented contexts, it may be important to identify the most zealous individuals—those who see the world in black and white. For example, suppose that we are interested in a new European apocalyptic sect similar to the Order of the Solar Temple, whose members committed mass suicide. Specifically, we would like to know how zealous are its members? In a case of this sort, measuring the degree of splitting within the texts (written or spoken) produced by the sect members is very important, and although it proved to be a challenge, we have shown that it is feasible to measure splitting in a text, and its relevance for the forensic context [23].

In sum, choosing the right approach and the right features is crucial. Now, by using specific examples, let me underline the problem of conducting an automatic personality analysis without due regard to the pragmatic aspect.

Lone-wolf perpetrators are a pressing issue for law enforcement agencies in the United States and in Europe. In a study conducted by [24], the researchers used "A unique dataset of 119 lone-actor terrorists and a matched sample of group-based terrorists" and compared the prevalence of mental illness (*Yes/No*) among lone-wolf terrorists and group terrorists. They found a significant difference between the two groups in this regard: among the lone-wolfs, the prevalence of mental illness was 32%, while among "group-based" terrorists it was 3%. The authors concluded that "...mental health professionals may have a role in *preventing* lone-actor terrorist attacks" (my emphasis) and that "...screening processes can be carried out by security agencies on patients that present similar antecedents and behaviors in medical evaluations." These scientifically invalid—and ethically dangerous—conclusions seem to ignore the simple lessons of reasoning, since the question is not whether there is a difference between lone-wolf and group-based terrorists, but whether mental illness is a significant risk factor and a relevant feature for intervention and prevention.

To address this question, one must ask what is the probability of someone engaging in acts of terrorism *given* their mental illness. The answer is almost nil. It might be inferred from the above study that people who suffer from mental illness pose a danger to society—but such an inference is scientifically invalid, ethically dangerous, and pragmatically irrelevant. Therefore, in the context of personality analysis, which is problem-oriented, one should clearly examine whether:

1. The findings are scientifically valid.
2. Pragmatically meaningful and usable.
3. Whether the implications are ethically justified.

3.5 Summary and Conclusions

The above critique is imperative for the reflective scientist, data engineer, or practitioner. However, critical reflections must not mask the achievements and future potential of computational personality analysis. Automatic personality analysis can have enormous benefits in improving our understanding of people in contexts ranging from screening for mental health problems, to the effective recruitment of human resources in companies. This field is still in its infancy, and there are several challenges to be addressed:

1. Most of the approaches to the automatic analysis of personality rely on *low-level features* (such as words), or their simple categorization. However, the complexity of human personality cannot be easily encompassed by low-level features alone. There is a need for more sophisticated methods that use deep syntactic-semantic analysis and infer personality dimensions through higher and more abstract features, that are extracted from the text.

Almost all the studies in the field rely on a tagged corpus, where texts are produced by individuals who are tagged according to their personality dimensions. In some cases, such corpora are *extremely difficult to obtain*—and even when they are, their artificial nature means that they lack ecological validity. In addition, their "shelf life" is limited, due to the contextual, dynamic, and changing nature of language.

2. Personality is a dynamic phenomenon that "lives" in time, and sometimes the most important information is identified by analyzing the behavior of personality dimensions along the timeline. When trying to identify whether the mental state of a teenager is moving toward a tipping-point of despair, for example, we must take the trajectory of the mental state into account.

In conclusion, most ML approaches to computational personality analysis adopt a "ready-to-wear" approach, whereby ML classifiers are trained, validated, and tested on a tagged corpus. However, as with any ready-to-wear approach, this approach is limited in its ability to provide the "client" with the best fit. The promise of computational personality analysis is huge [25], and addressing the challenge of building such a system in vivo requires reflectivity and sensitivity to various issues, such as the ones discussed above.

References

1. Corr PJ, Matthews G, editors. The Cambridge handbook of personality psychology. Cambridge, UK: Cambridge University Press; 2009.
2. Snyder M, Deaux K. The Oxford handbook of personality and social psychology. New York: Oxford University press; 2012.
3. Funder DC. The personality puzzle. 5th ed. New York: WW Norton & Co.; 2013.
4. Neuman Y. Personality from a cognitive-biological perspective. Phys Life Rev. 2014;11:650–86.
5. Neuman Y. Shakespeare for the intelligence agent toward understanding real personalities. Lanham, MD: Rowman & Littlefield; 2016.
6. Westen D, Shedler J, Bradley B, Defife JA. An empirically derived taxonomy for personality diagnosis: bridging science and practice in conceptualizing personality. Am J Psychiatr. 2012;169:273–84.
7. Freeman HD, Gosling SD. Personality in nonhuman primates: a review and evaluation of past research. Am J Primatol. 2010;72(8):653–71.
8. Streiner DL. Diagnosing tests: using and misusing diagnostic and screening tests. J Pers Assess. 2003;81:209–19.
9. Neuman Y, Cohen Y, Assaf D, Kedma G. Proactive screening for depression through metaphorical and automatic text analysis. Artif Intell Med. 2012;56:19–25.
10. Russ E, Shedler J, Bradley R, Westen D. Refining the construct of narcissistic personality disorder: diagnostic criteria and subtypes. Am J Psychiatr. 2008;165:1473–81.
11. Neuman Y. Artificial intelligence in public health surveillance and research. In: Luxton DD, editor. Artificial intelligence in behavioral and mental health care. Amsterdam: Elsevier/Academic Press; 2016. p. 231–54.
12. Neuman Y, Cohen Y, Neuman Y. How to (better) find a perpetrator in a haystack. J Big Data. 2019; https://doi.org/10.1186/s40537-019-0172-9.
13. Tausczik YR, Pennebaker JW. The psychological meaning of words: LIWC and computerized text analysis methods. J Lang Soc Psychol. 2009;29:24–54.
14. Fast E, Chen B, Bernstein MS. Empath: understanding topic signals in large-scale text. In: Proceedings of the 2016 CHI conference on human factors in computing systems. ACM; 2016. p. 4647–57.
15. Neuman Y, Assaf D, Cohen Y, Knoll J. Profiling school shooters: automatic text-based analysis. Front Psych. 2005; https://doi.org/10.3389/fpsyt.2015.00086.
16. Neuman Y, Cohen Y. A vectorial semantics approach to personality assessment. Sci Rep. 2014; https://doi.org/10.1038/srep04761.

17. Fernandes A, Dutta R, Velupillai S, Sanyal J, Stewart R, Chandran D. Identifying suicide ideation and suicidal attempts in a psychiatric clinical research database using natural language processing. Sci Rep. 2018; https://doi.org/10.1038/s41598-018-25773-2.
18. Neuman Y, Perlovsky L, Cohen Y, Livshits D. The personality of music genres. Psychol Music. 2016;44:1044–57.
19. Block J. A contrarian view of the five-factor approach to personality description. Psychol Bull. 1995;117:187–215.
20. Molenaar PC, Campbell CG. The new person-specific paradigm in psychology. Curr Dir Psychol Sci. 2009;18:112–7.
21. Psychodynamic diagnostic manual (PDM). Silver Spring, MD: Alliance of Psychoanalytic Organizations; 2006.
22. Bowins B. Psychological defense mechanisms: a new perspective. Am J Psychoanal. 2004;64:1–26.
23. Neuman Y, Assaf D, Cohen Y, Knoll JL. Metonymy and mass murder: diagnosing splitting through automatic text analysis. In: Arntfield M, Danesi M, editors. The criminal humanities: an introduction. New York: Peter Lang; 2016. p. 61–73.
24. Corner E, Gill P. A false dichotomy? Mental illness and lone-actor terrorism. Law Hum Behav. 2015;39:23–34.
25. Neuman Y. Computational personality analysis: introduction, practical applications and novel directions. New York: Springer; 2016.

Intelligent Virtual Agents in Behavioral and Mental Healthcare: Ethics and Application Considerations

4

David D. Luxton and Eva Hudlicka

4.1 Introduction

The past 10 years have witnessed rapid growth in the development and use of embodied intelligent virtual agents (IVAs) and nonembodied conversational agents (CAs), such as chatbots, and their application in behavioral and mental healthcare. IVAs and CAs have been successfully used as coaches to support behavior change (e.g., smoking cessation, starting exercise programs, nutrition), to provide support for caregivers, and to provide limited treatment functions, in areas where highly scripted protocols are applicable (e.g., elements of cognitive behavioral therapy, dialectical behavior therapy, or motivational interviewing) [1].

Along with significant new technological development and promising applications in this area, ethical issues have emerged regarding the most appropriate use of this technology. In addition to privacy issues, which are particularly critical in healthcare applications, there are issues regarding the safety of persons who interact with them. For example, how should an agent handle a situation where the user expresses suicidal ideation? Another ethical issue regards the possibility of deception, in situations when users may not know whether they are interacting with an IVA that is autonomous or one controlled by a human.

Our objective of this chapter is to summarize the state-of-the-art in the use of IVAs and CAs in behavioral care, and to discuss the range of applications of these agents, including their potential use in therapeutic gaming environments and virtual

D. D. Luxton (✉)
Department of Psychiatry and Behavioral Sciences, University of Washington, Seattle, WA, USA
e-mail: ddluxton@uw.edu

E. Hudlicka
Psychometrix Associates and Therapy 21st, Amherst, MA, USA
e-mail: hudlicka@ieee.org

F. Jotterand, M. Ienca (eds.), *Artificial Intelligence in Brain and Mental Health: Philosophical, Ethical & Policy Issues*, Advances in Neuroethics,
https://doi.org/10.1007/978-3-030-74188-4_4

or augmented reality applications. We then discusses the ethical issues associated with their use, including ethical considerations associated with artificial relationships, bias (e.g., cultural, demographic, or linguistic implicit biases introduced during development), and the black-box problem (lack of transparency regarding the agent's behavioral choices). We conclude with recommendations to begin to address these ethical issues.

4.1.1 Technical Overview

Conversational agents (CAs) are software programs that emulate human communication through verbal dialogue. The most common forms are *chatbots* that employ basic text interface for communication. Speech recognition, natural language processing, and speech synthesis technologies can also give CAs ability to conduct limited conversations with human users through verbal conversation.

Intelligent virtual agents (IVAs) add more realism to conversational emulation through computer-generated, animated, embodied, artificially intelligent virtual characters. IVAs can have the visual appearance of humans or any other form, ranging from simple cartoonish characters to highly detailed and lifelike three-dimensional forms [1–5]. The visual appearance and the interaction capabilities of IVAs vary greatly, and designers can customize them to match the human user's needs and preferences; that is, customize physical appearance and mannerisms, speech dialect, use of local colloquialisms, and other characteristics that match them to a user's cultural background, race/ethnicity, gender, or socioeconomic status [6].

Conversational agents are most commonly deployed on the Internet, or as personal assistants such as Amazon's Alexa, Apple's Siri, and Microsoft's Cortana. CAs can also be accessed on personal computers, kiosks, and mobile devices (i.e., smartphones, tablet computers, and smartwatches) [7], and IVAs are also used on the displays of some robots to make them interactive and more personable [8]. IVAs are typically deployed as components of tutoring or coaching systems, or as agents providing support or guiding the user through some therapeutic protocol in behavioral health technologies. The majority of the behavioral healthcare applications are still the research phase.

A significant development in CA and IVA technologies is the addition of capabilities supporting affective interaction [9–15], that is, recognition of the human user's emotions, expression of the agent's synthetic emotions to the human user, and affect-adaptive interaction with the user, including the development of "empathic" relationship between the agent and the user [2, 16–18]. *Affective-adaptive interaction* refers to a human–agent interaction where the agent is able to recognize (some of) the human's emotions, and respond appropriately, thereby exhibiting aspects of emotional and social intelligence humans expect from human interaction partners. The term "synthetic emotions" refers to models and expressions of emotions in machines. The term intends to highlight the fact that machines do not "experience" emotions as humans do, but can model (aspects of) affective processing and can display affective expressions using the channels available in the agent's embodiment, e.g., face, gestures, as outlined above. It is beyond the scope of this chapter to describe how synthetic emotions are modeled within agents. A detailed discussion of emotion modeling can be found in [1].

In the case of IVAs, embodiment provides specific expressive channels that are available to display affective expressions, which the human user perceives as a particular emotion. For example, a "talking head" agent has two available expressive channels: the face (to display facial expressions) and the head (to convey nonverbal information via head movement, such as nods or shaking). An agent with the embodiment of an upper torso augments these channels with hand gestures and upper body movements, and a fully embodied agent then adds body posture and full-body technologies mentioned above. Embodied agents thus have the capabilities to communicate affective and conversational states nonverbally, via the available channels. For example, in response to a human user's statement such as "I'm feeling depressed today" or "I've had a rough week," an affective virtual agent might display an expression of caring interest and empathy via facial expression (empathic caring), head movement (tilt), hand gestures (open arms), and torso movement (leaning forward).

Affective agents may also employ affective user modeling and explicitly model synthetic emotions within their architectures, to support affect-adaptive interaction and to dynamically produce affectively realistic agent behavior (e.g., [3, 10]) *Affective user modeling* refers to the ability of the agent to represent information about aspects of the user's affective behavior; for example, which events are likely to trigger which emotions, how are different emotions expressed by the user and how do they influence the user's behavior.

An essential component of affect-adaptive interaction is the ability of the agent to recognize the human interaction partner's emotions. Emotion recognition involves the collection of user data reflecting his/her emotional state, such as facial expressions, head movement, hand gestures, posture and body movement, as well as speech prosody and physiological data such as skin conductance. Pattern recognition classification algorithms and machine learning are then used to map the data onto a set of emotional states. Currently, emotion recognition techniques are able to recognize the "basic" emotions (e.g., fear, anger, sadness, and happiness) with accuracy rates approaching those of humans [19], and progress is being made in the ability to recognize nuanced and complex emotions, such as guilt, shame, and pride [20]. Nonetheless, recognition of naturalistic emotions (i.e., emotions displayed in non-laboratory settings and during unconstrained interactions) is more complicated and remains one of the significant challenges in emotion recognition research [19].

Together, these capabilities enable the agents to display aspects of emotional and social intelligence (e.g., awareness of social cues and user goals) and allow them to adapt to the changing states and needs of the human users. This, in turn, contributes to the development of what the human user perceives as a supportive, understanding, and empathic human–agent relationship. The existence of such has been shown to improve user engagement and thus make the agents more effective in their tasks, for example, behavior change coaching or providing support [1, 14, 17, 18, 21].

4.2 Practical Applications in Healthcare and Benefits

4.2.1 Use in Care Settings

Conversational agents provide many potential benefits, including the capability to support clinical care and provide health behavior coaching [22–25]. In terms of the more straightforward logistical benefits, CAs offer 24/7 availability and availability in remote areas where access to mental health professionals may be limited. Because IVAs have the ability to engage in naturalistic interactions with humans through dialog and nonverbal expression, interaction with them requires minimal training or no training at all on the part of the human user, again enhancing accessibility and usability. The ability to interact with human users who do not have prior training is especially beneficial for populations such as children, the elderly, and individuals with disabilities [1].

A major benefit of using conversational agents and virtual embodied agents is their ability to provide an interactive and engaging experience, and make users feel understood, while also reducing social anxiety [26–29]. For example, Bickmore and colleagues [30] tested a virtual nurse named "Elizabeth" that helped patients to understand hospital discharge information such as follow-up care and medication requirements. Displayed on a touchscreen, the virtual character used a synthetic voice to speak, and displayed animated nonverbal behavior (e.g., facial expressions, hand gestures, shifts in posture). Evaluation of user satisfaction revealed that patients preferred to receive the discharge planning information from the virtual nurse versus a doctor or nurse because it spent more time with them and never seemed rushed by other demands.

In another study, Mccue and colleagues [31] tested the use of CAs in treating patients with chronic pain and depression in inner city outpatient clinics and found that persons who interacted with a CA reported full compliance with the CA suggestions to reduce stress as well as high degree of compliance (89%) with healthy eating suggestions. Furthermore, 78% said they trusted the ECA "very much" and a significant subset of the participants (44%) indicated that they would prefer interaction with an EVA over interaction with a clinician. While results support the use of CAs and IVAs for these purposes, further validation is needed to assess for whom and when these agents are most appropriate.

4.2.2 IVAs in Serious Games

In addition to deployment as standalone agents or as components of an interactive system, IVAs can also be incorporated into computer games as non-playing characters. This is the case for both games designed for entertainment and serious games, including games developed for training and learning purposes, and to support psychotherapy (e.g., provide opportunities for the practice of specific skills, such as conversational skills to address social anxiety). Games have a unique ability to engage the players and to provide highly immersive learning, training, and therapeutic environments that can be customized to the user's specific learning needs or therapeutic goals [1].

Serious games represent the fastest-growing segment of the global gaming market [32]. Just as with games for entertainment, serious games typically include a game "storyline," which evolves across distinct physical contexts within the simulated game world [23]. Serious games also involve multiple nonplaying characters (NPCs) and different tasks that the player aims to achieve as she or he progresses through game levels. The skills to be learned and practiced by the player are embedded within the gameplay, and levels of the game provide progressively more challenging tasks. Depending on the type of training, coaching, or task to be learned, as well as on the age and abilities of players, the gameplay may focus on the "serious" task, or it may incorporate these tasks with segments of gameplay designed only for entertainment. The latter is more common for games aimed at children and younger users. Games provide an opportunity to create highly customized learning, training, and therapeutic environments and protocols.

Game developers can customize the storyline, gameplay levels, the NPCs, and the reward structure to provide an optimum training and learning experience for the user. As we noted previously in this chapter in regard to IVAs, the NPC appearance and behavior can be designed to match players' individual and cultural preferences as well as specific learning and training needs. Both of these technologies mentioned above take advantage of two innate human needs and capabilities: the desire and ability to "connect" (i.e., to attach) emotionally and the desire and need to "play."

One example of a serious game with embodied virtual agents (EVAs) is Ricky and the Spider (https://www.rickyandthespider.uzh.ch/en.html). Developed at the University of Zurich and intended for children ages six through 12 with OCD, the game integrates evidence-based CBT techniques. The game is intended to be played under the supervision of a human therapist and includes a psychoeducation component (i.e., information about OCD symptoms, techniques to reduce symptoms, and elements of controlled exposure therapy) and EVAs that model persons with OCD symptoms as well as provide therapeutic support. The agents help children to address their symptoms in a fun, interactive environment by leveraging their desire to game play.

In summary, IVAs are emerging as a useful technology to assist and augment tasks typically carried out by human care providers. Evidence of their effectiveness is also growing; however, reviews of studies testing clinical applications of IVAs have emphasized the limitations of the existing research and the need for trials that examine efficacy and safety [33–35]. Most published studies are quasi-experimental, involving the testing and evaluation of CAs by users and only a few randomized controlled trials (RCTs) of interventions delivered by IVAs have been reported in the literature (e.g., [36, 37]). As we noted earlier, further validation studies are needed to determine for whom and when these agents are most appropriate.

4.3 Ethical Issues

As can be expected, the introduction of advanced technologies such as IVAs also raises several ethical issues. These span the range from privacy concerns, through questions regarding client safety, to the much more complex matters of artificial empathy and artificial relationships. In this section, we discuss some of the categories of ethical issues that need to be addressed by both the clinical and the technical communities involved in developing and using IVA technologies in behavioral health.

4.3.1 User Safety

With increased use of intelligent autonomous or semi-autonomous systems, there is a potential risk of harm to people if a system does not adequately address situations when a user needs immediate crisis support or if other action related to safety is required. Consider the following scenarios where a person is seeking counseling from an online chatbot. A person seeking help for depression discloses they are experiencing suicidal thoughts and have a plan to end their life. A person seeking relationship counseling discloses that they would like to kill their spouse. A person completing a health assessment indicates that they are suddenly having radiating chest pains. What should the chatbot program be required to do?

Designers of autonomous and semi-autonomous systems that interact with people should consider giving agents the capability to detect these types of risks and take appropriate action. In some scenarios, the detection of threat could be automated by the system, which could then respond in an appropriate manner, for example, to immediately display available resources to the user (e.g., suicide prevention hotline) or to contact a specific person (relative, psychotherapist, or another provider). Procedures for keeping a human-in-the-loop, when feasible, represent another approach toward addressing this issue [6]. That is, when threat is detected, a human can review the information and then, if appropriate, contact the user and directly intervene or make an appropriate referral.

Presently, IVAs and online chatbots are generally considered to serve as "coaches" and not replacements for healthcare professionals. Therefore, many of the ethical and legal requirements that human healthcare professionals are expected to follow, such as duty-to-warn, may not seem to apply. However, in user safety scenarios such as those we describe in this section, and as these technologies become more accepted and commonplace, it is conceivable that the same ethics and legal requirements expected of human professionals will become increasingly important. It is, therefore, essential that developers and administrators of these systems provide adequate information to patients regarding the scope of use, risks, limits, and expectations.

4.3.2 Risks Associated with Overreliance on Technology

Another potential risk to users is overreliance on technology (CAs or IVAs) and the assumption the technology is adequately addressing their healthcare needs. While the use of CAs for counseling may be appropriate for some cases, their use may not be appropriate for more severe or comorbid health conditions when the user should seek the advice of a professional healthcare provider. Luxton [38] has highlighted the problem caused when virtual care systems are inadequately controlled based on the scope of their tested capabilities. Services accessible on the Internet or on mobile apps, for example, that claim to provide particular clinical services or benefits may not be adequate or appropriate to do the services they are purported to provide.

Here, again, the system developers and administrators (i.e., the company or government organization making the service available) should include an explicit warning to the user, and outline circumstances under which the help of a healthcare professional should be sought, rather than reliance on the technology. Furthermore, this problem could be at least partially addressed by requiring system developers and administrators to show users the "credentials" of the virtual care providers. These credentials may include information about how the technology was developed based on evidence-based clinical practices, how the system was tested, and how it is updated and monitored. Furthermore, end-users should also be provided with the means to voice any concerns regarding safety or quality of services provided by virtual care providers and to have those issues appropriately reviewed and resolved [38, 39].

4.3.3 Risks to Privacy

Just as with any other technology that collects and processes health data, the use of CAs has the potential to increase the risk for misuse of user information and data breaches. Given that our thoughts and emotions are the most personal and private aspects of our lives, the development and use of technologies that sense, infer, or track our emotions, therefore, create significant ethical challenges [1, 38]. User modeling and especially affective user modeling presents an ethical challenge because the models may contain the most guarded personal information about the users: the emotions they feel, including "undesirable" emotions and the events that trigger those emotions, including triggers that may be considered inappropriate by others [1]. This is especially the case in any behavioral health applications where the users may be addressing a particularly painful experience, or reveal issues, thoughts, and emotions that could have adverse repercussions if they were made public or disclosed to other parties (e.g., employers, insurance companies). Luxton [38] also emphasizes that the risks to privacy resulting from the use of virtual care providers extends to abuse by governments or other entities who wish to control individuals for political reasons or suppress dissent. Hackers also present a risk and could potentially exploit sensitive data of users.

Consider a situation where a member of a company's executive team is seeking treatment for substance abuse via an IVA due to privacy concerns and the possibility that sensitive information, if compromised, could harm her career and reputation, or potentially information could be used for malicious purposes by a competitor. Again, appropriate disclosures and descriptions of privacy measures should be provided to the users, and, of course, the software should employ state-of-the-art methods for ensuring privacy, including encryption and local storage of sensitive data. In situations where it might be necessary to disclose personal data, such as for insurance reimbursement or to share records with other medical providers, the user should be provided with complete description of the protocols governing such disclosures and have the choice of opting out.

4.3.4 Deception

Deception occurs when the users may not know whether they are interacting with a human who's controlling the IVA, or whether the IVA is autonomous [38, 39]. Miller [40] and Riek and Watson [41] have described this as "Turing deceptions," whereby a person is unable to determine whether they are interacting with a machine (i.e., software program) or not. This creates an ethical problem, especially when working with intellectually challenged and psychologically vulnerable persons [8] and represents an active area of research (e.g., [42]). For example, some types of patients, such as those with dementia or delusional or psychotic psychopathologies may be especially at risk of harm if they experience challenges discerning a machine from a real person [24]. In all healthcare situations, disclosure regarding the control of the machine's behavior (autonomous or human-controlled) and informed consent regarding the service to be provided are prudent ways to address this issue.

4.3.5 Artificial Relationships

Modern IVAs can have the appearance of empathetic concern that is highly interactive and human-like. Even when a patient is consciously aware that a care provider is an artificial agent (i.e., CA or IVA) running as a simulation on a computer, the patient can be expected to experience intense emotions during the interaction and may develop feelings about, and attachments to, the simulated agent. Even when disclosure is made that the system is "just a machine," some patients may believe that the machine is "alive." Relatedly, there is the issue of when the artificial relationship with an agent should end. We demonstrate these issues with the following example vignette.

Mrs. T is an 85-year-old widow, with mild to moderate vascular dementia, who recently moved to a nursing home from her residence. She reluctantly agreed to move at the strong urging of her three adult children, who all live more than 1500 miles away and, therefore, cannot visit frequently, and who were worried about her safety.

Three weeks after moving to the facility, Mrs. T developed symptoms of depression and was diagnosed with a depressive disorder. Although a low dose of an SSRI medication alleviated the symptoms, Mrs. T continued to experience intense periods of sadness and continued to isolate in her room and not take part in the social activities offered by the nursing home.

The nursing home facility where Mrs. T now lives is one of the sites for a National Institutes of Health-funded study exploring the effectiveness of IVAs as social companions for the elderly. Mrs. T's two sons and daughter, as well as her long-time primary care physician, were consulted regarding her possible participation in the study, and all agreed that this would be a good idea. Mrs. T was informed of the possibility to participate in the 6-month study, and also agreed, albeit somewhat reluctantly. Informed consent process was completed by study research assistance. The effects of her dementia resulted in some short-term memory loss, the condition was not so far advanced as to affect her judgment, and she was considered fully capable of understanding the informed consent form.

Informed consent was administered by one of the study research assistants, which included a detailed description of the study, and informed Mrs. T that the verbal interaction between her and the IVA companion would be recorded and analyzed by the study staff and that her mood data would be collected. The mood data collection consisted of a daily mood tracking and a weekly administration of a modified version of the PHQ-9.

The study involved three 1-h sessions with a social, relational agent acting as a companion. At study onset, Mrs. T acknowledged that she was fully aware that the IVAs are computer-generated, have no emotions or capacity for such, even though the virtual companion can display facial expressions corresponding to the basic emotions. It is also able to conduct a limited dialog in natural language, which revolves around simple questions regarding Mrs. T's well-being. The IVA's functionality includes the ability to convey empathic expressions (verbal and nonverbal) to Mrs. T when she expresses sadness or feelings of loneliness.

Three weeks following the enrollment in the study, Mrs. T's mood began to improve. She began to look forward to the visits with the IVA, and her enjoyment was evident by her observable behavior (facial expressions and verbalizations). She also became increasingly relaxed with the IVA, and began to share more of her thoughts and feelings with it.

Five weeks after enrollment, Mrs. T began to ask for more extended and more frequent visits. She also asked whether the IVA could remain with her. After the researchers told her that this was not possible due to study protocol, Mrs. T expressed disappointment, sadness, and some frustration. The study staff then began to observe that she began to withdraw and expressed angry feelings to IVA.

The study staff analyzing the verbal data also noted that she was telling the virtual companion that she was angry at her children for putting her in the nursing home. Both the frequency and the intensity of these feelings began to increase rapidly after Mrs. T was told that she could not spend more time with the virtual agent.

Although constructed for illustrative purposes, the situation represented by this vignette is well within the realm of possibilities given the rapid expansion of these technologies into behavioral healthcare. As can be seen from the example, numerous complex and as yet unresolved ethical issues arise as we begin to venture into the realm of empathic agents and artificial relationships with such agents. Significant research and changes in the administration of healthcare services will need to take place to begin to adequately address these issues.

4.3.6 Bias in Design

As we noted earlier, developers can design IVAs to have changeable physical appearance and mannerisms, speech dialects, use of local colloquialisms, and other characteristics that make them consistent with a user's cultural background and preferences. The goal of these variations in the agent's appearance and behavior is to help establish rapport with users, enhance engagement, and thereby help improve adherence to treatment and health outcomes [38]. The design of the IVA appearance and behavior is, however, susceptible to biases. For example, an IVA design may be based on a particular race or ethnicity that is not representative of the users who are intended to interact with it. Thus, designers must carefully consider the users' preferences and the context of use during the development and deployment process.

IVAs also employ a knowledge engine with information that it uses in conversation to enable it to recognize and reason with information. This knowledge base, and how an IVA may produce new knowledge through machine learning, is susceptible to bias that may not be as obvious as overt physical biases. *Algorithmic bias* occurs when a computer program makes systematic and repeatable errors that lead to unfair outcomes, such as when privileging one group of users over others [43]. The causes of algorithmic bias may include problems with missing data, small sample size and underestimation, persons (i.e., patients) not identified by algorithms, and misclassification and measurement error. Moreover, the values of the programmers and organizations collecting, choosing, or using data to train the algorithm may also introduce bias [43].

Take, for example, an IVA app intended by its developers to provide counseling for persons with depression across multiple countries around the world. Even when the well-intentioned programmers used translators to develop the IVA's natural language dialogue, they may not have built the underlying conversational and knowledge engine components with adequate attention to the cultural contexts and nuances applicable in the target areas of use. The developers may have also designed the system to learn the patterns and preferences of its users through interaction with them over time via machine learning. If particular behavioral characteristics of end users are not adequately factored into in the design of the system due to statistical underestimation, for example, the IVA system may adapt in a less than optimal way, leading to unsuitable counseling recommendations and low user satisfaction with the system. Thus, the deployment of this IVA app may result in inequitable and unfair outcomes for particular end user groups. It is therefore essential that designers consider and test for potential biases before deployment of systems, and monitor for them afterwards.

4.3.7 Black-Box Problem

The complexity of the AI algorithms that make CA and IVAs possible can make them seem like a mysterious black-boxes when the algorithms associated with their decision-making and behavior are not easily audited or understood by humans [6]. For example, the use of sophisticated, multi-layer neural networks and distributed knowledge bases can result in complexities that humans cannot easily visualize or grasp conceptually.

Liability risks may increase considerably with the use of highly autonomous CAs and IVAs because of the difficulty in predicting the actions of these systems across every situation. Consider a scenario where an IVA using highly complex data and neural networks tells a patient to use a particular medication. If the medication proves to cause harm to the person, and the person sues the company who developed the IVA, wouldn't you expect that the company be able to describe how the IVA system made its decision? If the developers cannot explain how the system works, then it can be said that it is operating as a black-box. Requirements for an audit trail to describe the decision process of autonomous systems has been proposed as one way to help address the "black-box" issue [38, 39].

4.3.8 Legal Responsibility and Liability

Consider again a scenario whereby a CA or IVA system fails to appropriately assess for or alert when a user discloses that they are suicidal. If a person dies, could the family sue the company who provided the service for failing to appropriately detect or respond to the risk? Should responsibility also be with the developers of the technology, not just the company using the service? What about scenarios where autonomous agents are making independent decisions about the care of patients? Should the developer, the administrator, or both be liable?

It is conceivable that the company or organization, such as a public healthcare provider, could be liable if harm comes to a patient while using the system. This could occur if the providers of the service failed to adequately disclose the risks and limitations involved with using the agent.

The issue of responsibility and liability becomes more complicated when the IVA is designed to be autonomous. A moral agency refers to when a person can discern right from wrong and can be held accountable for his or her actions. Sullins [44] proposes that intelligent machines (e.g., robots) are moral agents when there is a reasonable level of abstraction under which the machine has autonomous intentions and responsibilities. If this is the case, then the machine may also be seen as responsible at some level for the actions it makes or does not make. However, autonomous behavior may not be enough to hold an intelligent machine responsible for its actions. Another requirement is that the agent would have to have acted freely. This raises the question as to whether an agent that is built by and controlled by humans can have free will and act freely. Moreover, unlike human care providers, IVAs cannot accept appropriate responsibility for their actions nor do they have the moral consequences that humans do [38].

4.4 Recommendations

One way to begin to help address the current and emerging ethical challenges associated with the use of IVAs is to revise or develop new professional ethics codes and practical guidelines [8, 38]. While several mental healthcare professional organizations have included provisions regarding the use of current technology (see Luxton et al. [39] for review), most existing professional ethics codes and practice guidelines do not yet address the use of technologies that stand-in for human professionals.

In regard to both addressing user safety and responsibility and liability associated with CAs and IVAs, we recommend the following:

1. The agent or the responsible party must discuss with the user any potential risks to their emotional or physical well-being that might result from the interaction with the agent. This could include, for example, relying on the agent more than on human caregiver and becoming more attached to the agent than to family members.
2. The human user always has the right to know if s/he is interacting with a human, an autonomous agent, or a human controlling a virtual agent or chatbot.
3. The human user has the right to terminate the interaction with the virtual agent at any point, for any reason, and not to require any explanation.
4. The agent should be able to detect if the user is becoming too emotionally dependent or attached, broach this topic with the user, and discuss possible risks with the user.

While these recommendations are not inclusive of all ethical issues or address every possible application of CAs in behavioral healthcare, they do serve as a starting point. We suggest review of the recommendations of Luxton [38] regarding the design and use of virtual care providers in healthcare.

4.5 Summary and Conclusions

Now that we have entered into the age of virtual care providers, we must understand the benefits, risks, and ethical and legal requirements regarding their use. Developers of these systems must be aware of the legal and ethical issues, and integrate capabilities into the design and deployment of these systems that consider these. Healthcare providers who use these systems as part of their care should seek training regarding the benefits and the appropriate use of IVAs. The public also needs to be aware of the benefits and limitations of these systems. These requirements will necessitate the development of new educational curricula and may require specialized types of certification to ensure that the users' rights are adequately protected.

The use of CAs also has the potential for substantial cost savings for both healthcare providers and persons seeking care. CAs can be replicated and scaled to meet the growing needs for services, and unlike human care providers, CAs do not need years of training, or salaries that human providers require [38]. Luxton [21] notes that AI technology, including virtual care providers, could lead to significant reductions in the long-term costs from untreated behavioral and mental health conditions as well as improved productivity resulting from a healthier population. Luxton cautions, though, that healthcare insurance companies or governments could require the use of AI care providers without allowing consumers the choice to seek services from human care providers. Consequently, Luxton speculates that the helping professions could lose something inherent to the helping professions: *human-to-human* expression of empathy, care, and compassion in exchange for automated systems with simulated expressions of empathy.

We must emphasize that the discussion in this chapter in no way intends to imply that IVAs or serious games should replace human mental health providers or traditional face-to-face therapy. These technologies cannot function at the level of an experienced, empathic human therapist. Rather, they have a unique role in the delivery of behavioral healthcare, both supportive of and distinct from the functions of human healthcare providers, including enhancing dissemination of evidence-based treatment, making treatment more accessible, supporting treatment between sessions with a human provider (facilitating homework and skills practice), and adapting to individual needs and cultural preferences. We are hopeful that this technology will continue to serve those in need of services and improve the well-being of people around the world.

References

1. Hudlicka E. Virtual affective agents and therapeutic games. In: Luxton DD, editor. Artificial intelligence in behavioral and mental healthcare; 2015.
2. Prendinger H, Ishizuka M, editors. Life-like characters: tools, affective functions, and applications (cognitive technologies). Berlin Heidelberg: Springer; 2004.
3. Becker-Asano C. WASABI: affect simulation for agents with believable interactivity. IOS Press; 2008.
4. de Rosis F, Pelachaud C, Poggi I, Carofiglio V, de Carolis B. From Greta's mind to her face: modelling the dynamics of affective states in a conversational embodied agent. Int J Hum Comput Stud. 2003;59:81–118.
5. Lim MY, Aylett R. Feel the difference: a guide with attitude!. In: Proceedings of IVA. 2007. p. 317–30.
6. Luxton DD. Ethical challenges of conversational agents in global public health. Bullein of the World Health Organization. 2020;98:285–287. https://www.who.int/bulletin/volumes/98/4/19-237636.pdf
7. Luxton DD, June JD, Sano A, Bickmore T. Intelligent mobile, wearable, and ambient technologies in behavioral health care. In: Luxton DD, editor. Artificial intelligence in behavioral and mental health care. San Diego: Elsevier Academic Press; 2015.
8. Luxton DD, Riek L. Handbook of rehabilitation psychology. In: Brenner L, Reid-Arndt BS, Elliott, et al., editors. Artificial intelligence and robotics for rehabilitation. 3rd ed. Washington DC: American Psychological Association Books; 2019.
9. Aylett RS. Agents and affect: why embodied agents need affective systems. In: 3rd Hellenic conference on AI. Samos, Greece: Springer; 2004.
10. Becker-Asano C, Wachsmuth I. Affect simulation with primary and secondary emotions. In: Proceedings of IVA. 2008. p. 15–28.
11. Becker C, Kopp S, Wachsmuth I. Why emotions should be integrated into conversational agents. In: Nishida T, editor. Conversational informatics: an engineering approach. Wiley; 2007. p. 49–68.
12. Castellano G, et al. Long-term affect sensitive and socially interactive companions. In: 4th Intl. Workshop on human computer conversation, Bellagio, Italy. 2008.
13. Poggi I, Pelachaud C, de Rosis F, Carofiglio V, De Carolis B. A believable embodied conversational agent. In: Stock O, Zancanaro M, editors. Multimodal intelligent information presentation. Text, speech and language technology, vol. 27. Dordrecht: Springer; 2005.

14. Hudlicka E, Lisetti C, Hodge D, Paiva A, Rizzo A, Wagner E. Artificial agents for psychotherapy. In: Proceedings of the AAAI spring symposium on emotion, personality and social behavior. Menlo Park, CA: AAAI; 2008. TR SS-08-04. p. 60–4.
15. Hudlicka E, Payr S, Ventura R, Becker-Asano C, Fischer K, Leite I, Paiva A, von Scheve C. Social interaction with robots and agents: where do we stand, where do we go? In: Proceedings of the third international conference on affective computing and intelligent interaction, Amsterdam, Holland. 2009.
16. Paiva A, Leite I, Boukricha H, Wachsmuth I. Empathy in virtual agents and robots: a survey. ACM Trans Interact Intell Syst. 2017;7:3. https://doi.org/10.1145/2912150
17. Paiva A, Dias J, Sobral D, Aylett R, Sobreperez P, Woods S, Zoll C, Hall L. Caring for agents and agents that care: building empathic relations with synthetic agents. In: Intl. joint conference on autonomous agents and multiagent systems, International Joint NY, NY. 2004.
18. Bickmore T, Picard RW. Establishing and maintaining long-term human-computer relationships. ACM Trans Comput Hum Interact (TOCHI). 2005;12(2):293–327.
19. Zeng Z, Pantic M, Roisman GI, Huang TS. A survey of affect recognition methods: audio, visual, and spontaneous expressions. Patt Anal Mach Intell IEEE Trans. 2009;31(1):3958.
20. Gunes H, Schuller B. Automatic analysis of social emotions, invited chapter for social signal processing book. In: Vinciarelli A, Pantic M, Burgoon J, Magnenat-Thalmann N, editors. Cambridge University Press; 2017. p. 213–24.
21. Luxton DD. Artificial intelligence in psychological practice: current and future applications and implications. Profess Psychol Res Pract. 2014;45(5):332–9. https://doi.org/10.1037/a0034559.
22. Bickmore T. Relational agents: effecting change through human-computer relationships. Cambridge, MA: MIT; 2003.
23. Hudlicka E. Affective gaming in education, training and therapy: motivation, requirements, techniques. In: Felicia P, editor. Handbook of research on improving learning and motivation through educational games: multidisciplinary approaches. IGI Global; 2011.
24. Luxton DD, editor. Artificial intelligence in behavioral and mental health care. San Diego: Elsevier/Academic Press; 2015.
25. Schulman D, Bickmore T, Sidner CL. An intelligent conversational agent for promoting long-term health behavior change using motivational interviewing. AI and health communication—papers from the AAAI 2011 spring symposium (SS-11-01).
26. Bailenson JN, Yee N. Digital chameleons: automatic assimilation of nonverbal gestures in immersive virtual environments. Psychol Sci. 2005;16:814–9. https://doi.org/10.1111/j.1467-9280.2005.01619.x.
27. Gratch J, Wang N, Gerten J, Fast E, Duffy R. Creating rapport with virtual agents. In: Pelachaud C, Martin JC, André E, Chollett G, Karpouzis K, Pelé D, editors. Intelligent virtual agents. Berlin/Heidelberg: Springer; 2007. p. 125–38. https://doi.org/10.1007/978-3-540-74997-4_12.
28. Kang SH, Gratch J, Wang N, Watt JH. Does the contingency of agents' nonverbal feedback affect users' social anxiety? In: Proceedings of the 7th international joint conference on autonomous agents and multiagent systems, vol. 1. Liverpool: International Foundation for Autonomous Agents and Multiagent Systems; 2008. p. 120–7.
29. Lucas GM, Gratch J, King A, Morency LP. It's only a computer: virtual humans increase willingness to disclose. Comput Hum Behav. 2014;37:94–100. https://doi.org/10.1016/j.chb.2014.04.043
30. Bickmore TW, Pfeifer LM, Jack BW. Taking the time to care: empowering low health literacy hospital patients with virtual nurse agents. In: CHI '09 proceedings of the SIGCHI conference on human factors in computing systems.
31. McCue K, Shamekhi A, Bickmore T, Crooks D, Barnett K, Haas N, et al. A feasibility study to introduce an embodied conversational agent (ECA) on a tablet computer into a group medical visit. Annual Meeting of the American Public Health Association. 2015. Available at: https://apha.confex.com/apha/143am/webprogram/Paper329324.html.
32. Adkins SA. The 2017–2022 global game-based learning market serious game revenues spike to $8.1 billion by 2022. 2017. Available at: http://seriousplayconf.com/wp-content/

uploads/2017/07/Metaari_2017-2022_Global_Game-based_Learning_Market_Executive_Overview.pdf.
33. Laranjo L, Dunn AG, Tong HL, Kocaballi AB, Chen J, Bashir R, Surian D, Gallego B, Magrabi F, Lau AYS, Coiera E. Conversational agents in healthcare: a systematic review. J Am Med Inform Assoc. 2018;25(9):1248–58. https://doi.org/10.1093/jamia/ocy072.
34. Vaidyam AN, Wisniewski H, Halamka JD, Kashavan MS, Torous JB. Chatbots and conversational agents in mental health: a review of the psychiatric landscape. Can J Psychiatr. 2019;64(7):456–64. https://doi.org/10.1177/0706743719828977.
35. Gaffney H, Mansell W, Tai S. Conversational agents in the treatment of mental health problems: mixed-method systematic review. JMIR Ment Health. 2019;6(10):e14166. https://doi.org/10.2196/14166.
36. Fitzpatrick KK, Darcy A, Vierhile M. Delivering cognitive behavior therapy to young adults with symptoms of depression and anxiety using a fully automated conversational agent (Woebot): a randomized controlled trial. JMIR Ment Health. 2017;4(2):e19.
37. Craig TK, Rus-Calafell M, Ward T, Leff JP, Huckvale M, Howarth E, et al. AVATAR therapy for auditory verbal hallucinations in people with psychosis: a single-blind, randomised controlled trial. Lancet Psychiatry. 2018;5(1):31–40. https://doi.org/10.1016/S2215-0366(17)30427-3.
38. Luxton DD. Recommendations for the ethical use and design of artificial intelligent care providers. Artif Intell Med. 2014;62(1):1–10. https://doi.org/10.1016/j.artmed.2014.06.004.
39. Luxton DD, Anderson SL, Anderson M. Ethical issues and artificial intelligence technologies in behavioral and mental health care. In: Luxton DD, editor. Artificial intelligence in behavioral and mental health care. San Diego: Elsevier Academic Press; 2015.
40. Miller KW. It's not nice to fool humans. IT Profess. 2010;12(1).
41. Riek LD, Watson RNW. The age of avatar realism. IEEE Robot Automat. 2010;17(4):37–42.
42. Fiske A, Henningsen P, Buyx A. Your robot therapist will see you now: ethical implications of embodied artificial intelligence in psychiatry, psychology, and psychotherapy. J Med Internet Res. 2019;21(5):e13216. https://doi.org/10.2196/13216.
43. Gianfrancesco MA, Tamang S, Yazdany J, Schmajuk G. Potential biases in machine learning algorithms using electronic health record data. JAMA Intern Med. 2018;178(11):1544–7. https://doi.org/10.1001/jamainternmed.2018.3763.
44. Sullins JP. When is a robot a moral agent? In: Anderson M, Anderson SL, editors. Machine ethics. New York, NY: Cambridge University Press; 2011.

Machine Learning in Stroke Medicine: Opportunities and Challenges for Risk Prediction and Prevention

5

Julia Amann

5.1 Introduction

> "The essence of practicing medicine has been obtaining as much data about the patient's health or disease as possible and making decisions based on that. Physicians have had to rely on their experience, judgement, and problem-solving skills while using rudimentary tools and limited resources." [1]

Precision medicine aims to individualize prevention, diagnostics, and therapeutics by understanding differences in individuals' genetics, lifestyle, and environment [2]. Over the past years, we have been witnessing an unprecedented push toward a more data-driven approach in healthcare that promises to take precision medicine to the next level, in part through artificial intelligence (AI). Simply put, AI can be understood as a set of sophisticated computational methods that seek to mimic human cognitive functions, including visual perception, speech recognition, and decision-making [3, 4]. AI uses certain machine learning (ML) algorithms to "learn" features from large datasets [3] and recognize patterns that are often invisible to the human eye [5–7]. Capitalizing on the availability of big data and ever-increasing computational power and storage capacities [1, 8], these novel tools seek to improve population health and well-being and to reduce healthcare costs.

A surge in scientific publications documents the potential to harness artificial intelligence in healthcare to prevent, diagnose, and treat diseases [9]. One of the pressing disease areas in focus for AI researchers is stroke, a leading cause of disability and mortality worldwide [3, 8]. Researchers aim to develop applications to optimize stroke diagnosis, treatment, and rehabilitation [10–12], and they also use AI to better understand risk. Several well-established risk prediction models have

J. Amann (✉)
Health Ethics and Policy Lab, Department of Health Sciences and Technology, ETH Zurich, Zurich, Switzerland
e-mail: julia.amann@hest.ethz.ch

F. Jotterand, M. Ienca (eds.), *Artificial Intelligence in Brain and Mental Health: Philosophical, Ethical & Policy Issues*, Advances in Neuroethics, https://doi.org/10.1007/978-3-030-74188-4_5

been developed as tools for stroke prevention [13]. Prevention plays an instrumental role in reducing the global burden of stroke [14], and the strategic adoption and development of AI-driven prediction tools can contribute substantially to this mission [1, 13]. These new tools open welcome opportunities and introduce new questions for us, of course. We find ourselves only at the beginning of this exciting journey that will without a doubt confront us with novel ethical, societal, and regulatory challenges.

This chapter surveys the global burden of stroke and describes current practices for reducing stroke incidence and stroke mortality rates. In particular, the chapter reviews how ML applications are applied to stroke risk prediction and prevention and identifies important technological and methodological challenges for using AI in these contexts. The chapter concludes by drawing the readers' attention to some of the questions and ethical challenges that arise as clinicians widely adopt ML-based applications in practice.

5.2 Burden of Stroke

Stroke is one of the leading causes of disability and mortality worldwide [14–17]. Even though a decrease in stroke mortality and incident rates was observed from 1990 to 2016, absolute numbers show an increase in stroke-related mortality and disability [15, 16]. The absolute number of people affected by stroke almost doubled during this time [16] with incidence rates in low- to middle-income countries exceeding those observed in high-income countries [18]. Researchers estimate that in 2016, there were over 80 million people affected by stroke, many of them younger than 70 years of age [15, 16]. In 2017, Europe counted 1.5 million stroke diagnoses and nine million stroke survivors, with 1.2 million experiencing severe limitations in their activities of daily living [19]. That same year, 0.4 million people died because of stroke [19]. The increase in absolute numbers is largely attributed to population aging and growth [20, 21]. Yet, a noteworthy increase was also recorded in stroke incidence rates in younger age groups (15- to 49-year olds) [16].

The global increase in stroke incidents poses major challenges for healthcare systems, and these challenges extend beyond a patient's hospital stay. Patients who survive a stroke long to return to normality [22]. However, following hospital discharge, stroke survivors and their families must cope with the aftermath of stroke. People who suffered a stroke often experience more or less severe physical, cognitive, and emotional deficits that may limit their ability to perform certain activities in daily life [23, 24]. As a result, they remain at least partially dependent on an informal caregiver, usually a family member or partner [25]. Stroke survivors and informal caregivers commonly report physical, emotional, social, and financial challenges and concerns [26, 27]. They also face service deficiencies in health and social care, limited options for service offers outside of healthcare, and a paucity of options for continuity of care. All of this lays an additional burden on those affected by stroke, leaving them frustrated and under emotional strain [27].

In addition to the impact of stroke on individuals, societies are faced with the economic burden of stroke [28]. Healthcare utilization, informal care provision, and the loss of productivity in the workforce contribute to these rising costs [21, 29, 30]. A recent study analyzing stroke-related costs for 32 European countries estimates that total costs added up to €60 billion in 2017. This includes €27 billion (45%) incurred by healthcare systems, €5 billion (8%) incurred by social care systems, an estimated €16 billion (27%) for informal care costs, and €13 billion (20%) owed to lost productivity due to early death or absence from work [19]. While lower total costs to the healthcare system have been reported for the United States for 2014/2015 [31], per capita healthcare-related spending on stroke was higher in the USA compared to Europe [19]. Similar costs were reported for stroke-related healthcare costs per stroke survivor living in the USA and Europe [19].

5.3 Stroke Prevention: A Public Health Priority

As the global stroke burden increases, researchers and policymakers call for more efficient stroke prevention and management strategies and improved access to stroke services [16, 17, 32, 33]. In 2006, the World Health Organization (WHO) highlighted neurological disorders, including stroke, as a public health priority [34]. With its Global Status Report on Noncommunicable Diseases 2014, WHO aimed to unite and support nations in the fight against stroke and vascular diseases [32, 33].

There is common agreement that prevention is one, if not the most, promising strategy to reduce the burden of stroke [16, 35–37]. It is well established that there are non-modifiable (e.g., sex, gender, genetics) and modifiable (e.g., smoking cessation, physical inactivity) risk factors for stroke [38, 39]. Modifiable risk factors are the obvious targets of stroke prevention efforts. In an international case-control study, researchers found that ten risk factors (history of hypertension, current smoking, waist-to-hip ratio, diet risk score, regular physical activity, diabetes mellitus, binge alcohol consumption, psychosocial stress and depression, cardiac diseases, and ratio of apolipoproteins B to A1) were associated to 90% of the risk of stroke [39]. The authors concluded that lifestyle interventions targeting blood pressure reduction, smoking cessation, and the promotion of physical activity and a healthy diet could help to significantly reduce the burden of stroke.

There are two main approaches in stroke prevention [40]: population-wide prevention strategies and prevention strategies that target high-risk individuals. Population-wide strategies aim at modifying behavioral and lifestyle risk factors in the entire population to promote health maintenance [41]. In doing so, they can also contribute to preventing other diseases and chronic conditions (e.g., hypertension and diabetes mellitus) that constitute known stroke risk factors [14]. Recent advances in our ability to accurately assess individual risk for cardiovascular diseases have motivated some countries to prioritize risk-based screening approaches to identify individuals at risk [42, 43].

Despite a formal distinction between these two approaches, it is important to note that stroke risk is a continuum with no determined threshold at which certain interventions are automatically indicated. Therefore, it may not be appropriate to categorize individuals into low-, moderate-, and high-risk groups when communicating absolute cardiovascular risk [44]. To effectively reduce stroke incidence and mortality rates, efforts must be undertaken to educate the general population about known behavioral risk factors [14, 43]. In addition, inexpensive screening strategies should be adopted to assist clinicians in identifying and protecting high-risk individuals [14, 43].

5.4 The Advent of Data-Driven Risk Prediction Models

Early prediction of stroke risk is the cornerstone of stroke prevention [45]. Identifying individuals who could benefit most from specific therapeutics or interventions helps them get the care they need and simultaneously helps avoid unnecessary treatments for others [10, 46, 47]. To date, several well-established statistically derived risk prediction models have been developed to provide long-term risk prediction [42, 45, 48, 49]. Clinicians commonly rely on these models to assess long-term risk, because the models provide parameters that are easy to interpret, such as odds ratios, relative risks, and hazard ratios [50]. However, these traditional models are subject to several limitations. They can, for example, only include a small number of risk factors (predictors) and generally do not include image-based morphological characteristics [13, 50, 51] nor behavioral risk factors (except smoking) or independent genetic factors [43]. Moreover, traditional approaches rely on certain assumptions of linearity, thus forcing models to behave in a certain way [51]. Often, traditional models are not generalizable across different populations due to the specific characteristics of the cohorts they were derived from [13]. This may lead clinicians to over- or underestimate risk for their patients [52].

Researchers are now trying to use ML in cardiovascular diseases and stroke risk assessment to overcome some of the challenges associated with traditional risk prediction models. ML methods use computational algorithms to relate all or some predictor variables of a given set to an outcome variable [50]. Classification and regression are the two primary tasks performed by ML-based algorithms [13]. Put simply, classification tasks categorize input data into predefined labels or outcomes (e.g., event or no event), whereas regression tasks predict some real-valued output (e.g., real-valued percentage risk between 0% and 100%). Despite various commonalities, ML differs from traditional statistical approaches in some aspects [53–55]. Contrary to classical statistics, ML is a data-driven approach that does not rely on a predefined model and assumption of data normality [53, 56]. Moreover, unlike traditional statistics which are focused on the "typical patient," ML is capable of making inferences at the individual level, taking into account individual differences in the data [53]. ML is also inherently a multivariate approach that can be used to analyze complex and heterogeneous kinds of data and incorporate them into risk prediction models, making it a promising solution for stroke risk prediction [53, 54, 57].

Studies investigating the use of these techniques in cardiovascular diseases and stroke prediction indicate that ML-based approaches can boost prediction accuracy. A recently published review found that the most common ML-based algorithms used in cardiovascular risk assessment are support vector machines, artificial neural networks, linear and logistic regression, and tree-based algorithms, such as random forests and gradient tree boosting [13]. In their review, Jamthikar et al. further showed that ML-based algorithms performed better compared to traditional regression-based methods for risk assessment, and that including both image-based features and conventional cardiovascular risk factors drives prediction accuracy. Indeed, imaging plays a pivotal role in cardiovascular and stroke risk detection. Ultrasound, in particular carotid ultrasound screening, can also easily be performed in routine clinical practice—unlike other non-invasive techniques, such as computed tomography or magnetic resonance imaging [47]—making ultrasound an invaluable tool for stroke prevention. In line with these findings, Ambale-Venkatesh et al. [58] emphasized the importance of subclinical disease markers obtained from imaging, electrocardiography, and blood tests. The authors found that ML in conjunction with deep phenotyping (i.e., multiple evaluations of different aspects of a specific disease process) enhanced prediction accuracy of cardiovascular events compared to traditional risk scores.

Several other studies provide similar evidence. In a prospective cohort study using routine clinical data, for example, researchers compared four machine-learning algorithms (random forest, logistic regression, gradient boosting machines, neural networks) to an established algorithm (American College of Cardiology guidelines) for first cardiovascular event prediction over 10 years [46]. Their findings show that ML techniques outperformed the established algorithm, leading to a significantly more accurate risk prediction. Similarly, a team of researchers demonstrated that their hybrid ML approach to stroke prediction significantly reduced the false-negative rate in comparison to conventional approaches, while the overall error increased only slightly [59]. In addition to increasing prediction accuracy, authors also recognize the potential of ML-based approaches to help identify new potential risk factors and to generate a better understanding of the role of novel biomarkers [59, 60].

5.5 From Data-Driven Risk Prediction to Stroke Prevention

Accurate risk prediction allows clinicians and patients to act. Enabled by advances in AI technologies that can analyze vast volumes of health data in an efficient and accurate manner [4], precision medicine aims to provide treatment and prevention tailored to individuals' variability in genetics, environment, and lifestyle [1]. At present, doctors recommend lifestyle changes to their patients, advising them to change known, modifiable risk factors to prevent stroke. Yet, their advice often goes unheeded. We should eat healthy, refrain from smoking and eschew excessive alcohol consumption, exercise regularly, stay hydrated, and the list goes on and on. To adhere to all these health-promoting recommendations in a world full of competing priorities, temptation, and imposed restrictions (e.g., financial constraints, poor access) may be too much to ask and simply not a realistic goal for many people. Earlier work has shown that there are incongruities between what people know they should do and their actual health behavior. So even though interventions (e.g., public health campaigns) may help to improve people's knowledge, these interventions may ultimately fail to induce, and more importantly, sustain behavior change—a phenomenon commonly referred to as the knowledge-behavior gap [61, 62].

Precision medicine is a promising approach to bridge this gap. It enables physicians and researchers to predict more accurately which prevention strategies will be most effective for which groups of people [1]. Understanding their natural predisposition to stroke may, in turn, motivate individuals to take on a more active role in their own health to reduce their individual stroke risk [14, 63]. In this context, the potential of mobile monitoring devices with real-time feedback systems has been highlighted as a tool for stroke prevention [10, 60, 64–67]. However, despite the promise these novel technologies hold for enabling personalized risk assessment and promoting stroke prevention, achieving stroke prevention via these means will largely depend on patients' acceptance and uptake of the technology. Tran et al. investigated chronic patients' perceptions of wearable biometric monitoring devices and AI systems that enable remote measurement and analysis of patient data in real-time [68]. In addition to capturing the perceived benefits and dangers of using these new technologies, the authors also assessed patients' readiness for using them. Their findings indicated that only half of the patients who participated in the study viewed digital tools and AI in healthcare as an opportunity, while 11% even considered them a danger, fearing that these will lead to the replacement of humans. In light of these findings, it is not surprising that 35% of patients indicated that they would refuse to integrate such devices into their care. More research is needed to better understand individuals' underlying motivations and fears that influence their attitudes toward the use of mobile monitoring devices and AI in healthcare. It is currently also unclear how well these new tools will be received by healthcare professionals. So, while AI-powered technologies are evolving rapidly, providing unprecedented opportunities for precision medicine in stroke prevention, the integration of these technologies into clinical practice raises several questions.

A project that will shed light on some of these questions is PRECISE4Q, a project funded under the European Union's Horizon 2020 Research and Innovation Program [69–71]. PRECISE4Q aims to identify and quantify risk factors and individual risk factor patterns. To do so, it combines heterogeneous data from a variety of sources, including large retrospective longitudinal stroke registry data, biobank data, and insurance data. What distinguishes PRECISE4Q from many other efforts in the field is its hybrid modeling approach, which combines ML methods and theory-driven (mechanistic modeling) approaches to risk prediction. Within the course of the project, a Digital Stroke Patient Platform will be established to collect and integrate large-scale data sets. This platform will also feature novel hybrid model architectures, structured prediction models, complex deep learning and gradient boosting models, as well as Clinical Decision Support Systems (CDSS) for stroke risk assessment, treatment outcomes, rehabilitation programs, and a socio-economic planning tool. A thorough validation of the models is planned with clinical data generated by prospective clinical studies and retrospective analyses of health registries, cohort studies, health insurance data, and electronic health records. The CDSS envisioned by PRECISE4Q will allow clinicians to simulate how an individual's stroke risk will evolve and change under different circumstances over time. In other words, clinicians will be able to simulate how different risk factors (e.g., smoking) will contribute to disease occurrence and how the individual will respond to different possible interventions (e.g., lifestyle intervention, medication). This will assist them in providing individuals with tailored recommendations based on their natural predisposition. For individuals, this means that they will learn not only their individual stroke risk but also what they can do to reduce this risk.

Another promising avenue for future research is the use of natural language processing to automatically extract information on lifestyle modification assessment and/or advice in clinical practice from electronic health records [72–74]. Such analyses can provide an objective evaluation of current clinical practice and improve our understanding of the timing of lifestyle modification and patient, clinic, and provider characteristics that are associated with or predictive of lifestyle modification documentation [73]. Understanding how and when clinicians assess lifestyle modification and provide advice to patients holds important implications for the development of prevention strategies. These insights can inform the improvement of care delivery and documentation in practice. Combining tools aimed at understanding current clinical practice with sophisticated risk prediction models, such as the ones described earlier, constitutes an opportunity to deepen our understanding of stroke prevention.

5.6 Technological, Methodological, and Ethical Challenges

Machine learning holds great promise for stroke prevention, yet it is also subject to some challenges and limitations. There are three common areas of challenges that clinicians and researchers should be mindful of as they seek to maximize the advantages of ML in stroke prevention, and in healthcare more generally: (1) challenges in data sourcing; (2) challenges in application development; (3) challenges in deployment in clinical practice [75]. Given that patients' health and well-being are at stake, it is of critical importance to investigate the technological and methodological challenges that arise at each stage and to consider their potential real-life consequences. It is also important to note that challenges occurring at one stage may have consequences for the subsequent stages. Challenges and limitations at the stage of data sourcing, for example, inevitably affect application development and deployment in clinical practice.

5.6.1 Data Sourcing

High-quality big data is key to accurate predictions. To develop ML systems that can be deployed in clinical practice, a continuous supply of large datasets is needed initially to train, validate, and improve algorithms [3, 76]. Yet, inadequate access to well-established patient and population-based datasets constitutes a major challenge for many ML-based data scientists and developers [13]. These professionals lack access to data partly because effective data sharing is currently not sufficiently incentivized by the medical scientific community [3, 10, 13, 77]. International research collaborations can help to mitigate this challenge. In the long run, effective data sharing strategies also need to be in place to facilitate and incentivize data sharing across institutions.

Another challenge to data sourcing relates to data protection and privacy regulations. Personal data are often subject to protective regulations that may impede data sharing. The European General Data Protection Regulation (GDPR), for example, entails a comprehensive set of regulations for the collection, storage, and use of personal information that will affect AI implementation in healthcare in several ways [76, 78]. The GDPR requires that individuals give explicit and informed consent before any organization collects personal data. It also grants individuals the right to track what data organizations are collecting about them, and it empowers them to direct an organization to discard their data. While these regulations rightly aim to protect patient privacy, they of course also impose certain restrictions on researchers and clinicians who seek to utilize these data. At present, the long-term impact of the GDPR and similar regulations on the implementation of AI in healthcare remains to be seen.

Closely related to data sourcing, data harmonization across different sources can also be quite problematic for data scientists. Given that very few studies provide comprehensive datasets for large numbers of participants, collaborative efforts are currently underway in the scientific community to harmonize and synthesize heterogeneous data across studies [79]. However, data harmonization is a time-consuming task that demands significant technological and scientific investments [80, 81].

5.6.2 Application Development

As outlined in this chapter, there is substantial evidence to suggest that ML-based algorithms can provide robust and accurate models for cardiovascular and stroke risk assessment, and can often outperform traditional regression-based approaches. Yet, there are several potential challenges and pitfalls to be mindful of when it comes to developing apps based on these algorithms. One of the key challenges in application development is algorithmic bias, which leads to systemic and unfair discrimination against certain individuals or groups of individuals [82, 83]. Even if no discrimination is intended, we know that the way data is collected, selected, prepared, and used to train ML-based algorithms can introduce bias [82]. Datasets used to develop stroke risk prediction models may, for example, suffer from missing data, misclassification, and measurement error, which can lead researchers and clinicians to make inaccurate predictions for subgroups of patients [84]. In other words, bias can occur when data sources do not reflect the true epidemiology within a given demographic [75]. As an example, consider that cardiovascular disease is often underdiagnosed in women because their symptoms are described as atypical

[85, 86]. Using such data to train ML-based algorithms may further reinforce this trend.

It has also been shown that ML methods perform poorly on imbalanced datasets, as they will be biased towards the majority group [59, 87, 88]. In other words, insufficient training samples and imbalanced class distribution will limit predictive performance in cases of rare occurrences [89]. In the case of stroke risk prediction, this may, for instance, pose limitations when we aim to develop predictive models for younger populations since the vast majority of available records likely describe older age groups [89]. Even though several balancing techniques have been developed, it is still a challenge to detect and address this bias in ML models [88].

But what does persistent algorithmic bias mean in practice? Algorithmic bias can cause enormous harm and contribute to increasing existing health inequalities in the real world [83]. A prominent example is the case of racial bias in commercial algorithms used in the U.S. healthcare system. In their 2019 study, Obermeyer et al. [90] found evidence indicating that a widely used algorithm was significantly biased against black patients. Due to this racial bias, a significantly lower number of black patients were identified for extra care. The authors demonstrated that bias occurred because the algorithm predicted healthcare costs rather than illness, not accounting for the fact that unequal access to care means that healthcare spending is lower for black patients than for white patients. The study carried out by Obermeyer et al. [90] serves as a striking example of how ML-based algorithms can reinforce existing inequalities and cause harm. It also raises the question: how many biased algorithms are still out there operating day in, day out? Importantly, this kind of bias is by no means limited to the US or to US race demographics. Similar problems can just as well be embedded in European algorithms, hiding similar (or different) kinds of social disparity.

5.6.3 Deployment in Clinical Practice

Finally, the practical implementation of AI technologies in healthcare is not without its own challenges [76, 91]. Trust plays a fundamental role in the implementation process. To obtain acceptance, AI-powered tools must first gain healthcare providers' and patients' trust [92]. As a first important step to gaining trust, tools should comply with existing data protection requirements and be transparent as to how outcomes and recommendations are derived [75]. However, at present, many ML models are considered black boxes that do not explain how their predictions are derived in a way that humans can grasp [93]. Unlike well-established regression-based methods where a clear relationship can be observed between the input variables and the output variable, the internal workings of ML algorithms are not easy to interpret for most clinicians [10]. As a result, clinicians may be wary of ML-based algorithms and reluctant to adopt them in practice [13]. This may also have to do with the fact that clinicians owe their patients explanations as to how certain recommendations were derived. Patients may, in turn, be more likely to follow recommendations regarding stroke prevention if they receive a clear explanation of why certain prevention measures (e.g., exercise regime, medication) are preferable over others in their particular situation. Even though concepts like AI explainability, interpretability, and transparency have gained traction in the scientific community, there is a need for strengthening cooperation among medical practitioners and data scientists to tackle these issues in a collaborative manner [13].

There is also uncertainty regarding who can be held liable for adverse events that result from the use of ML-based algorithms. This uncertainty may, in turn, hamper trust and impede the adoption of these technologies in practice [75]. This point is also linked to clinical validation and efficacy. To foster trust in ML-based algorithms, data scientists and researchers have to show that their algorithms yield accurate predictions and that they can be integrated into clinical practice securely and efficiently for the benefit of patients [10]. In the case of stroke risk prediction and prevention, this means that novel ML-based approaches will have to compete against established models to win over clinicians' and patients' trust. Clinicians and patients, in turn, will have to exercise good judgment about what and whom to trust.

5.7 Conclusion

Novel ML-driven approaches to stroke risk prediction allow researchers to overcome some of the challenges frequently associated with traditional risk prediction models. Capitalizing on the advantages of ML, physicians, and researchers will also be able to predict more accurately which type of interventions will be most effective for which groups of people. This will, in turn, help them to provide patients with tailored recommendations based on their natural predisposition, empowering them to reduce their individual risk of suffering a stroke. Yet, while ML methods offer unprecedented opportunities for precision medicine in stroke prevention, several technological and methodological challenges remain. As outlined in this chapter, challenges can be grouped into three broad categories: (1) challenges in data sourcing, (2) challenges in application development, (3) challenges in deployment in clinical practice.

Having identified some of the opportunities and challenges of machine learning in stroke risk prediction and prevention, it is time to ask ourselves what impact these dynamics will have on individuals and the delivery of care, more generally. Even though it will certainly take some time before ML-based tools can (at least partially) replace established approaches for stroke risk assessment and prevention, we should already prepare for the questions that will arise as these applications are broadly adopted in practice: how will they impact the doctor-patient relationship? How will they affect public trust in the healthcare system? As great strides are made in precision medicine for stroke, how can we ensure everyone will benefit from these gains—what about low- to middle-income countries where stroke incidence rates exceed those observed in high-income countries? What about individuals who refuse to have their data collected and analyzed? These and several other questions raise important ethical concerns that require further investigation. Only by committing to ethical conduct, methodological rigor, and patient safety will we harness the full potential of data-driven predictive modeling in stroke.

Funding This work was supported by funding from the European Union's Horizon 2020 research and innovation program under grant agreement No. 777107 (PRECISE4Q).

Acknowledgments The author would like to thank Vince I Madai and Stephanie Bishop for critically reviewing this chapter.

References

1. Mesko B. The role of artificial intelligence in precision medicine. Exp Rev Precis Med Drug Develop. 2017;2(5):239–41. https://doi.org/10.1080/23808993.2017.1380516.
2. Huang BE, Mulyasasmita W, Rajagopal G. The path from big data to precision medicine. Exp Rev Precis Med Drug Develop. 2016;1(2):129–43.
3. Jiang F, Jiang Y, Zhi H, Dong Y, Li H, Ma S, et al. Artificial intelligence in healthcare: past, present and future. Stroke Vasc Neurol. 2017;2(4):230–43.
4. Patel UK, Anwar A, Saleem S, Malik P, Rasul B, Patel K, et al. Artificial intelligence as an emerging technology in the current care of neurological disorders. J Neurol. 2019:1–20.
5. Jha S, Topol EJ. Adapting to artificial intelligence: radiologists and pathologists as information specialists. JAMA. 2016;316(22):2353–4.
6. Pesapane F, Codari M, Sardanelli F. Artificial intelligence in medical imaging: threat or opportunity? Radiologists again at the forefront of innovation in medicine. Eur Radiol Exp. 2018;2(1):35.
7. Attia ZI, Noseworthy PA, Lopez-Jimenez F, Asirvatham SJ, Deshmukh AJ, Gersh BJ, et al. An artificial intelligence-enabled ECG algorithm for the identification of patients with atrial fibrillation during sinus rhythm: a retrospective analysis of outcome prediction. Lancet. 2019;394(10201):861–7.
8. Tran BX, Vu GT, Ha GH, Vuong Q-H, Ho M-T, Vuong T-T, et al. Global evolution of research in artificial intelligence in health and medicine: a bibliometric study. J Clin Med. 2019;8(3):360.
9. Ienca M, Ferretti A, Hurst S, Puhan M, Lovis C, Vayena E. Considerations for ethics review of big data health research: a scoping review. PLoS One. 2018;13(10):e0204937.
10. Saber H, Somai M, Rajah GB, Scalzo F, Liebeskind DS. Predictive analytics and machine learning in stroke and neurovascular medicine. Neurol Res. 2019;41(8):681–90.
11. Sakai K, Yamada K. Machine learning studies on major brain diseases: 5-year trends of 2014–2018. Jpn J Radiol. 2019;37(1):34–72.
12. Feng R, Badgeley M, Mocco J, Oermann EK. Deep learning guided stroke management: a review of clinical applications. J Neurointervent Surg. 2018;10(4):358–62.
13. Jamthikar A, Gupta D, Khanna NN, Araki T, Saba L, Nicolaides A, et al. A special report on changing trends in preventive stroke/cardiovascular risk assessment via B-mode ultrasonography. Curr Atheroscler Rep. 2019;21(7):25.
14. Feigin VL, Norrving B, George MG, Foltz JL, Roth GA, Mensah GA. Prevention of stroke: a strategic global imperative. Nat Rev Neurol. 2016;12(9):501.
15. Feigin VL, Abajobir AA, Abate KH, Abd-Allah F, Abdulle AM, Abera SF, et al. Global, regional, and national burden of neurological disorders during 1990–2015: a systematic analysis for the Global Burden of Disease Study 2015. The Lancet Neurol. 2017;16(11):877–97.
16. Feigin VL. Anthology of stroke epidemiology in the 20th and 21st centuries: assessing the past, the present, and envisioning the future. Int J Stroke. 2019;14(3):223–37.
17. Feigin VL, Krishnamurthi RV, Parmar P, Norrving B, Mensah GA, Bennett DA, et al. Update on the global burden of ischemic and hemorrhagic stroke in 1990-2013: the GBD 2013 study. Neuroepidemiology. 2015;45(3):161–76.
18. Feigin VL, Lawes CM, Bennett DA, Barker-Collo SL, Parag V. Worldwide stroke incidence and early case fatality reported in 56 population-based studies: a systematic review. Lancet Neurol. 2009;8(4):355–69.

19. Luengo-Fernandez R, Violato M, Candio P, Leal J. Economic burden of stroke across Europe: a population-based cost analysis. Eur Stroke J. 2020;5(1):17–25.
20. Roth GA, Forouzanfar MH, Moran AE, Barber R, Nguyen G, Feigin VL, et al. Demographic and epidemiologic drivers of global cardiovascular mortality. N Engl J Med. 2015;372(14):1333–41.
21. Di Carlo A. Human and economic burden of stroke. Oxford University Press; 2009.
22. Graven C, Sansonetti D, Moloczij N, Cadilhac D, Joubert L. Stroke survivor and carer perspectives of the concept of recovery: a qualitative study. Disabil Rehabil. 2013;35(7):578–85.
23. Forsberg-Wärleby G, Möller A, Blomstrand C. Psychological Well-being of spouses of stroke patients during the first year after stroke. Clin Rehabil. 2004;18(4):430–7.
24. Hill V. Live well after stroke: methods of a community-based, occupational therapist–led, life management intervention. Ann Phys Rehabil Med. 2018;61:e514.
25. Redfern J, Gordon C, Cadilhac D. Longer-term support for survivors of stroke and their carers. Stroke Nurs. 2019;2:323–45.
26. Wray F, Clarke D. Longer-term needs of stroke survivors with communication difficulties living in the community: a systematic review and thematic synthesis of qualitative studies. BMJ Open. 2017;7(10):e017944.
27. Pindus DM, Mullis R, Lim L, Wellwood I, Rundell AV, Aziz NAA, et al. Stroke survivors' and informal caregivers' experiences of primary care and community healthcare services–a systematic review and meta-ethnography. PLoS One. 2018;13(2):e0192533.
28. Rajsic S, Gothe H, Borba H, Sroczynski G, Vujicic J, Toell T, et al. Economic burden of stroke: a systematic review on post-stroke care. Eur J Health Econ. 2019;20(1):107–34.
29. Mozaffarian D, Benjamin E, Go A, Arnett D, Blaha M, Cushman M, et al. Heart disease and stroke statistics-2016 update: a report from the American Heart Association. Circulation. 2016;133(4):e38.
30. Saka Ö, McGuire A, Wolfe C. Cost of stroke in the United Kingdom. Age Ageing. 2009;38(1):27–32.
31. Benjamin EJ, Muntner P, Bittencourt MS. Heart disease and stroke statistics-2019 update: a report from the American Heart Association. Circulation. 2019;139(10):e56–e528.
32. Mendis S, Davis S, Norrving B. Organizational update: the world health organization global status report on noncommunicable diseases 2014; one more landmark step in the combat against stroke and vascular disease. Stroke. 2015;46(5):e121–e2.
33. Mendis S, Armstrong T, Bettcher D, Branca F, Lauer J, Mace C, et al. Global status report on noncommunicable diseases 2014. World Health Organization; 2014.
34. Aarli J, Tarun D, Janca A, Muscetta A. Neurological disorders: public health challenges. World Health Organization; 2006.
35. Meschia JF, Bushnell C, Boden-Albala B, Braun LT, Bravata DM, Chaturvedi S, et al. Guidelines for the primary prevention of stroke: a statement for healthcare professionals from the American Heart Association/American Stroke Association. Stroke. 2014;45(12):3754–832.
36. Goldstein LB, Bushnell CD, Adams RJ, Appel LJ, Braun LT, Chaturvedi S, et al. Guidelines for the primary prevention of stroke: a guideline for healthcare professionals from the American Heart Association/American Stroke Association. Stroke. 2011;42(2):517–84.
37. World Health Organization. Prevention of cardiovascular disease. World Health Organization; 2007.
38. Boehme AK, Esenwa C, Elkind MS. Stroke risk factors, genetics, and prevention. Circ Res. 2017;120(3):472–95.
39. O'donnell MJ, Xavier D, Liu L, Zhang H, Chin SL, Rao-Melacini P, et al. Risk factors for ischaemic and intracerebral haemorrhagic stroke in 22 countries (the INTERSTROKE study): a case-control study. Lancet. 2010;376(9735):112–23.
40. Rose G. Sick individuals and sick populations. Int J Epidemiol. 1985;14(1):32–8.
41. Feigin VL, Krishnamurthi R, Bhattacharjee R, Parmar P, Theadom A, Hussein T, et al. New strategy to reduce the global burden of stroke. Stroke. 2015;46(6):1740–7.
42. Parmar P, Krishnamurthi R, Ikram MA, Hofman A, Mirza SS, Varakin Y, et al. The Stroke Riskometer(TM) App: Validation of a data collection tool and stroke risk predictor. Int J Stroke. 2015;10(2):231–44.

43. Feigin VL, Brainin M, Norrving B, Gorelick PB, Dichgans M, Wang W, et al. What is the best mix of population-wide and high-risk targeted strategies of primary stroke and cardiovascular disease prevention? J Am Heart Assoc. 2020;9(3):e014494.
44. Feigin VL, Norrving B, Mensah GA. Primary prevention of cardiovascular disease through population-wide motivational strategies: insights from using smartphones in stroke prevention. BMJ Glob Health. 2017;2(2):e000306.
45. Diener A, Celemin-Heinrich S, Wegscheider K, Kolpatzik K, Tomaschko K, Altiner A, et al. In-vivo-validation of a cardiovascular risk prediction tool: the Arriba-pro study. BMC Fam Pract. 2013;14:7. https://doi.org/10.1186/1471-2296-14-13.
46. Weng SF, Reps J, Kai J, Garibaldi JM, Qureshi N. Can machine-learning improve cardiovascular risk prediction using routine clinical data? PLoS One. 2017;12(4):e0174944.
47. Jamthikar A, Gupta D, Khanna NN, Saba L, Araki T, Viskovic K, et al. A low-cost machine learning-based cardiovascular/stroke risk assessment system: integration of conventional factors with image phenotypes. Cardiovas Diagn Ther. 2019;9(5):420.
48. D'agostino RB, Vasan RS, Pencina MJ, Wolf PA, Cobain M, Massaro JM, et al. General cardiovascular risk profile for use in primary care. Circulation. 2008;117(6):743–53.
49. Nobel L, Mayo NE, Hanley J, Nadeau L, Daskalopoulou SS. MyRisk_Stroke calculator: a personalized stroke risk assessment tool for the general population. J Clin Neurol. 2014;10(1):1–9.
50. Goldstein BA, Navar AM, Carter RE. Moving beyond regression techniques in cardiovascular risk prediction: applying machine learning to address analytic challenges. Eur Heart J. 2017;38(23):1805–14.
51. Kakadiaris IA, Vrigkas M, Yen AA, Kuznetsova T, Budoff M, Naghavi M. Machine learning outperforms ACC/AHA CVD risk calculator in MESA. J Am Heart Assoc. 2018;7(22):e009476.
52. Garg N, Muduli SK, Kapoor A, Tewari S, Kumar S, Khanna R, et al. Comparison of different cardiovascular risk score calculators for cardiovascular risk prediction and guideline recommended statin uses. Indian Heart J. 2017;69(4):458–63.
53. Vieira S, Pinaya WHL, Mechelli A. Introduction to machine learning. In: Machine learning. Elsevier; 2020. p. 1–20.
54. Rajkomar A, Dean J, Kohane I. Machine learning in medicine. N Engl J Med. 2019;380(14):1347–58.
55. Beam AL, Kohane IS. Big data and machine learning in health care. JAMA. 2018;319(13):1317–8.
56. Olesen AE, Grønlund D, Gram M, Skorpen F, Drewes AM, Klepstad P. Prediction of opioid dose in cancer pain patients using genetic profiling: not yet an option with support vector machine learning. BMC Res Notes. 2018;11(1):78.
57. Ngiam KY, Khor W. Big data and machine learning algorithms for health-care delivery. Lancet Oncol. 2019;20(5):e262–e73.
58. Ambale-Venkatesh B, Yang X, Wu CO, Liu K, Hundley WG, McClelland R, et al. Cardiovascular event prediction by machine learning: the multi-ethnic study of atherosclerosis. Circ Res. 2017;121(9):1092–101.
59. Liu T, Fan W, Wu C. A hybrid machine learning approach to cerebral stroke prediction based on imbalanced medical dataset. Artif Intell Med. 2019;101:101723.
60. Li X, Liu H, Du X, Zhang P, Hu G, Xie G, et al, editors. Integrated machine learning approaches for predicting ischemic stroke and thromboembolism in atrial fibrillation. In: AMIA annual symposium proceedings. American Medical Informatics Association; 2016.
61. Petosa R. Using behavioral contracts to promote health behavior change: application in a college level health course. Health Educ. 1984;15(2):22–7.
62. Lira M, Kunstmann S, Caballero E, Guarda E, Villarroel L, Molina J. Cardiovascular prevention and attitude of people towards behavior changes: state of the art. Revista Medica de Chile. 2006;134(2):223–30.
63. Garrido P, Aldaz A, Vera R, Calleja M, de Alava E, Martín M, et al. Proposal for the creation of a national strategy for precision medicine in cancer: a position statement of SEOM, SEAP, and SEFH. Clin Transl Oncol. 2018;20(4):443–7.

64. Kökciyan N, Chapman M, Balatsoukas P, Sassoon I, Essers K, Ashworth M, et al. A collaborative decision support tool for managing chronic conditions. Stud Health Technol Inform. 2019;264:644–8.
65. Kario K. Perfect 24-h management of hypertension: clinical relevance and perspectives. J Hum Hypertens. 2017;31(4):231–43.
66. Li KHC, White FA, Tipoe T, Liu T, Wong MC, Jesuthasan A, et al. The current state of mobile phone apps for monitoring heart rate, heart rate variability, and atrial fibrillation: narrative review. JMIR Mhealth Uhealth. 2019;7(2):e11606.
67. Lowres N, Neubeck L, Salkeld G, Krass I, McLachlan AJ, Redfern J, et al. Feasibility and cost-effectiveness of stroke prevention through community screening for atrial fibrillation using iPhone ECG in pharmacies. Thromb Haemost. 2014;111(06):1167–76.
68. Tran V-T, Riveros C, Ravaud P. Patients' views of wearable devices and AI in healthcare: findings from the ComPaRe e-cohort. NPJ Digit Med. 2019;2(1):1–8.
69. PRECISE4Q Consortium. PRECISE4Q: predictive modelling in stroke. 2020. www.precise4q.eu. Accessed 25 Mar 2020.
70. Frey D. Schlaganfallbehandlung: Künstliche Intelligenz als Game-Changer. kma-Das Gesundheitswirtschaftsmagazin. 2018;23(11):32–4.
71. CORDIS EU Reserach Results. Personalised medicine by predictive modeling in stroke for better quality of life. 2020. https://cordis.europa.eu/project/id/777107. Accessed 23 Mar 2020.
72. Shoenbill K, Song Y, Craven M, Johnson H, Smith M, Mendonca EA. Identifying patterns and predictors of lifestyle modification in electronic health record documentation using statistical and machine learning methods. Prev Med. 2020;136:106061.
73. Shoenbill K, Song Y, Gress L, Johnson H, Smith M, Mendonca EA. Natural language processing of lifestyle modification documentation. Health Informatics J. 2020;26(1):388–405.
74. Liu F, Weng C, Yu H. Natural language processing, electronic health records, and clinical research. In: Clinical research informatics. London: Springer; 2012. p. 293–310.
75. Vayena E, Blasimme A, Cohen IG. Machine learning in medicine: addressing ethical challenges. PLoS Med. 2018;15(11):e1002689.
76. He J, Baxter SL, Xu J, Xu J, Zhou X, Zhang K. The practical implementation of artificial intelligence technologies in medicine. Nat Med. 2019;25(1):30–6.
77. Blasimme A, Fadda M, Schneider M, Vayena E. Data sharing for precision medicine: policy lessons and future directions. Health Aff. 2018;37(5):702–9.
78. McCall B. What does the GDPR mean for the medical community? Lancet. 2018;391(10127):1249.
79. Fortier I, Doiron D, Burton P, Raina P. Invited commentary: consolidating data harmonization—how to obtain quality and applicability? Am J Epidemiol. 2011;174(3):261–4.
80. Fortier I, Raina P, Van den Heuvel ER, Griffith LE, Craig C, Saliba M, et al. Maelstrom research guidelines for rigorous retrospective data harmonization. Int J Epidemiol. 2017;46(1):103–5.
81. PRECISE4Q Consortium. How to tackle the challenges of Data Integration. In: PRECISE4Q: predictive modelling in stroke. 2020. https://precise4q.eu/how-to-tackle-the-challengesof-data-integration. Accessed 24 May 2021.
82. Aysolmaz B, Iren D, Dau N, editors. Preventing algorithmic bias in the development of algorithmic decision-making systems: a Delphi study. In: Proceedings of the 53rd Hawaii International Conference on System Sciences. 2020.
83. Wong P-H. Democratizing algorithmic fairness. Philos Technol. 2019:1–20.
84. Luxtona DD. Ethical implications of conversational agents in global public health. Bull World Health Organ. 2020;98:285–7.
85. Baron AA, Baron SB. High levels of HDL cholesterol do not predict protection from cardiovascular disease in women. Prev Cardiol. 2007;10(3):125–7.
86. Lau ES, Sarma A. Utility of imaging in risk stratification of chest pain in women. Curr Treat Options Cardiovasc Med. 2017;19(9):72.
87. Wu Y, Fang Y. Stroke prediction with machine learning methods among older Chinese. Int J Environ Res Public Health. 2020;17(6):1828.

88. Krawczyk B. Learning from imbalanced data: open challenges and future directions. Prog Artific Intell. 2016;5(4):221–32.
89. Hung C-Y, Chen W-C, Lai P-T, Lin C-H, Lee C-C, editors. Comparing deep neural network and other machine learning algorithms for stroke prediction in a large-scale population-based electronic medical claims database. In: 2017 39th annual international conference of the IEEE engineering in medicine and biology society (EMBC). IEEE; 2017.
90. Obermeyer Z, Powers B, Vogeli C, Mullainathan S. Dissecting racial bias in an algorithm used to manage the health of populations. Science. 2019;366(6464):447–53.
91. Higgins D, Madai VI. From bit to bedside: a practical framework for artificial intelligence product development in healthcare. Adv Intell Syst. 2020;2(10) https://doi.org/10.1002/aisy.202000052.
92. Siau K, Wang W. Building trust in artificial intelligence, machine learning, and robotics. Cutt Bus Technol J. 2018;31(2):47–53.
93. Rudin C. Stop explaining black box machine learning models for high stakes decisions and use interpretable models instead. Nat Mach Intell. 2019;1(5):206–15.

Respect for Persons and Artificial Intelligence in the Age of Big Data

6

Ryan Spellecy and Emily E. Anderson

6.1 Introduction

Calls to revise the US system of human research protections to adapt to changes in the practice of medicine and biomedical research are not new. Pragmatic clinical trials [1], community engaged research [2], and learning health systems [3] all pose unique ethical challenges that could not have been envisioned when the regulatory system was developed in the 1970s. Similarly, the use of "big data" in medical research has received increased attention in recent years as conducting such research has become cheaper and easier due to advances in computing technology. "Big data" is a somewhat vague term and used in a variety of different ways [4–8]. Here we focus specifically on the use of existing patient data combined across institutions in research that is subject to the Common Rule (e.g., federally funded or conducted at research institutions that choose to uniformly apply the Common Rule), and the application of artificial intelligence (AI) to that big data, particularly in research related to mental health.

The relationship between big data and AI in medical research is essential in some regards. IBM defines big data in terms of the three "V"s of volume, variety, and velocity [8]. Volume is obvious, and we will discuss the application of this to medical data below. Velocity refers to the speed at which data can be created and accessed and is relevant to big data and AI in medical research. However, the most salient "V" for this topic is variety. Daniel O'Leary describes

R. Spellecy (✉)
Center for Bioethics and Medical Humanities, Medical College of Wisconsin, Milwaukee, WI, USA
e-mail: rspellecy@mcw.edu

E. E. Anderson
Neiswanger Institute for Bioethics, Stritch School of Medicine, Maywood, IL, USA
e-mail: emanderson@luc.edu

F. Jotterand, M. Ienca (eds.), *Artificial Intelligence in Brain and Mental Health: Philosophical, Ethical & Policy Issues*, Advances in Neuroethics,
https://doi.org/10.1007/978-3-030-74188-4_6

the multiple sources and varieties of data, including wearable electronic devices that monitor our health, social media, phones, and radio frequency identification (RFID) [9]. AI's influence on variety becomes essential when we discuss "unstructured data," meaning data that are simply amassed and lack patterns or structure. To remedy this, "Under situations of large volumes of data, AI allows delegation of difficult pattern recognition, learning, and other tasks to computer-based approaches [10]."

This intersection of AI and big data raises ethical challenges, specifically for privacy, confidentiality, and informed consent [11–13]. Privacy and confidentiality concerns are only magnified when the focus of research is on mental illness or other stigmatized health conditions. Therefore, we argue, this intersection necessitates novel thinking around how to best fulfill the Belmont principle of respect for persons when conducting such research, particularly in the absence of any specific guidance from current research regulations for research using AI.

6.2 Big Data and AI in Biomedical Research

Opportunities abound for health researchers to aggregate and analyze data originally collected for non-research purposes. Perhaps the best example of the creation of very large, inter-institutional data sets for health research is PCORnet. PCORnet is an initiative funded by PCORI (Patient-Centered Outcomes Research Institute) that seeks to combine 11 Clinical Data Research Networks (CDRNs), 18 Patient Powered Research Networks (PPRNs), and one coordinating center to create a database of 100 million covered lives [9]. The CDRNs are health system-based networks, created by linking the clinical data warehouses of large institutions, while the PPRNs are operated and governed by patients and their partners. For example, the Mental Health Research Network (MHRN), a CDRN, combines data from13 health systems that serve approximately 12.5 million patients across 15 states (17% of whom have a mental health condition) [14]. The MoodNetwork PPRN aims to enroll at least 50,000 patients with major depressive disorder and bipolar disorder to provide longitudinal data through medical records and surveys and potentially participate in prospective comparative effectiveness studies [15].

The data in repositories such as PCORNET exists in both structured and unstructured formats, and AI can enable researchers to create structure where it is lacking. A unique challenge for big data research on electronic medical records (EMR) lies in data mining the typed "free text" notes that clinicians enter into the EMR. Notes have great research potential, but placing them in a structure that allows researchers to analyze them requires AI, as it would be impossibly complex and time consuming for people to do this unassisted. As O'Leary notes "Natural language, natural visual interpretation, and visual machine learning will become increasingly important forms of AI for big data." Natural language in particular can enable AI to comb through free text notes in the EMR, provide structure, and ultimately enable their use in research.

6.3 Big Data Health Research and the Belmont Principles

Perhaps most obviously, PCORnet and similar research that relies on large data sets poses challenges to informed consent. Whether or not this was the original intention of its authors, the Belmont framework for ethical human subjects research and the resulting federal regulations elevate individual prospective informed consent above all other ethical considerations, prioritizing the principle of respect for persons above beneficence and justice while also making respect for persons synonymous with informed consent. Therefore it is no surprise that discussions of the ethics of health research using big data tend to focus on the challenges of informed consent. Simply stated, the key ethical challenge of big data health research lies in balancing respect for persons with the potential benefits. However, when respect for persons is inappropriately and narrowly conceived as individual prospective informed consent, as many have argued, [16, 17] this sets up big data research for ethical failure. Given the size of some data networks, the problem of informed consent in the context of big data seems intractable. Consent from all subjects is not merely "impracticable" (to use a regulatory term), it is impossible. In minimal risk research, when informed consent is not possible, the alternative is usually simply to waive consent and be done with it. However, we argue that a waiver, even when ethically permissible, does not demonstrate respect for persons, if we interpret the spirit of the principle of respect for persons to require extra protections not just for those with diminished autonomy, such as persons with mental illnesses that impair their capacity to consent, but for those who cannot give truly informed consent due to practical constraints. Here, we would like to stimulate discussion and analysis of other processes and measures that might be capable of demonstrating respect for persons in research more broadly defined, when informed consent is not possible. In order to do this, we must first discuss privacy and confidentiality in the context of health research that relies on big data, as the most likely and potentially most severe harms in such research would be from an informational breach.

6.4 Privacy and Confidentiality

In research, privacy is commonly understood as pertaining to information about which an individual has a reasonable expectation that access is controlled by the individual, whereas confidentiality is commonly understood as pertaining to information that an individual has entrusted to another, with an *understanding* that the information will only be used for certain purposes. The primary risks to subjects in research that uses large data networks are those potential harms that might result from a breach of confidential information. There is also the risk of dignitary harm as a result of a perceived invasion of privacy.

When patients discuss symptoms with their therapists, agree to take medications, or discuss mental health diagnoses, they generally believe that this information will not be shared with anyone except other health care providers and third-party payers. However, in reality, this information is frequently accessed by researchers, most commonly in chart review studies. Such research is done with Institutional Review Board (IRB) approval in a manner consistent with the Code of Federal Regulations that govern research, and without the patient's consent. In ethical terms, the patient provides her physician with information, often sensitive information, *in confidence* with the *understanding* that it will be kept *private*, and that information (de-identified) is then shared with a researcher the patient has never even heard of without her consent. In a strict sense, the patient's privacy has been violated, her confidentiality has been breached. This scenario occurs innumerable times every day across the United States…and importantly, it is allowable by the federal regulations as long as other protections are in place to. It is important to recognize that we do violate privacy and we do breach confidentiality when we engage in such research, and it is deemed ethically acceptable when appropriate steps are taken to minimize the possibility of data breaches (e.g., by not collecting and storing identifiers).

While under HIPAA, any disclosure of mental health therapy notes require patient authorization, there are certain parts of an electronic health record related to mental health that are NOT considered therapy notes, such as prescriptions and medication monitoring, modalities of treatment, results of clinical tests, and summaries of treatment plans, symptoms, and progress. Additionally, persons with mental illness also confide details of their symptoms, diagnoses, and medications to non-mental health practitioners, and these details, which may have serious legal or employment ramification if breached, may end up in various places in the medical record, including as free text notes that are not protected as therapy notes.

Because of potential dignitary harms of invasion of privacy, the sensitive nature of mental illness, and the significant harms that could result from a breach of confidential mental health related information, we will later propose that, as an ethical requirement, researchers should think proactively about how to engage patients and communities to conduct AI research in mental health that uses data from electronic health records.

6.5 Notification and Broad Consent: Ethically Insufficient

Research suggests that people generally favor the use of their data for health research [18]. This is used to justify the use of both notification (informing without getting consent) and broad consent (getting consent without fully informing) practices. While many institutions seek legal and ethical cover by including a notification to patients that data might be used for research, this is ethically insufficient for two reasons. First, these notices are often buried in a HIPAA privacy notice in the clinical consent, and so are read by few patients. Second, such notices offer no opt out option. Let's say the patient did read and understand such a notice in her hospital's privacy policies and simply decides that it is consistent with her values to allow her data to be used for research. In this case, then there is no violation of confidentiality or privacy. She need not even have full information concerning the research to be conducted, nor even information beyond that her data may be used for future unspecified research (with standard protections). Perhaps one of her deep moral convictions is that people should help others, she sees such research as an instance of helping others, and so agrees to allow her data to be used for research. Perhaps she is passionate about reducing stigma associated with mental illness and as a result, believes strongly in sharing her mental health information with researchers. The specific ethical reason is not important. Competent adults do not need full and comprehensive information to make autonomous, informed choices [19]. What is important, ethically, is whether or not the patient or subject consents to the data use.

There have been novel proposals to reevaluate the means by which we obtain consent, including blanket, broad, and dynamic consent [19–22]. However, these proposals still fall short of the traditional aims of informed consent in the context of big data, as potential research subjects cannot know at the time of consent the studies to which they are in essence consenting. Instead, they consent to vague categories of research.

6.6 A Broader Conceptualization of Respect for Persons and a Balance with Other Principles

No Belmont principle is absolute, and the Belmont report exhorts us to identify the relevant ethical principles and *balance* them against one another. There are great potential benefits to big data health research, especially when balanced against the very small chance of potential harms that may result from an unintended informational breach. The ethical consideration typically becomes whether or not the benefit of such data research outweighs the affront to respect for persons. This balancing, however, is not straightforward as respect for persons and beneficence in this case appeal to fundamentally different ethical concerns.

We seek to reframe the issue: If the risks are minimal and consent impracticable, and appropriate confidentiality protections are in place, waiving informed consent can only be ethically permissible if research demonstrate some effort towards demonstrating respect for persons through one or more of the strategies we suggest below.

The Federal Regulations that govern research recognize the ethical tension between respect for persons and beneficence. 45 CFR 46.116(d) contains a provision for waiving informed consent for certain types of research, such as research on data from clinical records. To grant such a waiver, an IRB must document that four conditions are met, the second of which is that the waiver or alteration will not adversely affect the rights and welfare of the subjects. The difficulty with this recommendation is it makes perfect sense to consider whether or not waiving the requirement for informed consent will adversely affect the *welfare* of potential research subjects. This is a simple application of beneficence, and is a simply risk/benefit calculation. Of course, in big data research on mental illness, stigma and the unique harms that could result from breach of confidential information, including legal and employment harms, must be taken into account, but the likelihood and magnitude of these harms can be anticipated and weighed against potential benefits of the research. What does not make sense is whether or not the *rights* of a potential subject will be adversely affected. While benefit and harm are terms that are amenable to considerations of degrees, rights are not. Rights are simply either violated or not. They cannot be "adversely affected." If an individual is denied his right to vote, we do not say that his rights have been adversely affected, as though his right was decreased. It has been violated. Rights simply do not admit of degrees as harms and benefits do. So, to attempt to balance one Belmont principle (beneficence) that admits of degrees and can be increased, decreased, balanced, or ignored, with another Belmont principle (respect for persons) that, at least in this case, is simply upheld or not, is futile.

We could simply state that in evaluating such research, we recognize the great potential for benefit and the minimal potential for harm, and so are justified in waiving informed consent. However, it does not follow that our ethical responsibilities with regard to respect for persons have been completed. This is the key to the concern noted above, that thinking about respect for persons only in terms of informed consent will lead us to waive consent and be done with it. We argue that, in the era of AI and big data, we can and should conceptualize respect for persons as something broader than the right to self-determination through informed consent. Doing so develops respect for persons as something that admits of degrees, and the ethical obligation of researchers and IRBs then becomes thinking through how to better balance respect for persons with beneficence, rather than simply determine if/when consent can be waived. To do this, we should explore other potential means of demonstrating respect for persons that do not rely solely on informed consent and shift focus away from rights-based thinking. We can look to other models of research for means of doing this, primarily through patient and other stakeholder engagement and apply this to AI research on mental health.

6.7 Beyond Informed Consent

We would like to suggest ways that researchers might demonstrate respect for persons, not through individual informed consent but through patient and other stakeholder engagement. These suggestions are heavily influenced by the traditions of community engaged (CEnR) and community-based participatory research (CBPR), research conducted under the emergency exception for informed consent (EFIC) regulations, as well as some of the recent work in biobanking. Ranging from least to most "engaged," we recommend notifying potential research participants that research using their personal health data may occur; sharing information about research results with the public; consulting with individuals who represent the interests of potential research participants; and including public members in research oversight activities at the institutional level.

CEnR and CBPR in mental health research are not new. These strategies have been used in mental health research for some time, and guidance exists on the overall ethical approach to mental health research [23, 24], as well as utilizing such methods in mental health research in specific populations [23, 25]. What presents a novel challenge is utilizing these methods in mental health research that leverages big data an AI. Identifying representative stakeholders from large data sets being used for multiple types of studies may be challenging. Additionally, when engaging stakeholders in clinical trials of medication or community-based intervention research the research is easier to explain and the aims more tangible than in AI research. However, as AI becomes more integrated with our daily lives, citizens are becoming more interested in the potential harms and benefits and may be likely to want to engage in research partnerships.

Notification There is tremendous value in letting the community of potential research participants know what research you are planning to do. Time and time again institutional transparency has proven to have great extrinsic as well as intrinsic value. Many academic medical centers use a variety of strategies to let their patients know about the kinds of research that is going on and the fact that their medical records may be accessed appropriately for certain kinds of research. One common example is Research Match, a registry developed and utilized by a consortium of Clinical and Translational Sciences Award (CTSA) institutions [26]. Such practices should be implemented more widely, as they promote public awareness of research. To be sure, the regulatory conditions for waiving consent in minimal risk research have a requirement for notification "after participation," but they are vague about which studies require notification, and it's unclear why notification should wait until "after participation." Notification should be considered across many more types of research, although evidence is needed regarding the most effective and respectful forms of notification.

Information There is also tremendous value in letting people know the results of research—positive or negative. Initiatives are underway to improve dissemination of results of federally funded research and of clinical trials that are federally funded or conducted to gather data for applications to the FDA (NIH Open Access Policy, clinicaltrials.gov). However, the research community could be doing a lot better at this, including ensuring that results are published where lay people are likely to read them. This practice is common in CBPR and CEnR, and can maintain and improve community/academic partnerships [27–29]. For example, Dirks et al. report on how community member involvement in disseminating the results of research with a decision-support tool to aid in depression management can broaden reach and increase acceptability of the information [30].

Consultation Asking potential research participants "What do you think about what we're planning to do?", which is qualitatively different from simply notifying, not only demonstrates respect for persons, it can also improve the relevance of research questions and findings (Note: this really only demonstrates RFP when researchers actually listen.).

EFIC research invokes a specific regulatory framework (the "Final Rule," 21 CFR 50.24) that allows researchers to conduct research that is greater than minimal risk yet waives the requirement for informed consent, as long as certain requirements of the Final Rule are met. This research is conducted in settings in which consent is not possible (i.e., heart attack victims who are unable to provide consent due to their condition and cannot be proactively consented as it is unknown who will suffer a heart attack), yet the research is essential to advancing healthcare knowledge. This is important, as this is the only research that is greater than minimal risk that the regulations permit without informed consent, and so the additional safeguards become key [31–33]. It is also important because it establishes a precedent for regulating other means of demonstrating respect for persons beyond prospective written informed consent. One of those safeguards is that the researchers, with approval and oversight from the IRB, must conduct community consultation for the research study. The point of community consultation is not to gain community consent or proxy consent. Rather, the point is to consult with the community and learn whether or not the community thinks that this kind of research ought to be done in their community and what changes, if any, should be made to the research plan to make it more acceptable to the community. This provides an ethical model for using data for research purposes without consent. If it can be established that, just as in EFIC research, consent is not feasible and the benefit is great, a preferable, ethical alternative to doing nothing at all would be to engage the community in conversations about the research or use of data.

Several different models of community consultation have been employed in EFIC studies. Models include querying a convenience sample, random digit telephone surveys, targeted focus groups, large community meetings/public forums, community advisory boards, or some combination of these methods [34–40]. Each of these approaches has distinct advantages and disadvantages, and the appropriateness of each approach will vary depending on the purpose of the research.

Even better is to obtain ongoing consultation through mechanisms like community advisory boards (CABs), which are common in community engaged research (CEnR) [41]. CEnR is an approach to research that "provides communities with a voice and role in the research process beyond providing access to research participants" and may include working with communities to identify research priorities, systematically studying the views of community members regarding research protocols prior to implementation, community advisory and review boards, hiring community members as part of the research team, and including community members as co-investigators [24] Unlike EFIC, CEnR is not a regulatory model but rather an approach to research that follows a set of principles aimed at fulfilling ethical and process goals, such as establishment of an equitable, sustainable partnership between academic researchers and community partners (CTSA Principles of Engagement). Funders of AI research might consider requiring consultation or other forms of engagement when individual informed consent is not practical.

In CEnR, stakeholder engagement is not meant to be a replacement for individual informed consent, but when done correctly can demonstrate respect for persons—as well as for communities *qua* communities. The challenge is how best to engage key stakeholders in the research process in a manner that is not merely ad hoc or after the fact, but one that does so in the spirit of respect for persons. Developing plans and guidelines *before* engaging stakeholders will fail to involve community members in decisions regarding data use and will not foster a sense of ownership. A successful CEnR partnership requires engaging the community early in and frequently throughout the process eliciting input on all aspects (i.e., identifying concerns and needs of the community, employing community members as members of the research team, forming a community advisory council, etc.), and actually incorporating community input into the research design, implementation, analysis, and dissemination.

6.8 Include Participant Representatives in Project Leadership

In CEnR, relationships are ideally bi-directional, that is, respect for all parties is encouraged, acknowledging that the researcher has as much to learn from the community as the community does from the researcher. In some research, particularly when individual prospective informed consent is not possible, including community partners as part of the research leadership team can be essential in understanding how to best approach the community regarding the use of their data in research.

Specific to the use of AI in big data research in mental health, CEnR can help to truly engage the public in the oversight and goals of such research. Additionally, by engaging communities in a bi-directional manner in such research, the research will be improved by addressing the research priorities of communities as well as making the research more transparent, which can help ameliorate some of the well-publicized concerns that the public has about AI research [42, 43].

Increase Public Participation in Institutional Research Oversight This suggestion is a bit different from others in that it refers to including the lay members of the public in research activities at the institutional level and is therefore not necessarily something that can be implemented in a specific study. Many calls have been made throughout the past several decades for more public members on IRBs [44]. Such calls note that while the regulations require non-affiliated and non-scientist members to serve on the IRB, a non-scientist might be an administrative assistant from the institution, and a non-affiliated might be a retired physician. Such examples likely fall short of providing a community voice for the lay public. Such an effort also broadens respect for persons to communities and not just individuals.

Many if not all of these activities can also be justified on other ethical grounds beyond respect for persons, but thinking of them in terms of what they can do to demonstrate respect for persons is helpful for big data health research studies in which individual informed consent would be impossible to obtain from every potential participant.

6.9 Stakeholder Engagement in AI Mental Health Research

AI research specific to mental health is beginning to grow but most is still in the proof-of-concept phase. Such research uses not only EMR data but also data from patient reported outcomes, brain imaging, novel monitoring systems such as smart phones, and social media. Much of this research has been aimed at improving diagnostic clarity, identifying mental illness earlier, personalizing treatment, or identifying patients at increased risk of suicide [45]. The extent to which patient stakeholders have been engaged in this research is unclear.

How might the lessons of community engagement be applied to AI research that uses big data in mental health? CABs are a ready example. Local and national mental health advocacy organizations could be contacted to provide representation on advisory boards for mental health research with big data. Such CABs could also be asked to weigh in on research priorities, use of confidential patient data, and broader community engagement strategies. . For example, in the MoodNetwork, patient stakeholders from a variety of different advocacy groups were involved in developing the network website, patient surveys, and recruitment materials [46].

A unique strength of the mental health community is the number of advocacy groups, many of which are already actively engaged in research. Groups like the National Alliance on Mental Illness (NAMI) and Mental Health America serve persons with mental illnesses and their loved ones daily, and could be ideal groups to engage in the guidance and oversight of this research. Additionally, there are organizations that advocate on behalf of specific diseases or populations. Dry Hooch is a veteran-led organization that provides, among other things, peer counseling for veterans in mental health and substance use disorders.

Researchers could use CEnR strategies to work *with* Dry Hooch to apply AI research on issues of veterans' mental health. In such a scenario, veterans themselves would identify the mental health issues of relevance and importance to them, in conjunction with their academic partners. Additionally, veterans would have a seat at the table in the design of the research and data analysis, lending credibility to the use of AI in big data research on veterans' mental health and building trust in the research enterprise.

6.10 Conclusion

The use of AI in conducting research on big data for purposes other than the reason for which was initially collected poses unique ethical challenges. Fortunately, there are examples particularly from EFIC research and CEnR that provide models for conducting this important work in a manner that adheres to the highest ethical standards. Engaging patient and other community stakeholders in meaningful, bi-directional, sustainable partnerships can help researchers demonstrate respect for research participants, even in the absence of direct interaction with individual participants.

In discussions of ethics of AI and big data health research, we encourage less focus on the technical aspects of informed consent and more imagination regarding ways to demonstrate respect for persons. This can be accomplished through implementation of some or all of the engagement strategies we have outlined. The engagement strategies presented here also promote other ethical aims, such as the requirement of social or scientific value, the incorporation of more and diverse public voices into the process of independent review, and the elevation of the values of good stewardship of research resources and transparency and public accountability.

References

1. Califf RM, Sugarman J. Exploring the ethical and regulatory issues in pragmatic clinical trials. Clin Trials Lond Engl. 2015;12(5):436–41.
2. Ross LF, Loup A, Nelson RM, Botkin JR, Kost R, Smith GR, et al. Human subjects protections in community-engaged research: a research ethics framework. J Empir Res Hum Res Ethics. 2010;5(1):5–17.
3. Faden RR, Kass NE, Goodman SN, Pronovost P, Tunis S, Beauchamp TL. An ethics framework for a learning health care system: a departure from traditional research ethics and clinical ethics. Hast Cent Rep. 2013;Spec No:S16–27.
4. Boyd D, Crawford K. Critical questions for big data. Inf Commun Soc. 2012;15(5):662–79.
5. Gandomi A, Haider M. Beyond the hype: big data concepts, methods, and analytics. Int J Inf Manag. 2015;35(2):137–44.
6. Hashem IAT, Yaqoob I, Anuar NB, Mokhtar S, Gani A, Ullah Khan S. The rise of "big data" on cloud computing: review and open research issues. Inf Syst. 2015;47:98–115.
7. Wu X, Zhu X, Wu G, Ding W. Data mining with big data. IEEE Trans Knowl Data Eng. 2014;26(1):97–107.
8. Zikopoulos P, de Roos D, Parasuraman K, Deutsch T, Giles J, Corrigan D. Harness the power of big data the IBM big data platform. 1st ed. New York; Singapore: McGraw-Hill Education; 2012. 280 p.
9. Collins FS, Hudson KL, Briggs JP, Lauer MS. PCORnet: turning a dream into reality. J Am Med Inform Assoc JAMIA. 2014;21(4):576–7.
10. O'Leary DE. Artificial intelligence and big data. IEEE Intell Syst. 2013;28(2):96–99.
11. Ienca M, Ferretti A, Hurst S, Puhan M, Lovis C, Vayena E. Considerations for ethics review of big data health research: a scoping review. PLoS One. 2018;13(10) Available from: https://www.ncbi.nlm.nih.gov/pmc/articles/PMC6181558/.
12. Rothstein MA. Ethical issues in big data health research: currents in contemporary bioethics. J Law Med Ethics J Am Soc Law Med Ethics. 2015;43(2):425–9.
13. Vayena E, Blasimme A. Health research with big data: time for systemic oversight. J Law Med Ethics. 2018;46(1):119–29.
14. Building capacity for stakeholder engagement in the mental health research network [Internet]. 2018 [cited 2020 Mar 30]. Available from: https://www.pcori.org/research-results/2018/building-capacity-stakeholder-engagement-mental-health-research-network.
15. Mood patient-powered research network - phase I [Internet]. 2014 [cited 2020 Mar 30]. Available from: https://www.pcori.org/research-results/2013/mood-patient-powered-research-network-phase-i.
16. May T, Craig J, Spellecy R. Viewpoint: IRBs, hospital ethics committees, and the need for "translational informed consent". Acad Med. 2007;82(7):670–4.
17. Secko DM, Preto N, Niemeyer S, Burgess MM. Informed consent in biobank research: a deliberative approach to the debate. Soc Sci Med. 2009;68(4):781–9.

18. Kim J, Kim H, Bell E, Bath T, Paul P, Pham A, et al. Patient perspectives about decisions to share medical data and biospecimens for research. JAMA Netw Open. 2019;2(8):e199550.
19. Blasimme A, Moret C, Hurst SA, Vayena E. Informed consent and the disclosure of clinical results to research participants. Am J Bioeth AJOB. 2017;17(7):58–60.
20. Grady C, Eckstein L, Berkman B, Brock D, Cook-Deegan R, Fullerton SM, et al. Broad consent for research with biological samples: workshop conclusions. Am J Bioeth. 2015;15(9):34–42.
21. Spellecy R. Facilitating autonomy with broad consent. Am J Bioeth AJOB. 2015;15(9):43–4.
22. Wendler D. Broad versus blanket consent for research with human biological samples. Hast Cent Rep. 2013;43(5):3–4.
23. Maiter S, Simich L, Jacobson N, Wise J. Reciprocity: an ethic for community-based participatory action research. Action Res. 2008;6(3):305–25.
24. Dubois JM, Bailey-Burch B, Bustillos D, Campbell J, Cottler L, Fisher CB, et al. Ethical issues in mental health research: the case for community engagement. Curr Opin Psychiatry. 2011;24(3):208–14.
25. Stacciarini J-MR, Shattell MM, Coady M, Wiens B. Review: community-based participatory research approach to address mental health in minority populations. Community Ment Health J. 2011;47(5):489–97.
26. Harris PA, Scott KW, Lebo L, Hassan N, Lighter C, Pulley J. ResearchMatch: a national registry to recruit volunteers for clinical research. Acad Med J Assoc Am Med Coll. 2012;87(1):66–73.
27. Chen PG, Diaz N, Lucas G, Rosenthal MS. Dissemination of results in community-based participatory research. Am J Prev Med. 2010;39(4):372–8.
28. Kamaraju S, Olson J, DeNomie M, Visotcky A, Banerjee A, Asan O, et al. Community breast health education for immigrants and refugees: lessons learned in outreach efforts to reduce cancer disparities. J Cancer Educ [Internet]. 2018 [cited 2019 Sep 26]. Available from: https://doi.org/10.1007/s13187-018-1412-y.
29. Lopez EDS, Brakefield-Caldwell W. Disseminating research findings Back to partnering communities: lessons learned from a community-based participatory research approach. Metrop Univ. 2005;16(1):59–76.
30. Dirks LG, Avey JP, Hiratsuka VY, Dillard DA, Caindec K, Robinson RF. Disseminating the results of a depression management study in an urban Alaska native health care system. Am Indian Alsk Native Ment Health Res Online. 2018;25(1):62–79.
31. Spellecy R. Unproven or unsatisfactory versus equipoise in emergency research with waived consent. Am J Bioeth. 2006;6(3):44–5.
32. Derse AR. Emergency research and consent: keeping the exception from undermining the rule. Am J Bioeth. 2006;6(3):36–7.
33. Kipnis K, King NMP, Nelson RM. Trials and errors: barriers to oversight of research conducted under the emergency research consent waiver. IRB Ethics Hum Res. 2006;28(2):16–9.
34. Baren JM, Anicetti JP, Ledesma S, Biros MH, Mahabee-Gittens M, Lewis RJ. An approach to community consultation prior to initiating an emergency research study incorporating a waiver of informed consent. Acad Emerg Med Off J Soc Acad Emerg Med. 1999;6(12):1210–5.
35. Contant C, McCullough LB, Mangus L, Robertson C, Valadka A, Brody B. Community consultation in emergency research. Crit Care Med. 2006;34(8):2049–52.
36. Dix ES, Esposito D, Spinosa F, Olson N, Chapman S. Implementation of community consultation for waiver of informed consent in emergency research: one Institutional Review Board's experience. J Investig Med Off Publ Am Fed Clin Res. 2004;52(2):113–6.
37. Kasner SE, Baren JM, Le Roux PD, Nathanson PG, Lamond K, Rosenberg SL, et al. Community views on neurologic emergency treatment trials. Ann Emerg Med. 2011;57(4):346–354.e6.
38. Kremers MS, Whisnant DR, Lowder LS, Gregg L. Initial experience using the Food and Drug administration guidelines for emergency research without consent. Ann Emerg Med. 1999;33(2):224–9.
39. Lowenstein DH, Alldredge BK, Allen F, Neuhaus J, Corry M, Gottwald M, et al. The prehospital treatment of status epilepticus (PHTSE) study: design and methodology. Control Clin Trials. 2001;22(3):290–309.

40. Mosesso VN, Brown LH, Greene HL, Schmidt TA, Aufderheide TP, Sayre MR, et al. Conducting research using the emergency exception from informed consent: the Public Access Defibrillation (PAD) Trial experience. Resuscitation. 2004;61(1):29–36.
41. McRae AD, Bennett C, Brown JB, Weijer C, Boruch R, Brehaut J, et al. Researchers' perceptions of ethical challenges in cluster randomized trials: a qualitative analysis. Trials. 2013;14(1):1.
42. van Dijck J. Datafication, dataism and dataveillance: big data between scientific paradigm and ideology. Surveill Soc. 2014;12(2):197–208.
43. Ekbia H, Mattioli M, Kouper I, Arave G, Ghazinejad A, Bowman T, et al. Big data, bigger dilemmas: a critical review. J Assoc Inf Sci Technol. 2015;66(8):1523–45.
44. Anderson EE. A qualitative study of non-affiliated, non-scientist institutional review board members. Account Res. 2006;13(2):135–55.
45. Graham S, Depp C, Lee EE, Nebeker C, Tu X, Kim H-C, et al. Artificial intelligence for mental health and mental illnesses: an overview. Curr Psychiatry Rep. 2019;21(11):116.
46. Sylvia LG, Hearing CM, Montana RE, Gold AK, Walsh SL, Janos JA, et al. MoodNetwork: an innovative approach to patient-centered research. Med Care. 2018;56(Suppl 10):S48–52.

Part II

AI for Digital Mental Health and Assistive Robotics: Philosophical and Regulatory Challenges

Social Robots and Dark Patterns: Where Does Persuasion End and Deception Begin?

7

Naveen Shamsudhin and Fabrice Jotterand

7.1 Introduction

Technologists, at both academic and corporate research centers, in collaboration with behavioral psychologists, neuroscientists, and sociologists, are developing socially interactive robots, seemingly emotionally and socially intelligent, for long-term interactive and assistive relations with humans. To allow for a seamless illusion of mutually empathic interaction to be experienced by the human user, these robots are being endowed with the technical capabilities to multi-modally *read* human cognitive and emotional "states" and appropriately respond in both verbal and non-verbal manners using its various sensors and actuators. Unfortunately, even with state-of-the-art socially interactive robots, humans are only able to engage for a short period of time or for sporadic contacts, before losing interest [1]. For successful long-term social interaction, robots need to be programmed with the ability to perceive and interpret their environment based on past experiences, identify and model mental states of humans. Most importantly, these robots need to understand, model, and exhibit the rich dynamics of human social and cultural communicative behavior and norms. When the nuances and paradigms of human to human communication are ported to robots, we must confront and accept some of the uncomfortable facets of human nature and psychology. Our cognitive biases—including our tendency to anthropomorphize, our persuasive styles of communication, and our stereotypes such as those on gender and culture, among many others—are being

N. Shamsudhin (✉)
Multi-Scale Robotics Lab, ETH Zurich, Zurich, Switzerland

F. Jotterand
Center for Bioethics and Medical Humanities, Medical College of Wisconsin, Milwaukee, WI, USA

Institute for Biomedical Ethics, University of Basel, Basel, Switzerland

F. Jotterand, M. Ienca (eds.), *Artificial Intelligence in Brain and Mental Health: Philosophical, Ethical & Policy Issues*, Advances in Neuroethics,
https://doi.org/10.1007/978-3-030-74188-4_7

used by robot programmers to design the future of long-term human–robot interactions.

Similar to the situation faced by social roboticists, the pressure to increase user engagement time, a key metric of product performance, is being felt by designers at non-robotic technology companies as well. For many of these companies, the revenue stream is directly related to the amount of time users engage with their services. Over the last two decades, companies and digital designers have increasingly employed design techniques, which are now collectively known as *dark patterns*. This approach exploits knowledge of human and social psychology to create habit-forming, privacy-intrusive products, which are not always in the best interest of the users, and detrimentally affect the user's mental and physical well-being. Look no further than your pockets, for a habit-forming exemplar: the smartphone and its app ecosystem.

In this work, we argue that the adoption and embodiment of these dark pattern strategies by roboticists is imminent and potentially harmful. While some of these dark patterns may be benign, others raise serious ethical and legal concerns. Compared to other digital technologies, such as web or smartphone interfaces, dark pattern harboring robots are potent social actors, and could more effectively compromise user trust and autonomy, and erode the overall well-being (mental and physical) of its users. This work further documents evidence of dark patterns surrounding commercially available social robots.

This chapter is structured into four sections. Section 7.2 explores the state of the art in socially interactive robotics, their application areas, and the challenges it faces. Section 7.3 highlights the use of anthropomorphic principles in social robot design to sustain long-term human–robot interaction. Section 7.4 introduces dark patterns with examples and considers cases of their deployment in social robots. Section 7.5 contains the normative analysis and discussions and finally Sect. 7.6 offers concluding remarks with suggestions for the development of social robots while avoiding the adoption of the use of dark patterns.

7.2 Robots in the Wild

7.2.1 Robots in the Realm of Personal Human Experience

Robots are increasingly entering the realm of personal human experience. Over 25 million service robots are estimated to have been sold in 2019 worldwide, vastly outnumbering industrial robot sales (see Fig. 7.1). Personal service robots which make up the majority of sales are mostly household robots designed to perform singular and useful tasks, such as vacuuming, mopping, lawn-moving (e.g., iRobot's Roomba™), or for entertainment and hobby purposes (e.g., Sony's aibo™). Assistance robots for personal healthcare applications, such as for the elderly and for the handicapped, currently only constitute a small fraction, numbering a few thousand units annually. Professional service robots, which require a properly trained operator for its functioning, can be found in a variety of applications areas

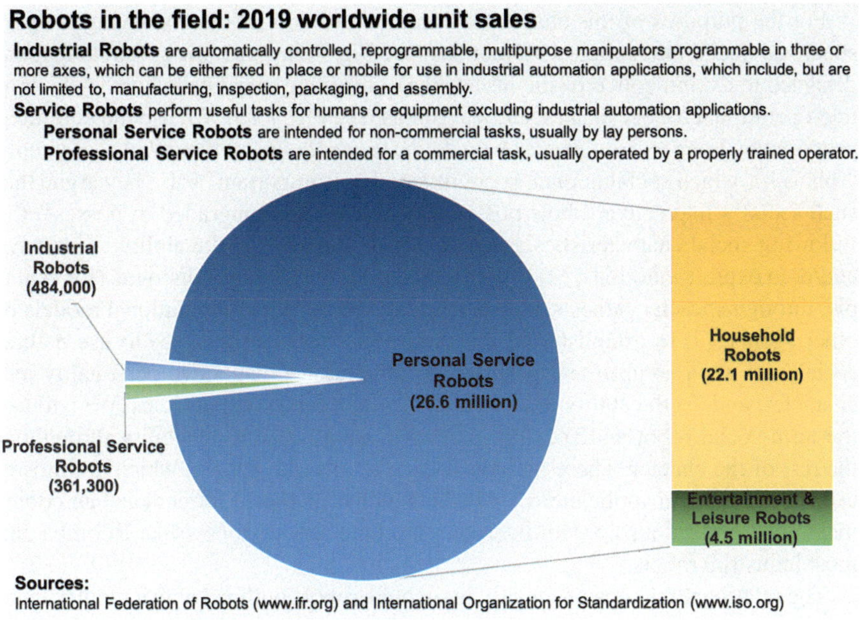

Fig. 7.1 Estimated worldwide annual robot unit sales [2]

such as in healthcare, in transportation, agriculture, and more popularly in public relations at reception desks of hotels and banks as well as at museums, malls, and supermarkets to guide and make recommendations to visitors and consumers (e.g., SoftBank Corporation's Pepper™). Within the next decade, our homes and workplaces, spaces of rich human and social experiences, will be crowded with a multitude of interconnected robotic devices. Hundreds of start-ups and corporations are busy at work attempting to bring to market more such service robots, that can deliver shopping to homes, prepare meals in the kitchen, deliver them to your table, perform a dance routine when you are bored, and recite bed-time stories to children. Alas, the all-in-one humanoid butler robot that can perform all the tasks above without malfunctioning is still confined to science fiction for the moment, and at best, decades away from reality, as the technical and social challenges are very complex.

7.2.2 Socially Interactive Robots

Not all robots need to be social. Most industrial robots work in environments cordoned off for humans, and do repetitive tasks, such as painting and welding, and do not need social skills for interacting with humans. Once programmed by a human operator, they work repetitively in their stationary workspaces. Other industrial robots, that are mobile and move around factory floors picking and delivering goods, have the ability to detect objects in their vicinity and avoid collision, a primitive form of environmental awareness.

For the purposes of this chapter, we are interested in robots that engage in personalized social interaction with humans (see Fig. 7.2a), and not in robots that are designed to exhibit collectivistic insect-like or bird-like social behaviors [4, 5] or to tele-operational robots such as surgical robots (see Fig. 7.2b). We use the definition proposed by Fong and colleagues who define "socially interactive robots" as simply "robots for which social human–robot interaction is important" [6]. They argue that such socially interactive robots possess (or will soon be upgraded to possess) the following social characteristics associated with humans: (1) the ability to perceive and/or to express emotions; (2) to communicate with high-level dialogue (for example, through speech synthesis); (3) to learn and recognize computational models of other agents; (4) to establish and maintain social relationships; (5) to use natural social cues such as gaze and gesture; (6) to exhibit a distinctive personality and character; and (7) the ability to learn and develop social competencies. We will use the term social robot and socially interactive robot interchangeability throughout the rest of the chapter. These include service robots like Pepper, which are mostly used in commercial applications, and aibo which is marketed for personal entertainment and companionship. However, they exclude service robots like Roomba and most industrial robots.

The relational features of socially interactive robots outlined above qualify them as social actors. In 1998, Fogg proposed a functional triad diagram [3], where he placed various technologies on a triangular map categorizing them as either tools,

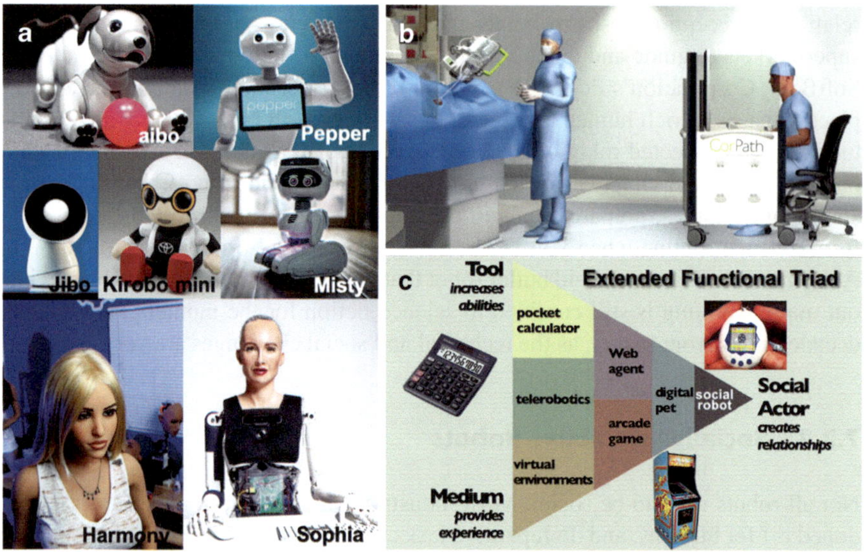

Fig. 7.2 (**a**) Social robots may be completely anthropomorphic (Sophia, Harmony), or zoomorphic (aibo), or hybrid entities (Pepper, Jibo, Kirobo Mini, Misty). (**b**) Tele-operated robots for medicine or surgery are also designed with a human–robot interaction in mind, but they are not considered social robots. (**c**) Extending the functional triad of technologies proposed in Fogg so as to include social robots that have emerged post-1999 [3]

mediums, or social actors or somewhere in between. Social robots are not mere tools nor mediums, but they are social actors capable of creating and "maintaining" relationships with humans. In light of these considerations, we can now extend his functional triad (see Fig. 7.2c) by placing social robots at the social actor vertex, displacing the digital pets (like Tamagotchi) which were the state of the art at the time of his proposal.

7.2.3 The Challenge of Long-Term Interaction with Social Robots

Though social robots are used in research labs, and pilot studies have been conducted in a multitude of application areas, only very few have seen large-scale commercialization with practical use and adoption, such as aibo, a robotic dog for entertainment and companionship applications [7, 8], and Pepper for business, retail, and home applications [9].

One of the biggest design challenges, if not the biggest, facing the introduction of social robots into the wild, is the challenge of sustaining long-term human–robot interaction and engagement over days, weeks, and years, beyond the initial period of novelty [1, 10]. Currently most human–robot social interactions are short-lived, except for the structured interactions used in elderly and differentially abled care [11]. They usually last a few minutes to utmost an hour of one-off interaction, or sporadic interactions such as an information robot at an airport or at the bank lobby. We note that aibo is a notable exception to this norm, which thanks to its clever (but potentially deceptive) design principles [8, 12] has users that use it on a day-to-day basis for several years [13].

To increase the length and the depth of these interactions between humans and robots, social robots must communicate with humans much like humans communicate with each other, or in ways that humans are familiar with and find natural, such as with similarity in form and appearance, personalized interaction based on history of previous interactions, usage of behavioral norms like maintaining eye gaze and nodding, etc. There must be an establishment of trust, and an ability to connect at a "deeper and meaningful" level than just an exchange of information. To mimic the features of a human to human interaction, social robots would need to be able to express emotions such as empathy and compassion or demonstrate higher level of rationality and communication skills. Social roboticists have found various design strategies that collectively help in this effort which are detailed in Sect. 7.3.

7.3 Social Robot Design

7.3.1 Anthropomorphism in Social Robot Design

Roomba is the most successful robot on the market found in millions of homes around the world. Roomba does a single task, that is, vacuuming floors. It is not a socially interactive robot, even though it operates in our private social space, our homes. What is surprising is that even though Roomba has no models of social intelligence programmed into it, nor looks neither like a human or an animal, and has no ability to express human social cues, a large number of its owners give it names, ascribe personality traits to it, feel that it deserves credit for doing a good job at cleaning, and also get sad when it is damaged or has to be replaced.

The human response to Roomba is an excellent robotic example of anthropomorphism. Anthropomorphism generally refers to the attribution of human traits (such as cognition, personality, emotions, and intentionality) and human behavior to non-human entities such as animals, cartoon characters, and technological devices. From the point of view of Dennett [6, 14], for complex systems for which we may not have complete knowledge about its full physical characteristics nor its internal design and functions, we tend to ascribe an intentionality and agency to the system and assume that the actions of the systems arise from its internal beliefs and desires. Robots do seem to fall under the category of such complex systems.

Anthropomorphism is considered as a cognitive bias, a limitation or as a hindrance. For social robotics design, however, it has been a very useful and indispensable design ingredient and has been extensively investigated and implemented to shape human–robot interactions [15, 16]. Applied anthropomorphism in robot design functions "as a mechanism through which social interaction can be facilitated. Thus, the ideal use of anthropomorphism is to present an appropriate balance of illusion (to lead the user to believe that the robot is sophisticated in areas where the user will not encounter its failings) and functionality (to provide capabilities necessary for supporting human-like interaction)" [6]. Achieving this delicate balance between the illusion of mutually reciprocative social and affective relation and delivering the desired functionality is the golden aim of social roboticists.

Anthropomorphic design principles guide not only the external physical form of social robots but also its external behavior to amplify their sense of animation or "being alive," its likeability or attractiveness, and ultimately trust and a sense of mutual acceptance with the user. Physical design parameters include the size of the robot, its appearance such as the humanoid or zoomorphic form (Fig. 7.2), facial features such as having an eye, its apparent gender, etc., and behavioral parameters include robot posture, its change in eye gaze and facial expressions, and empathic responses like head nodding and action mirroring. A large two-meter social robot, purely by virtue of its size, might appear menacing, but if it was designed with the proportions and behavioral characteristics of a baby or with an overall cute aesthetics, it can evoke a positive emotional response in users [17].

7.3.2 Social Robot Design Is Not Neutral

Design is not neutral. Ask any practitioner. Stephen P Anderson [18], for instance, writes in his influential article "Towards an Ethics of Persuasion" that "[a]ll design influences behavior, even if we're not intentional about the desired behaviors" whereas Colin Gray points out, "[d]esign is inherently a persuasive act, where the designer creates intentional change in the world that either directly or indirectly induces behavioral or social change." [19]. Cynthia Breazeal, MIT professor and social robot entrepreneur remarks that "[s]ocial robots are designed to interact with people in a socio-emotional way during interpersonal interaction" [20]. She also points out that there are several application-specific design goals in building social robots. One is to have a research platform to "gain a scientific understanding of social intelligence and human sociality," while design goals for successful commercial social robots could be framed as follows: "The commercial success of these robots hinges on their ability to be part of a person's daily life. As a result, the robots must be responsive to and interact with people in a natural and intuitive manner" [21]. Softbank's innovation department, for instance, made no qualms about reporting that the design goal of its robot Pepper was to "intrigue and attract consumers" [9].

The short-term overarching goal for the success of social robots is creating an illusion to the user of a personalized, natural, and reciprocative empathic and social interaction with the robot, while the long-term goal of developing a strong artificial intelligence (AI) is at works [16]. With its carefully designed form and behavior, the social robot design intent can be alternatively framed as persuasion of the user into believing, at least temporarily, that the robot is human-like, has life-like properties, can be trusted, and there is value in the creation and maintenance of this human–robot relationship. Clearly, social robot design is not neutral and has a persuasive intent. The design of social robots is an example of persuasive design.

7.3.3 Social Robots as Exemplars of Persuasive Design

BJ Fogg defined Persuasive Technologies as "a computing system, device, or application intentionally designed to change a person's attitudes or behavior in a predetermined way" [3]. We use the term Persuasive Design to imply the design of persuasive technologies (see Fig. 7.3). Social robots are exemplars of persuasive design. With their combination of physical embodiment, presence, and mobility, their anthropomorphic design features and use of affective computing, they are capable of persuading us into the illusion of natural, reciprocative empathic communication. They are social actors, far more effective than conventional persuasive technologies [22, 23]. They not only can deploy fact or expert-based logical and rational arguments using high-level language to persuade but can also use techniques that appeal to the emotions or the use of social cues for change, i.e., nonverbal communication such as bodily language and gestures.

Social robots have applications in healthcare; in informing, educating and persuading people about and into positive personal behavioral change or for the social good. While purely information or fact-giving is less effective as we know from human to human communication (for example in doctor–patient relationships), adding a persuasive component can increase effectiveness or likelihood of behavior change such as increased treatment adherence [24]. Think of a social robot that reminds or informs the patient to take medication versus a robot that can persuade

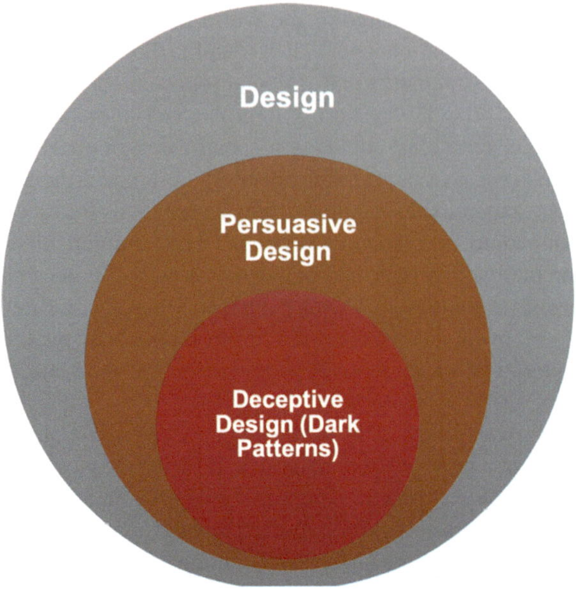

Fig. 7.3 A conceptual map of design. All design has a persuasive effect, whether the designer intends for persuasion or not. Persuasive design is a conscious strategy to persuade the user to change his or her beliefs, behavior, or attitudes. Deceptive designs are persuasive designs that employ deceptive techniques that undermine user autonomy and manipulate users into actions that are not in their best interest. These techniques are also known as *dark patterns*. The most abusive forms of dark patterns may be harmful, unethical, and/or illegal such as designing for addiction or for physical harm. This diagram encompasses design of all technologies including that of social robots

patients to take medication by providing arguments from analogy, from popular practice, or from expert opinion [25]. It is clear that the latter one will be more effective.

It should be noted that in Fogg's definition above, he does not specify or mention the mechanisms or cognitive processes through which the change is introduced in the receiver of the persuasive message by the technology. Attitude and behavior changing technologies are potent but at the same time social robot designers must be careful as to the methods they use in persuasion. It is tempting to deploy deceptive design to amplify behavior change so as to better benefit the designer rather than the user. The presence of dark patterns in a persuasive design may not always be evident [26]. To qualify as a dark pattern (1) the intent to persuade must be necessary, (2) the resulting user action is not in the user's best interest and is in the interest of the persuader, and (3) the methods employed are deceptive and covert. We could imagine a drug company designing a persuasive social robot which is given to patients who receive the drug sold with the drug to remind patients to take that drug but would increase consumption by manipulating slightly the dosage, or would vary the dosage to sub-populations to carry out trials without the knowledge of the patients.

7.4 Dark Patterns Meet Robotics

7.4.1 What Are Dark Patterns?

Dark Patterns is an umbrella term referring to a collection of practices used by designers, that exploit individual and social psychology and behavior to create user interfaces and user experiences that deceive or manipulate users (even create addiction) into doing things that they might not generally want to do. The user action thus performed is not in the user's best interest and benefits the designer or the company in question. The collection of dark patterns constitute deceptive design (see Fig. 7.3). Dark patterns may deceive users into making unwanted purchases, adopting excessive data mining configurations, creating unwanted negative habits [27], resulting in their incurring financial loss, increased intrusion of privacy, and addiction, respectively, while the designer gets increased revenue, increased user reach, and access to valuable user datasets. The phrase *dark patterns* was coined in 2011 by design specialist Harry Brignull with the registration of www.darkpatterns.org where he maintains a repository of such patterns found on e-commerce websites. Since then, more such instances of dark patterns in digital media and tools—websites, in video games [26], in mobile applications [28], in social media [29]—have been identified, compiled, and classified by vigilant netizens (see hashtags #darkpattern, #darkpatterns on Twitter), designers [19], community groups (for [30]), academic researchers [28, 31], and governmental organizations [32].

To illustrate examples of frequently encountered dark patterns on websites and apps, we have compiled a few of them in Fig. 7.4. Figure 7.4a shows screenshots of website interface elements of a search result on www.booking.com, one of the most widely used online hotel booking service. A structured combination of design elements such as colored text, pop-ups, icons, buttons, and numbers populate each hotel search result, creating a feeling of scarcity and fear-of-missing-out (FOMO) which encourages impulsive bookings. Why do we need to be informed with the message "You missed it!" in bold red text with an exclamation mark? It is fairly clear that most of these interface elements are irrelevant to the actual needs or interests of the user, and is in place only to benefit the company with increased bookings. Please note that such forms of dark patterns are very frequent on other e-commerce websites and is not restricted to www.booking.com.

Figure 7.4b displays a collection of mobile app and website screenshots highlighting visual dark patterns deployed by Facebook and Instagram to increase user engagement on their platforms. Collectively, the design elements such as "likes," push notifications, and the bright red spots known as badges, loading animations, along with other non-visual dark patterns such as notification sounds and the algorithms that use intermittent variable reward strategies and tailor the content shown to the user on the News Feed, to name a few of many, have adversely impacted user psychology. Users show symptoms of attachment and addiction [27] and reduced self-esteem and depression due to negative social comparisons [33, 34]. In addition, these design elements have seduced users into accepting privacy configurations that give away more user data than a user would generally want to share [28]. This

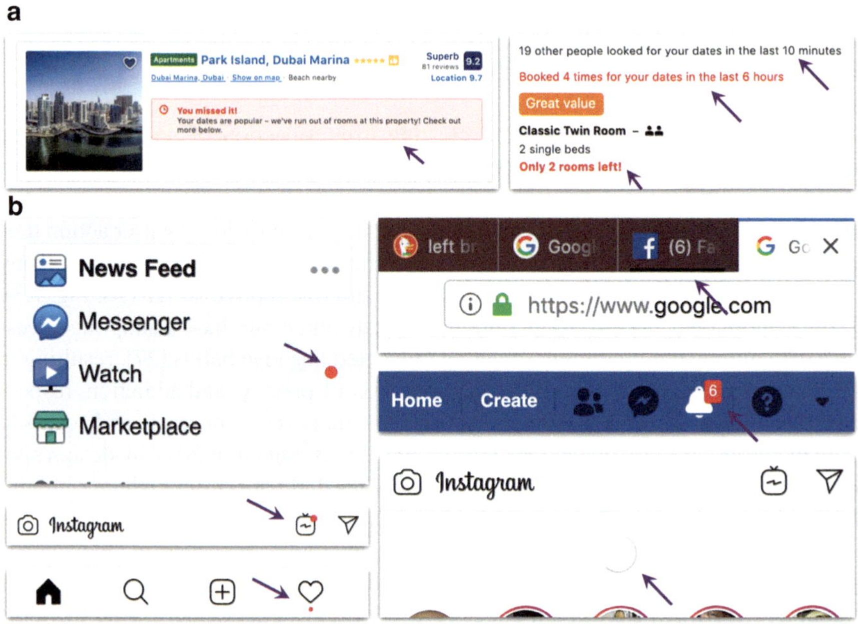

Fig. 7.4 Examples of design elements of dark pattern strategies found on popular website and smartphone applications. Purple arrows point towards visual and dynamic elements, which include texts in high-contrast color, push notifications, "likes," bright red colored badges, and loading animations that collectively contribute to a dark pattern strategy: (**a**) Screenshot of a single search result on www.booking.com that encourages impulsive purchasing. (**b**) Design elements on the web and app versions of Facebook and Instagram that increase engagement time but potentially increase user anxiety, attachment, and addiction

specific privacy-centered dark pattern is frequently named as *Privacy Zuckering* [19]. Such dark pattern deployment is not restricted to the Facebook group of companies alone, but are also used by the other four technology giants, namely Alphabet (Google), Apple, Amazon, and Microsoft. Collectively they deliver and serve digital services to practically everyone in the world with an internet connected smartphone or a computer.

7.4.2 Pervasiveness of Dark Patterns

Dark patterns have become so pervasive that it has almost become standard industry practice. In the last couple of years, various books and tutorials for designers have been published that serve as guides to create irresistible, tempting, evil, habit-forming and seductive digital products (adjectives have been borrowed from the book titles) [35–38]. Digital products, both software and hardware, compete for the attention of its users. Human attention has become an extremely valuable and scarce commodity in this information-rich digital world. With the multitude of distractions around, companies employ tactics to increase the time spent on their software, from games, to digital newspapers, and to social media websites. Hoover and Eyal rightly remark that "[a]s infinite distractions compete for our attention, companies are learning to master novel tactics to stay relevant in users' minds." [37].

Dark patterns aim at encouraging users to maximize user engagement time (at the loss of user's time), to make faster and more expensive purchase decisions (at the financial loss for user), to give up their personal identifiable data (resulting in privacy loss for user), and ultimately to undermine their own sense of personal identity (at the risk of technology redefining the user's sense of self). Both purchases and engagement time drive company profits through various means such as through third-party revenues like targeted advertisements, selling user data to data brokers, or by hooking customers to buy costly add-ons. The demands for massive year-on business growth demanded by shareholders and venture capitalists who fund these companies, have forced teams of designers, engineers, and marketers with the power of data and psychology to design products with dark patterns. What may have started as harmless ways of using knowledge of user psychology and behavior to improve user experience and engagement, has turned dark.

7.4.3 When Social Robots Meet Dark Patterns

The adoption and embodiment of dark pattern strategies into social robots seems imminent. We point out that the main challenge faced by social roboticists and other digital product and service companies is essentially the same: how to increase user engagement. Social robots, as pointed out, are potent social actors that can inform, educate, entertain, motivate, persuade, and provide companionship, while at the same time are engineered to maintain a delicate balance between illusion and trust.

Recently, a few academic studies have demonstrated the potential to program robots to manipulate or deceive humans. One such study [39] involved a mobile robot with speech synthesis capabilities that attempted to gain access to an unauthorized building. Using a well-known technique called piggybacking or tailgating, physically disguised and verbally addressing itself as a food delivery robot, the robot was able to convince authorized people to let the robot into the secure facility. Another recent study showed the ability of robots to extract valuable information from people by "lying" about their motivation [40]. These dark pattern practices used by robots can be also defined as robot *social engineering* [41], a term well-known in computer security, where humans manipulate humans to get hold of valuable information.

Are dark patterns already being deployed in commercially available social robots? To explore the presence and potential for dark patterns in robots beyond research labs, and to aid and focus our discussion, we use examples of two social robots available on the market, namely, (1) aibo, an entertainment and companionship robot, and (2) Pepper, an enterprise and home multi-application robot. The general concerns and implications that we raise here span the whole class of social robots and its development. Where and when appropriate, the identified robot dark patterns will be matched with the previously developed classification schemes developed for digital products and services [19, 28].

7.4.3.1 aibo

aibo™ is a robotic dog, developed by Sony Corporation and has the status of the world's first mass-market consumer robot for entertainment applications. Since its introduction in 1998, it has been a widely successful product having sold more than 150,000 units. The latest edition (model ERS1000) is available for sale in the United States at a price of $ 2899.99, and comes with a compulsory 3 years AI cloud service. It is also available for sale in Japan, but our analysis only covers the version offered in the USA. aibo can respond to voice commands and can perform up to 30 pre-programmed "tricks" (sequences of movements). Additional tricks can also be user programmed. It communicates non-verbally through bodily movements and gestures, noises, body temperature changes and through its multi-LED display in its eyes. It can recognize up to one hundred different user faces through the 20.7 mega pixel camera in its nose, which can also be set up to automatically take pictures a few times a day.

Dark patterns by definition, are deceptive and are not overt. To look out for potential dark patterns in the aibo ecosystem, we bring to consideration here three citations found on Sony's aibo website (https://us.aibo.com) and in its online marketing material; (1) "Your aibo can't wait to meet you" (see Fig. 7.5a), (2) "[aibo is] capable of forming an emotional bond with members of the household while providing them with love, affection, and the joy of nurturing and raising a companion." and (3) "Aibo keeps on growing and changing, constantly, updating it's data in the cloud. Over time, your approach to nurturing your aibo will gradually shape it's personality—it could be a doting partner, a wild fun-loving companion, or anywhere in between."

Phrase (1) projects onto us, what one of Sony's former lead robotics researcher Frédéric Kaplan calls the first of two clever design principles involved in the creation of aibo, "its apparent autonomy in the choice of its goals," that is the engineered illusion of aibo as a free creature [8]. Phrase (1) is featured on a website banner along with the picture of an *expectant* aibo in a kneeling position and a bright red colored "Store" button (see Fig. 7.5a). Here instead of using the scarcity effect or FOMO typically seen in e-commerce websites (Fig. 7.4a), the interface design of the aibo online store pushes our *darwinian buttons* of anthropomorphism, which attributes an intentional stance or agency to aibo, to make the purchase decision. Clearly, aibo, the robot, is not waiting for you to make that purchase.

Phrase (2) urges us to project love and affection to aibo and to nurture its growth as it is capable of emotionally bonding with humans. The *joy of nurturing* aibo that user's experience has clearly been a commercial success for Sony judging by aibo's sales figures, and this has come from the second design principle behind aibo, which in Kaplan's words is a kind of "[a]ffective blackmail. The owner must feel guilty if he doesn't take care of his pet" [8, 12]. The user is pressured to be responsible to provide care so that aibo develops and matures. As Phrase (2) and Kaplan confirm, the user's actions shape its personality, with a tricky feedback loop; the more the

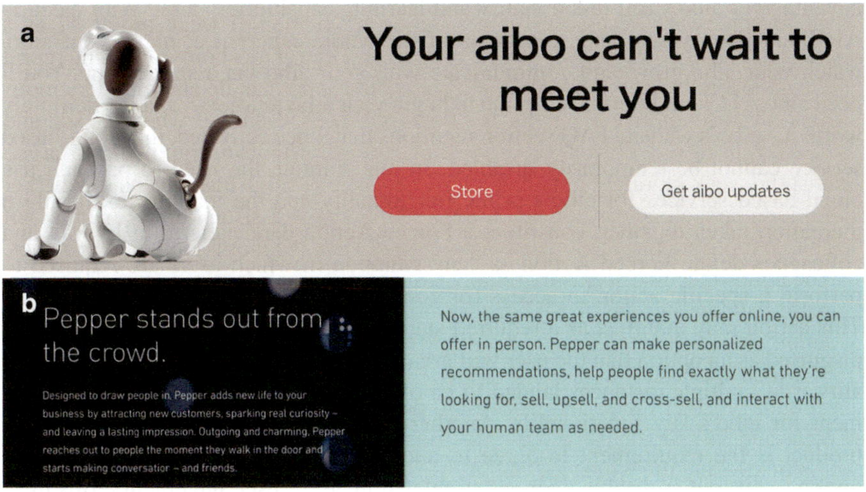

Fig. 7.5 Marketing material as seen on the websites of aibo and Pepper. The texts extensively play with anthropomorphism to convince viewers to purchase the robots. Additionally, the non-neutrality of the social robot design is also evident in the descriptive text accompanying Pepper

user interacts with the pet, the more crucial it becomes for the user to make sure that the pet does not get sad, grow improperly, or even "die." Such affective feedback loops have been used in the design of other digital toys for companionship, such as the extremely popular Tamagotchi.

The two key design principles behind aibo, paraphrased here as, the creation of the notion of aibo's apparent autonomy in the mind of its users, and designing aibo for affective blackmail of its users, are not transparent to the user and has only appeared in two relatively obscure technical publications [8, 12]. These design principles combined with the marketing claims detailed above, both of which play to our inherent anthropomorphic biases, is suggestive of a dark pattern strategy spread across departments of a multi-national corporation, that ultimately serves its economic interests and undermines the agency of users. To the author's best knowledge, there has been no attempt by Sony to clarify or make transparent as to how aibo's affective computing principles, as described by Kaplan, operate to its users. The aibo ecosystem now includes a variety of purchasable hardware (such as play toys and food bowls) and software add-ons (such as virtual coins). Isn't there a risk of creating an extreme form of emotional dependency through the implicit threat of aibo's improper growth? Shouldn't there be an approach of more transparency and explanation of aibo's internal algorithms to the users? Not doing so, but cloaking them could constitute a dark pattern [42].

Phrase (3) provides clues to another dark pattern. aibo's "growth and personality" development, a key feature of the companion robot, does not work unless users use Sony's cloud package and consent to the continuous upload of data to Sony's and third-party servers, which may include personally identifiable images, audio, and video. According to Sony, the cloud connectivity "is necessary to take advantage of aibo's full functionality and learning capabilities" and "With the aibo AI Cloud Plan, you get the essentials of the whole experience: name your aibo, watch your aibo grow, and communicate with your aibo via a special app. You'll need aibo AI Cloud Plan subscription to begin your aibo journey—and it's definitely worth it." aibo's website FAQ section mentions that, once activated, the cloud-based service cannot be temporarily disabled. In our opinion, the need to pay and use cloud-based services for aibo's prime functionality and the inability to disable it thereafter, taken together, constitute a Forced Action dark pattern [19]. Gray and colleagues define Forced Action as "any situation in which users are required to perform a specific action to access (or continue to access) specific functionality. This action may manifest as a required step to complete a process, or may appear disguised as an option that the user will greatly benefit from." aibo's cloud functionality is at multiple instances lauded to be of great use to the user and is a requirement for aibo's key functionality. Another example of Forced Action within the product is the requirement to agree to automatic software upgrades without the option to opt-out or disable them, as stipulated in aibo's user agreement terms [43].

Clocking a word count close to 12,500 spread across 23 pages, aibo's privacy statement and user terms of agreement [43, 44], are likely not read in depth or in detail when users first start using aibo, as studies with privacy policies of other digital products and services show [28, 45]. Included within aibo's privacy statement is a clause which requires a blanket agreement by the user to the processing of personal data, including recordings from the microphone and camera, to improve other unnamed products in Sony's offering, and also its transfer to unspecified third parties for service improvement and for advertising. Yet another clause in the user agreement [43] specifies that user content rights have to be granted to "Sony, its parents, subsidiaries, affiliates, successors, licensees and assigns, a non-exclusive, worldwide, perpetual, sublicensable, royalty-free license to use, host, store, modify, reproduce, distribute, create derivative works, publish, publicly perform and publicly display your User Content with respect to the Services. You hereby waive any moral rights you may have in and to any of your User Content, even if the User Content or a derivative work is altered or changed in a manner not agreeable to you."

The combined use of mandatory and non-opt-out cloud data processing, and the embedding of several blanketing and privacy-intrusive and data collection maximizing clauses, in a user unfriendly legal terminology within the privacy and user agreement policies, are suggestive of a privacy dark pattern strategy [28]. There are no technical or legal restraints that prevents human employees at Sony or at their third-party contract partners from accessing and viewing personal intimate data of aibo's users.

The privacy dark strategy deployed on aibo, including its non-opt out nature of aibo's cloud services and data processing, would be illegal in the European Union, under the new General Data Protect Regulation (GDPR). As of the time of writing this chapter, aibo (ERS-1000) has not been available for sale in the European Union. We also note that aibo is not offered for sale in the US state of Illinois, as its facial recognition system violates Illinois's data protection laws [44].

While aibo is marketed as an entertainment product, it has also been used in therapy of cognitively vulnerable populations to improve their psychological status and overall well-being [46], e.g., individuals with dementia, (Toshimitsu [47]), in social facilitation [48], and for loneliness reduction or companionship [49]. These cognitive vulnerable groups, which also includes contexts of children's education [50], may also be at risk of increased violation of privacy because of their inability to read or comprehend the complex terminology of the privacy policy or user agreement terms.

7.4.3.2 Pepper

Pepper is a socially proactive 1.2 meter tall humanoid robot, developed by Softbank Robotics, initially designed for business-to-business (B2B) applications with the goal to "intrigue and attract consumers" [9]. Introduced in 2014, approximately 20,000 units have been sold and are in operation in malls, retail stores, banks, schools, in elderly care and medical facilities worldwide. Only in Japan can it be found in homes, where it is available as a consumer product. Pepper is extremely versatile in its technical capabilities and also complex from a product point of view. Pepper available for enterprise applications adopts a robotics-as-a-platform model, where Pepper is sold or leased through regional third-party robotics companies to business customers who in turn operate these robots in retail, banking, educational, hospitality, and healthcare environments.

Pepper, as a mobile social robot platform operational in multiple public and commercial spaces worldwide, has the potential to harbor dark patterns. The presence of multiple actors, i.e., Softbank and its third-party service providers, the regional robot provider, the business/retail client and its employees, app developers, etc. in the Pepper ecosystem is a serious source of data privacy concern. One of the suppliers of Pepper in Europe, Humanizing Technologies, provides a content management system which can monitor every interaction that a user has made with Pepper. On their website the company writes that "We are tracking all behaviors and apps that are installed on your Pepper robot…This data helps you understand which behaviors of Pepper trigger emotions of your customers." Pepper utilizes an emotion engine powered by a third-party provider (Affectiva) to detect and classify subtle differences in facial expressions and voice. The partnership between Softbank and Affectiva raises further privacy concerns, as it is not transparent regarding how the data is handled and stored between these partners. Furthermore, researchers have reported the presence of several security flaws in Pepper's software which can allow a malicious third party to take control and command the robot [51].

Pepper can acquire personal persuasion histories about users, which contain information about the types of influences an individual is especially susceptible to and under what circumstances [52]. Personal persuasion profiles (including a variety of human signatures such as voice data, facial and affective data) of customers acquired by mobile social robots like Pepper in spaces like supermarkets and banks, open up opportunities for dark pattern deployment.

"Pepper gathers data over the course of conversations, learning people's tastes, traits, preferences, and habits to help personalize responses and better address needs. Pepper also collects new info to help you better understand both your customers and your business." appears on Softbank's website. Imagine a scenario where you enter a hardware store, when Pepper the sales assistant, recognizes you with its face recognition algorithms and your store visit history, and combining it with knowledge of your internet search history accessed through online data brokers, would guide you to the exact product you are looking for even without you saying a word, perhaps even recommending you a higher-priced product with better margins for the store. This is upselling, one of the touted opportunities for using Pepper (see Fig 7.5b).

7.5 Discussion

Social robots with its anthropomorphic design can potentially develop far more intimate and emotional relationships with its users than any other technology. Social robots are being developed to beneficially support humans in a wide range of contexts such as for entertainment, in education, for physical and emotional assistance and support, for elderly care, etc. Empirical data, such as from social robot intervention studies in psychosocial health contexts, indicate generally positive outcomes [53]. Other generally positive reviews point out reduced agitation and anxiety, increased social interaction, reduced loneliness, better use of medications amongst the elderly [54], more relaxation, more smiling, lesser pain, and openness and better communication within hospitalized children [55], etc. While the beneficial narrative of social robots is generally touted by robot researchers and designers, we question if the relationships formed between users and social robots particularly embedded with dark patterns negatively impact user's mental health and well-being?

We have seen the rise in user engagement on typical digital devices and associated services, such as smartphones and apps. Deployment of dark patterns utilizing behavioral psychology has a significant role to play in this upswing of user engagement. The time spent, however, may not always be in the user's best interest and could have a negative impact on various parameters of a user's psychosomatic well-being (quality and duration of sleep, levels of self-esteem, state of anxiety, etc.), hence raising issues about mental health as a matter of public health [29, 34, 56]. The extent of the public recognition of (potential) harm can be seen in the two acts introduced in 2019 by lawmakers in the USA in a bid to legally stop a sub-set of dark patterns deployed by large technology and social media companies, namely the Deceptive Experiences To Online Users Reduction (DETOUR) Act "to prohibit the usage of exploitative and deceptive practices by large online operators and to promote consumer welfare in the use of behavioral research by such providers." [57] and the Social Media Addiction Reduction Technology (SMART) Act "to prohibit social media companies from using practices that exploit human psychology or brain physiology to substantially impede freedom of choice, to require social media companies to take measures to mitigate the risks of internet addiction and psychological exploitation, and for other purposes" [58].

Realizing these negative user effects and fears of upcoming legal consequences, some companies have regulated themselves by rolling out screen time or engagement time monitoring add-ons to their digital products (such as Apple's Screen Time and Google's Digital Wellbeing). Some technology users are also intentionally staying away from tech products for periods of time, a phenomenon known as digital detox [59, 60], in the pursuit of well-being. While it can be said that the intensity of the harm on well-being is dependent on individual user susceptibility and usage context [26], aren't highly engaging interfaces directly effecting our ability to disengage, or in other words attenuating our personal autonomy of action?

Building upon the conceptual map of design (Fig. 7.3), we introduce a framework (illustrated in Fig. 7.6) to analyze design (of robots and in general all technologies) in terms of their effect on user autonomy which captures the gradative effects of design from being harmless to being harmful, unethical, and illegal.

Design (or the resultant technology) can deny user autonomy across four specific dimensions: (1) temporal—if we are not able to stop ourselves from spending valuable time on the design artifact, (2) monetary—if we are not able to stop ourself from spending money on the service, (3) data privacy—if we are not able to put a plug on giving up data, and consequently our privacy away, and (4) personal identity—if the technology undermines our sense of personal identity. An autonomy affirming technology would allow the user to freely stop using and paying for the product or service at will. Such a technology would also allow adequate informing and the freedom of choice in determining the optimum level of privacy as desired by the user.

It is important to remember that the process of development and market launch of a commercially viable or a "socially successful" social robot is a complex and expensive endeavor which requires a large amount of private and venture capital, even though the preliminary effort might have been funded by public funding in science and technology. Gray and colleagues write "complex entanglement among designer responsibility, organizational pressures, and neoliberal values often politicizes and prioritizes the profitability of design above other social motivations" [19].

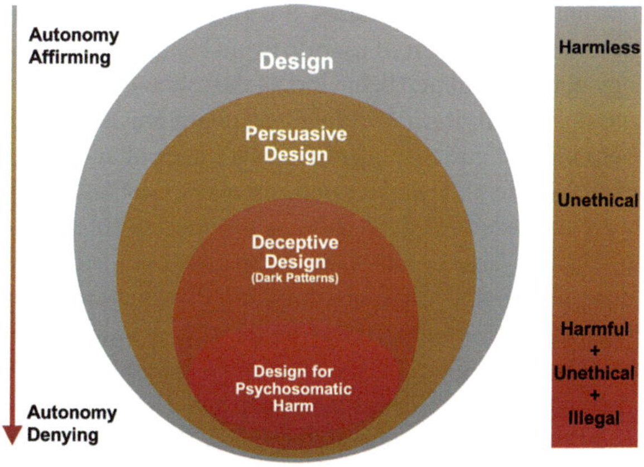

Fig. 7.6 Figurative schematic of our proposed framework for analysis

It is worth noting that out of the six organizers of the recent workshop titled *Social Robots in the Wild* which took place at the thirteenth ACM/IEEE International Conference on Human-Robot Interaction only one was an academic researcher while the rest were employees of technology companies developing social robots. Considering that deceptive design or the use of dark patterns in digital products is fairly common industry practice, if not a standard practice, the rise of deceptive design in commercial social robots is imminent. While empirical work on the negative effects of social robots on the psychosomatic health and well-being is an open field of research [61–63], we believe that prior research on the harm to mental health (in thematic areas such as dependency, addiction, loss of autonomy, loneliness, and anxiety) of users of well-established digital products and services (computers, smartphones, video games, social media, etc.) lays strong groundwork in this direction.

7.6 Conclusions

The dark patterns, and the underlying human cognitive biases and social factors that they exploit, we have identified in social robots are by no means exhaustive. We note that our investigation has been conducted only through publically available information and we have not purchased or used any of the aforementioned robots. It can be very well assumed that during actual usage, more potential dark patterns may emerge. We hope that these preliminary findings will form the basis of rigorous investigation and research in the future to develop a dark pattern typology that spans both software (web/app) based technologies and physical intelligent technologies like social robots. Dark pattern categorization in social robots requires a classification scheme quite different from the ones that have been developed for web or mobile applications, primarily because of the multi-modality of HRI, the physical embodiment, and their social context.

While users are increasingly becoming aware of dark patterns on websites and mobile applications, dark patterns embedded in robots are a relatively new phenomenon and hence harder to detect. With robot designers deliberately overplaying with our anthropomorphic biases, a certain degree of alertness and critical awareness is necessary on the part of users to avoid being misled by neither the marketing terminology nor the exhibited behaviors of the robot. Public information and education is paramount. Identifying, collecting, and compiling these dark patterns in social robots will help make it easier to identify recurrences of such patterns in other robotic products both for the public and for policy makers. Our suggestion would be to create a website with croudsourced dark patterns detected in robots and other embodied technology products. Websites like www.darkpatterns.org and www.darkpatterns.uxp2.com have done a remarkable job at compiling dark patterns in web-based tools and services. Additionally, a metric could be derived to indicate the number of appearances and severity of usage of dark patterns in a single product. It would be valuable if such a dark pattern score could be incorporated in the evaluation criteria of the upcoming Quality Mark from the Foundation for Responsible Robotics [64].

We urge social robot designers and their employers to avoid the usage of dark patterns and deceptive design. We encourage the adoption of an ethical design process by incorporating approaches such as Privacy by Design [65], Value Sensitive Design [66], Humane Design [30], and the IEEE's Ethically Aligned Design [67].

Social robots have immense potential for positive interventions in a society and to support human flourishing and well-being, and consequently, we point out to all social robot engineers and designers, an excellent ethical heuristic to prevent dark patterns creeping into their design process [68, 69]:

> The creators of a persuasive technology should never seek to persuade a person or persons of something they themselves would not consent to be persuaded to do.

References

1. Robinson NL, Turkay S, Cooper LAN, Johnson D. Social robots with gamification principles to increase long-term user interaction. 2019. p. 5.
2. International Federation of Robotics. Executive summary world robotics 2019 service robots. 2019. https://ifr.org/downloads/press2018/Executive_Summary_WR_Service_Robots_2019.pdf.
3. Fogg BJ. Persuasive technologies. Commun ACM. 1999;42(5):26–9. https://doi.org/10.1145/301353.301396.
4. Krieger MJB, Billeter J-B, Keller L. Ant-like task allocation and recruitment in cooperative robots. Nature. 2000;406(6799):992–5. https://doi.org/10.1038/35023164.
5. Kube CR, Zhang H. Collective robotics: from social insects to robots. Adapt Behav. 1993;2(2):189–218. https://doi.org/10.1177/105971239300200204.
6. Fong T, Nourbakhsh I, Dautenhahn K. A survey of socially interactive robots. Robot Auton Syst. 2003;42(3–4):143–66. https://doi.org/10.1016/S0921-8890(02)00372-X.
7. Hung L, Liu C, Woldum E, Au-Yeung A, Berndt A, Wallsworth C, Horne N, Gregorio M, Mann J, Chaudhury H. The benefits of and barriers to using a social robot PARO in care settings: a scoping review. BMC Geriatr. 2019;19(1):232. https://doi.org/10.1186/s12877-019-1244-6.
8. Kaplan F. Free creatures: the role of uselessness in the design of artificial pets. In: Proceedings of the 1st edutainment workshop. 2000.
9. Pandey AK, Gelin R. A mass-produced sociable humanoid robot: pepper: the first machine of its kind. IEEE Robot Automat Magaz. 2018;25(3):40–8. https://doi.org/10.1109/MRA.2018.2833157.
10. Leite I, Martinho C, Paiva A. Social robots for long-term interaction: a survey. Int J Soc Robot. 2013;5(2):291–308. https://doi.org/10.1007/s12369-013-0178-y.
11. Carros F, Meurer J, Löffler D, Unbehaun D, Matthies S, Koch I, Wieching R, Randall D, Hassenzahl M, Wulf V. Exploring human-robot interaction with the elderly: results from a ten-week case study in a care home. In: Proceedings of the 2020 CHI conference on human factors in computing systems. Honolulu, HI: ACM; 2020. p. 1–12. https://doi.org/10.1145/3313831.3376402.
12. Kaplan F. Artificial attachment: will a robot ever pass Ainsworth's strange situation test? In: Proceedings of humanoids. 2001. p. 125–32.
13. Kertész C, Turunen M. Exploratory analysis of sony AIBO users. AI Soc. 2019;34(3):625–38. https://doi.org/10.1007/s00146-018-0818-8.
14. Dennett DC. The intentional stance. MIT press; 1989.
15. Damiano L, Dumouchel P. Anthropomorphism in human–robot co-evolution. Front Psychol. 2018;9:468. https://doi.org/10.3389/fpsyg.2018.00468.

16. Duffy BR. Anthropomorphism and the social robot. Robot Auton Syst. 2003;42(3–4):177–90. https://doi.org/10.1016/S0921-8890(02)00374-3.
17. Dale JP, Goggin J, Leyda J, McIntyre AP, Negra D. The aesthetics and affects of cuteness. 1st ed. New York: Routledge; 2016. https://doi.org/10.4324/9781315658520.
18. Anderson SP. Towards an ethics of persuasion. UX Magazine, Dec 13, 2011. https://uxmag.com/articles/towards-an-ethics-of-persuasion.
19. Gray CM, Kou Y, Battles B, Hoggatt J, Toombs AL. The dark (patterns) side of UX design. 2018. p. 14.
20. Breazeal C. Social robots for health applications. In: 2011 annual international conference of the IEEE engineering in medicine and biology society. Boston, MA: IEEE; 2011. p. 5368–71. https://doi.org/10.1109/IEMBS.2011.6091328.
21. Breazeal C. The vision of sociable robots. In: Designing sociable robots. MIT Press; 2002.
22. Ham J, Midden CJH. A persuasive robot to stimulate energy conservation: the influence of positive and negative social feedback and task similarity on energy-consumption behavior. Int J Soc Robot. 2014;6(2):163–71. https://doi.org/10.1007/s12369-013-0205-z.
23. Siegel M, Breazeal C, Norton MI. Persuasive robotics: the influence of robot gender on human behavior. In: 2009 IEEE/RSJ international conference on intelligent robots and systems. St. Louis, MO: IEEE; 2009. p. 2563–68. https://doi.org/10.1109/IROS.2009.5354116.
24. Winkle K, Lemaignan S, Caleb-Solly P, Leonards U, Turton A, Bremner P. Effective persuasion strategies for socially assistive robots. In: 2019 14th ACM/IEEE international conference on human-robot interaction (HRI). IEEE; 2019. p. 277–85.
25. Rincon JA, Costa A, Novais P, Julian V, Carrascosa C. A new emotional robot assistant that facilitates human interaction and persuasion. Knowl Inf Syst. 2019;60(1):363–83. https://doi.org/10.1007/s10115-018-1231-9.
26. Zagal JP, Björk S, Lewis C. Dark patterns in the design of games. In: Proceedings of foundations of digital games. 2013. p. 8.
27. Schüll ND. Addiction by design: machine gambling in Las Vegas. Princeton University Press; 2014.
28. Bösch C, Erb B, Kargl F, Kopp H, Pfattheicher S. Tales from the Dark side: privacy dark strategies and privacy dark patterns. In: Proceedings on privacy enhancing technologies. 2016. p. 237–54.
29. Harris T. How technology is hijacking your mind — from a magician and Google design ethicist. Medium Magazine. May 18, 2016. https://medium.com/thrive-global/how-technology-hijacks-peoples-minds-from-a-magician-and-google-s-design-ethicist-56d62ef5edf3.
30. Center for Humane Technology. Human Design Guide (Alpha Version). 2019. https://humanetech.com/designguide.
31. Mathur A, Acar G, Friedman MJ, Lucherini E, Mayer J, Chetty M, Narayanan A. Dark patterns at scale: findings from a crawl of 11K shopping websites. ArXiv:1907.07032 [Cs]. 2019. https://doi.org/10.1145/3359183.
32. Forbrukerrådet (Consumer Council of Norway). Deceived By Design how tech companies use dark patterns to discourage us from exercising our rights to privacy. 2018. https://www.forbrukerradet.no/.
33. Appel H, Gerlach AL, Crusius J. The interplay between facebook use, social comparison, envy, and depression. Curr Opin Psychol. 2016;9:44–9.
34. Woods HC, Scott H. #Sleepyteens: social media use in adolescence is associated with poor sleep quality, anxiety, depression and low self-esteem. J Adolesc. 2016;51:41–9. https://doi.org/10.1016/j.adolescence.2016.05.008.
35. Nodder C. Evil by design: interaction design to lead us into temptation. Wiley; 2013.
36. Lewis C. Irresistible apps - motivational design patterns for apps, games, and web-based communities. Apress; 2014.
37. Hoover R, Eyal N. Hooked - how to build habit-forming products. Portfolio; 2014.
38. Anderson SP. Seductive interaction design: creating playful, fun, and effective user experiences. New Riders; 2011.
39. Booth S, Tompkin J, Pfister H, Waldo J, Gajos K, Nagpal R. Piggybacking robots: human-robot overtrust in university dormitory security. In: Proceedings of the 2017 ACM/IEEE

International conference on human-robot interaction - HRI '17. Vienna: ACM Press; 2017. p. 426–34. https://doi.org/10.1145/2909824.3020211.

40. Sanoubari E, Seo SH, Garcha D, Young JE, Loureiro-Rodriguez V. Good robot design or machiavellian? An in-the-wild robot leveraging minimal knowledge of passersby's culture. In: 2019 14th ACM/IEEE international conference on human-robot interaction (HRI). Daegu: IEEE; 2019, p. 382–91. https://doi.org/10.1109/HRI.2019.8673326.
41. Postnikoff B, Goldberg I. Robot social engineering: attacking human factors with non-human actors. In: Companion of the 2018 ACM/IEEE international conference on human-robot interaction - HRI '18. Chicago, IL: ACM Press; 2018, p. 313–14. https://doi.org/10.1145/3173386.3176908.
42. Chromik M, Eiband M, Völkel ST, Buschek D. Dark patterns of explainability, transparency, and user control for intelligent systems. In: Joint proceedings of the ACM IUI 2019 Workshops. 2019. p. 7.
43. Sony Electronics Inc. Aibo end user agreement. Sony Electronics Inc. 2019. https://us.aibo.com/terms/pdf/terms-ai.pdf.
44. Sony Electronics Inc. Aibo privacy policy. 2019. https://us.aibo.com/terms/pdf/aibo-privacy.pdf.
45. Obar JA, Oeldorf-Hirsch A. The biggest lie on the internet: ignoring the privacy policies and terms of service policies of social networking services. Inf Commun Soc. 2020;23(1):128–47. https://doi.org/10.1080/1369118X.2018.1486870.
46. Abdi J, Al-Hindawi A, Ng T, Vizcaychipi MP. Scoping review on the use of socially assistive robot technology in elderly care. BMJ Open. 2018;8(2):e018815. https://doi.org/10.1136/bmjopen-2017-018815.
47. Hamada T, Okubo H, Inoue K, MaruyamaJ, Onari H, Kagawa Y, Hashimoto T. Robot therapy as for recreation for elderly people with dementia - game recreation using a pet-type robot. In: RO-MAN 2008 - The 17th IEEE international symposium on robot and human interactive communication. Munich: IEEE; 2008. p. 174–79. https://doi.org/10.1109/ROMAN.2008.4600662.
48. Kramer SC, Friedmann E, Bernstein PL. Comparison of the effect of human interaction, animal-assisted therapy, and AIBO-assisted therapy on long-term care residents with dementia. Anthrozoös. 2009;22(1):43–57. https://doi.org/10.2752/175303708X390464.
49. Kanamori M, Suzuki M, Oshiro H, Tanaka M, Inoguchi T, Takasugi H, Saito Y, Yokoyama T. Pilot study on improvement of quality of life among elderly using a pet-type robot. In: Proceedings 2003 IEEE international symposium on computational intelligence in robotics and automation. computational intelligence in robotics and automation for the new millennium (Cat. No.03EX694), vol. 1. Kobe: IEEE; p. 107–12. 2003. https://doi.org/10.1109/CIRA.2003.1222072.
50. Decuir JD, Kozuki T, Matsuda V, Piazza J. A friendly face in robotics: sony's AIBO entertainment robot as an educational tool. Comput Entertain. 2004;2(2):14. https://doi.org/10.1145/1008213.1008236.
51. Giaretta A, De Donno M, Dragoni N. Adding salt to pepper: a structured security assessment over a humanoid robot. In: Proceedings of the 13th international conference on availability, reliability and security - ARES 2018. Hamburg: ACM Press; 2018. p. 1–8. https://doi.org/10.1145/3230833.3232807.
52. Kaptein, Maurits, and Dean Eckles. "Selecting effective means to any end: Futures and ethics of persuasion profiling." International conference on persuasive technology. Springer, Berlin, Heidelberg, 2010.
53. Robinson NL, Cottier TV, Kavanagh DJ. Psychosocial health interventions by social robots: systematic review of randomized controlled trials. J Med Internet Res. 2019;21(5):e13203. https://doi.org/10.2196/13203.
54. Pu L, Moyle W, Jones C, Todorovic M. The effectiveness of social robots for older adults: a systematic review and meta-analysis of randomized controlled studies. The Gerontologist. 2019;59(1):e37–51. https://doi.org/10.1093/geront/gny046.
55. Moerman CJ, van der Heide L, Heerink M. Social robots to support children's Well-being under medical treatment: a systematic state-of-the-art review. J Child Health Care. 2019;23(4):596–612. https://doi.org/10.1177/1367493518803031.

56. Clayton RB, Leshner G, Almond A. The extended iSelf: the impact of iPhone separation on cognition, emotion, and physiology. J Comput-Mediat Commun. 2015;20(2):119–35. https://doi.org/10.1111/jcc4.12109.
57. Warner M. Detour - deceptive experience to online users reduction act. 2019. https://www.congress.gov/116/bills/s1084/BILLS-116s1084is.pdf.
58. Hawley J. Social media addiction reduction technology (SMART) act. 2019. https://www.congress.gov/116/bills/s2314/BILLS-116s2314is.pdf.
59. Melton J, Verhulsdonck G, Shah V, Dunn P. Intentional non-use of the internet in a digital world: a textual analysis of disconnection narratives. In: 2019 IEEE international professional communication conference (ProComm). 2019. p. 65–6. https://doi.org/10.1109/ProComm.2019.00016.
60. Syvertsen T. Digital detox: the politics of disconnecting. Emerald Group Publishing; 2020.
61. Lacey C, Caudwell C. Cuteness as a 'dark pattern' in home robots. 2019. p. 8.
62. Nash K, Lea JM, Davies T, Yogeeswaran K. The bionic blues: robot rejection lowers self-esteem. Comput Hum Behav. 2018;78:59–63. https://doi.org/10.1016/j.chb.2017.09.018.
63. Sandoval EB. Addiction to social robots: a research proposal. In: 2019 14th ACM/IEEE international conference on human-robot interaction (HRI). Daegu: IEEE; 2019. p. 526–27. https://doi.org/10.1109/HRI.2019.8673143.
64. Foundation for Responsible Robotics. Assessment Principles of the FFR Quality Mark (blog). 2020. https://responsiblerobotics.org/quality-mark/assessment-principles/.
65. Danezis G, Domingo-Ferrer J, Hansen M, Hoepman J-H, Le Metayer D, Tirtea R, Schiffner S. Privacy and data protection by design-from policy to engineering. ArXiv Preprint ArXiv:1501.03726; 2015.
66. Friedman B, Hendry DG. Value sensitive design: shaping technology with moral imagination. MIT Press; 2019.
67. The IEEE Global Initiative on Ethics of Autonomous and Intelligent Systems. Ethically aligned design, 1st ed. IEEE; 2019.
68. Berdichevsky D, Neuenschwander E. Toward an ethics of persuasive technology. Commun ACM. 1999;42(5):51–8. https://doi.org/10.1145/301353.301410.
69. Fogg, Brian J. "Persuasive computers: perspectives and research directions." Proceedings of the SIGCHI conference on Human factors in computing systems. 1998. https://doi.org/10.1145/274644.274677.

8 Minding the AI: Ethical Challenges and Practice for AI Mental Health Care Tools

Nicole Martinez-Martin

8.1 Introduction

The use of artificial intelligence (AI) for mental health applications raises questions regarding the potential impact on fiduciary obligations in the therapeutic relationship, oversight, bias, and data protection. Health technologies utilizing AI present particular challenges for regulation. AI technologies that address mental and behavioral health may be used in different domains, from healthcare to education and consumer uses, and in some domains, there are not regulations or practices that provide protections for users' health data. There are also ways that bias can enter into AI tools, such as during the data collection and preparation stages. It is therefore necessary to consider how to utilize AI for mental health applications so that the resulting tools do not reflect and reinforce existing social problems and inequality. At the same time, AI can present opportunities for addressing existing inequalities and discrimination in mental health care. There is the potential for misuse of data and health information gathered from individual users, and users may not be sufficiently aware of negative repercussions from sharing their data. Finally, AI tools will likely impact the fiduciary obligations generally expected in the therapeutic relationship and it will be necessary to carefully consider likely areas of concern in order to prepare processes for integrating these tools appropriately into mental health care. This article chapter will engage emerging recommendations for best practices in this area, along with areas for empirical ethics research.

N. Martinez-Martin (✉)
Stanford Center for Biomedical Ethics, Stanford, CA, USA
e-mail: nicolemz@stanford.edu

F. Jotterand, M. Ienca (eds.), *Artificial Intelligence in Brain and Mental Health: Philosophical, Ethical & Policy Issues*, Advances in Neuroethics,
https://doi.org/10.1007/978-3-030-74188-4_8

8.2 Artificial Intelligence

"Artificial intelligence" generally refers to the use of machines to perform tasks that resemble cognitive functions that we associate with human intelligence, such as learning or solving problems [1]. Artificial intelligence (AI) can take different forms, as software or hardware, including intelligent autonomous agents, distributed networks, or robotics [2]. Although the term "machine learning" (ML) is sometimes used interchangeably with AI, ML more specifically applies to approaches that train computers to "learn"—recognizing patterns in massive datasets, including complex data interactions, and in the algorithms that are used within AI applications [3]. Machine learning has been used for data mining, image recognition, natural language programming, statistical learning methods, and neural networks, among other applications [4]. The ability of ML to detect patterns and connections that the humans programming the model would not have necessarily known to look for can bring significant benefits to scientific research. For example, ML can be used in order to analyze large quantities of data, such as electronic health records, in order to detect patterns and associations that may be relevant to patient health and outcomes [5]. These patterns can, in turn, be used for the purpose of predictive analytics and decision-making models, in ways that outperform traditional clinical prediction models [6]. ML has also been used to examine social media and websites in order to determine patterns in health-related behaviors [7]. ML and neural networks have been applied to constructing expert systems and clinical decision support systems, which are systems meant to provide and supplement the type of knowledge and skills generally supplied by human experts [8]. By incorporating ML, clinical decision support systems can provide recommendations without needing preprogrammed knowledge. As will be discussed more below, while the benefits of using ML to analyze massive datasets are considerable, the reasons or reasoning underlying the output of ML can be difficult to scrutinize, sometimes even for those who set up the ML system [9]. This is one reason that ML can present a challenge for regulation and oversight.

Natural language processing (NLP) is a subfield of AI that has a number of applications in mental health care [10]. NLP uses computational techniques to examine and classify language. NLP can be used for analysis of social media or vocal data to identify patterns relevant to mental health and behavior [11]. For example, NLP can be used as part of scanning clinical text for identifying symptoms of severe mental illness [12, 13], or for providing and analyzing psychotherapy encounters [14]. NLP can also be used to construct chatbots or virtual humans who can interact with people through text or voice [15]. AI techniques, such as ML and NP, have also been used for virtual reality and augmented reality technologies in order to make the virtual environments more engaging and interactive for the participant [16]. In looking at the different types of AI, one can note features that contribute to the challenging aspects of ethical applications of AI for mental health. The use of massive data sets presents areas of tension with data protection and privacy. The difficulty in knowing the reasoning or potential for bias in algorithms generated by ML can make evaluation of these technologies more difficult. Furthermore, when it comes to chatbots and VR in particular, AI-generated mental health technologies will have an impact on the therapeutic relationship in ways that go beyond traditional healthcare tools. Some of that impact could be positive, such as providing useful options for patients who may prefer sharing their feelings and information with a non-human. Potential negative repercussions include insufficient data protection or lack of clarity regarding liability for mistakes, which can also undermine trust overall in AI approaches to mental health care. The impact on the therapeutic relationship will need to be studied in order to better understand the benefits and burdens of AI mental health tools.

8.3 AI Mental Health Applications

AI is being integrated into a number of technologies for mental health care, from computing methods that can use massive data sets to assist in clinical decision-making, diagnosis, and treatment, to apps and wearables that can be used by patients and consumers for mental health assistance, and public health applications that assist in identifying behavioral health risks and solutions [17]. Some applications of AI involve boosting the capabilities of existing techniques and treatments, such as utilizing AI for deep brain stimulation approaches that respond dynamically to the needs of the patient [18]. It should be noted that the contexts for these different applications influence the types of ethical challenges encountered for that use. In the USA, for example, there are statutes that provide some protection for health information; however, these statutes and regulations generally apply to health information generated within the context of healthcare institutions and healthcare providers [19]. Even though some consumer AI applications can generate information about a person's mental health or behavior, that information may not have the same privacy protections that health information in the healthcare domain would be afforded [20]. Thus, as AI is being increasingly applied to mental health, it is important to note that many of these applications may be used in domains to which different privacy and user protection concerns are relevant, such as healthcare, consumer, or government institutions.

AI tools are being incorporated in the construction of expert systems [21, 22]. In clinical contexts, AI-informed expert systems may be used for such purposes as suggesting appropriate medications for a patient [23]. Predictive analytics are being increasingly utilized in healthcare environments, with AI often utilized for analyzing the data [24] These expert systems have traditionally been used in order to derive clinical rules or recommendations from the large amounts of data available in health systems, but, with the advent of more sophisticated AI, have become more focused on assisting with choices of differing probabilistic pathways [23]. Some have raised concerns that these decision-making tools will eventually replace the role of physicians, but the general goal is to make clinicians more effective with these tools. Providing sufficient training and support so that clinicians can utilize the information and findings provided by these AI-enhanced tools effectively remains a challenge [25]. In developing these decision-making tools, it is important to take account for how they may be influenced and affected by the context in which they are placed. In other words, the treatment decisions that are recommended by an AI tool will, in turn, impact the clinical environment, thus becoming another factor that will need to be accounted for in the analyses performed by the AI tools [26]. It is therefore important to carefully consider how the expert system will be implemented, so that it can be appropriately aligned with its environment and stakeholders.

AI has also become useful for development of technologies that provide simulations for therapeutic purposes. Autonomous conversational agents can be used to engage with a person, respond to text or vocal queries, and even provide some aspects of therapeutic interactions [16]. Chatbots can be used to respond to basic text queries regarding mental health needs in order to inform or direct the user to resources or services, and also for more complex interactions meant to provide aspects of therapy [27]. For example, Woebot is a conversational agent meant to address people with depression that incorporates tools drawn from cognitive behavioral therapy and can assist in monitoring mood, find learning videos and resources, and walk the user through "self-directed" therapy [28]. There is also an increasing role for robotics technology, incorporating AI, for mental health purposes. Robotics can be useful in cases where there may not be a person who can fill the role, such as robotics that can serve a companionship and support role (e.g., assisting users with getting exercise) for a patient [29]. Robots may be particularly useful in cases where the user may have reasons to prefer not to interact with a human for the therapy service. For example, robots have shown promise in assisting people on the autism spectrum develop skills, such as play [30] or social interaction [31]. With both chatbots and robotic technology, one of the ethical challenges relates to the possibility of blurred boundaries in user interactions with the bot, where users may lose sight of the fact that they are sharing information with a technology that can collect and pass along that data. It will be key to ensure that users are adequately informed about how privacy and confidentiality apply to the interactions, and how the design of the technologies may be used to address these types of concerns (such as switches or signals to the user when information is being recorded) [32].

Virtual reality (VR) is a technology that allows a user to experience a computer-generated simulated environment and interact with virtual persons or beings in that environment [33]. VR has become a tool for addressing a variety of mental health concerns, from use in PTSD treatments to assisting with diagnosis [34]. VR can also be used as a way to provide a virtual therapy space for real-time therapeutic interactions [35]. Augmented reality (AR) refers to the combination of VR with the world around someone by placing computer-generated images into the live video. AR has been used to help train mental health clinicians, remind psychiatric patients to take medications, and assist children who have autism learn to recognize facial emotions [33].

Mobile mental health applications also have been incorporating features through the use of AI. "Digital phenotyping" is a term commonly used to refer to approaches in which smartphones and mobile sensors are used to gather personal data from users, which is then analyzed in order to assess the user's cognitive and mental state, as well as make predictions [36, 37]. The data collected could be physiological functions, such as pulse, location information, tapping and keyboard interactions, or voice features [38]. Some approaches to digital phenotyping include analysis of social media posts and other internet use in order to assess behavioral health risks [39]. For clinical uses, the user would generally be asked to give informed consent and download an app onto their phone, which would passively collect the relevant personal data as the user goes about their usual daily activities. Beyond clinical usage, there are a range of institutions and organizations that may utilize digital phenotyping tools, such as educational institutions interested in assessing risk of suicide or of a student dropping out, the military assessing behavioral risks of recruits, insurance companies using such tools to set rates, employers, or consumer digital phenotyping for marketing purposes [40, 41].

As noted above, ethical concerns will differ depending upon the context of the application (e.g., different regulations and guidelines for data protection generally apply in healthcare contexts as opposed to consumer contexts). For uses that take place outside of healthcare, it is particularly important to examine the potential repercussions of inferences that can be drawn from an individual's personal data [42].

8.4 Ethical Challenges

Ethical challenges related to safety, effectiveness, or privacy are familiar areas of concern for new health technologies. Of course, AI tools in mental health care will raise varying ethical concerns according to their function. A conversational agent, for example, will likely raise concerns regarding how users interact with it therapeutically, that are different than concerns regarding how predictive analytics impact mental health care. Generally speaking, AI has some features that can pose difficulties for the traditional frameworks for addressing such ethical issues. The use of ML to generate algorithms, which puts the "reasoning" behind decisions into a proverbial "black box" can make it particularly challenging to examine and review the reasons behind the outputs generated by the algorithms. Thus agencies, such as the FDA, which is responsible for oversight of medical devices in the USA, have had to consider how to appropriately evaluate the accuracy and applications of AI technologies [43, 44]. A second issue, the use of these technologies in domains outside of healthcare, also impacts accountability and oversight. Information gathered in a healthcare setting would generally need to follow HIPAA privacy protections and involve informed consent procedures, which include protecting health and identifying information [45]. Digital phenotyping tools that could generate information and predictions about behavior and mental health, but are for consumer use, generally have fewer protections for user data or need for informed consent, often confined to notice about data practices on associated "terms and conditions" page. In some cases, the terms and conditions are misleading, not letting know the companies who may be receiving the data [46].

In the current big data environment, information that previously might be considered mundane or uninteresting, such as a grocery purchase or location at a particular moment, can be combined with other information and be transformed into health information [47]. Yet the paradigm for protection of health information is still based on traditional frameworks in which healthcare institutions and physicians are envisioned as the main domain for healthcare information [47]. Moreover, the massive amounts of data and techniques used for ML are often characterized as providing more objective results, but need to be carefully scrutinized for ways that bias may enter into the findings [48, 49]. Finally, while some argue that AI tools should just be seen as the same kind of device as any previous methods of assessing health risks, there are indications that people may regard AI tools as more objective than the human clinician or even as a third party involved in the clinical interaction [50]. For that reason, AI tools will likely influence the therapeutic relationship [51]. Because many ethical obligations are rooted in the therapeutic relationship, it is important to empirically study how AI impacts the therapeutic relationship in order to address any repercussions for associated ethical duties.

8.5 Therapeutic Relationship

The therapeutic relationship or alliance refers to the relationship that develops between a patient and the mental health care provider in order to achieve the goals for the patient [52]. In mental health care, the therapeutic relationship can involve the patient providing sensitive and emotionally charged personal information. The mental health care provider has professional obligations to protect the patient from harm and provide a foundation for achieving desired treatment outcomes [53]. For this reason, ethical values such as trust and confidentiality are key to the therapeutic relationship [54]. When it comes to the use of AI technologies for mental health care, there are many questions that may impact the therapeutic relationship. How might continuous monitoring affect trust? How do clinicians manage the massive amounts of data in order to extract meaningful information and communicate it to patients? Is the technology experienced as a "third party" in the clinical relationship? How will clinicians evaluate and incorporate findings from AI tools into their professional judgment and how patients will respond in terms of perceived stigma or bias in the predictions? There will be a need for empirical research to investigate the impact on trust in the clinician or digital tool, as well as how physicians rely and communicate health information and how patients view the competence of physicians and devices. Elderly and people with severe mental illness may face particular challenges in understanding the risks and benefits of using AI technology, or have different views regarding prioritizing ethical values, such as privacy [55]. When it comes to AI technologies such as conversational agents or robots that are used to interact with patients, designers and healthcare obligations must consider how the devices will affect these ethical obligations associated with the therapeutic relationship. Conversely, when these applications are used outside of healthcare institutions, are there ethical obligations generally found in the therapeutic relationship that should be addressed—for example, if a website analyzes its users' behavior, are there any duties to warn or direct users to resources that should be instituted [56].

For clinical use of these tools, organizations such as the American Psychiatric Association have been proactive in trying to establish recommendations for appropriate integration of these tools into clinical practice [57].

8.6 Safety and Effectiveness

Oversight for safety and efficacy of health technologies utilizing AI is still evolving. Regulation of health devices based on machine learning presents challenge because the reasons for particular results or findings may not be accessible for evaluation. In the USA, medical devices that utilize AI are subject to regulation by the Federal Drug Administration (FDA). The FDA has made significant efforts in recent years to establish effective approaches to regulate digital health technologies, including those that incorporate AI. The FDA has announced a Digital Health Program and a Pre-certification Program for manufacturers, which involves a shift from a product-based approach to a more process-based approach and does not address the issue of evaluating specific machine learning devices [58]. Professional organizations for computer science and AI have also discussed the need for designing AI systems that include mechanisms for a clinician or other user to receive more explanation of the bases of the results or findings that they have received [59].

Going forward, one significant issue for clinical applications of AI will be embedding established clinical standards in the ways that the tool is designed and used. ML approaches require large datasets and population sizes in order to produce validated models for expert systems and predictive analytics, and so issues of data sharing are important to consider. As systems and tools based on AI are increasingly integrated into healthcare, professionals will need to consider what the appropriate applications of AI for mental health care are, as well as the scope and limitations of the systems. In particular, interdisciplinary collaboration is needed for assessing if, when, and how AI applications are implemented, and different end users (clinicians, healthcare administrators, or patients) should be included in the development process in order to support ethical design and use of these tools [60]. As AI-based systems and tools are placed into different contexts and used among different populations, professionals using the system will need information on how the tools may best be used among different populations. Systems may need safeguards in place in order to ensure that the technologies are being used in the manner and for the population in which they have been validated. Of course, for technologies such as mental health apps or digital phenotyping, that may be used outside of healthcare institutions or, particularly, by consumers, it can be more difficult to establish lines of accountability and oversight that can ensure appropriate understanding and scope of use of the tools. In those instances, regulations that protect user data and require more robust user consent can help to inform users and require consent for use of their data.

Accountability for AI systems also involves questions regarding which entities are responsible for monitoring how the systems are functioning and being used, as well as liabilities for problems. If an AI tool causes harm or is not working as expected in a particular context, who is responsible for reporting and to whom they report? Furthermore, there will need to be consideration of how technologies such as expert systems or digital phenotyping may need to capability of monitoring risks of harm to patients or other users. In mental health contexts, where patients may disclose information that indicates a potential to harm self or others, how should tools monitor and assess such information and to whom will they need to report? These questions have come up in relation to conversational agents, in terms of whether these agents need to be programmed to provide resources or alert others if suicide risk is found [61]. Digital phenotyping is an area where, even if there is not direct disclosure from the patient about harming self or others, inferences could potentially be drawn from user data that leads to a prediction of harm [62]. Design of such tools need to incorporate consideration of monitoring and reporting potential harms, and institutions utilizing such systems need plans about how predictive tools and monitoring of potential for harm will be undertaken. Depending upon the jurisdiction, laws regarding duty-to-warn and other requirements will need to be taken into account.

8.7 Bias/Fairness

An important issue related to effectiveness and scope of use is methods for addressing the potential for bias in ML tools. The potential for bias can be viewed in terms of the potential for bias in the data used to construct the algorithms and bias in the algorithms themselves, as well as the potential for bias in how the algorithms may be used within a particular local context [63]. Because massive datasets are used in order to train ML systems to identify patterns in the data, the accuracy of the resulting algorithm depends on the quality of data in those training and validation sets [64]. Furthermore, if the dataset do not accurately reflect the population to the technology will be applied, then the bias in the data will be seen in the outcomes generated by the ML algorithm [65]. Thus ML systems could unfortunately both reflect and reinforce biases that are found in society. In terms of mental health applications of AI, social factors such as race, gender, and class can influence many aspects of mental health diagnosis outcomes. If the data used to generate an algorithm does not contain a representative sample, then the findings of the algorithm can be skewed. One of the reasons that it can be important to design "explainability" of the algorithm's reasoning into a tool is that algorithms may not have sufficient information (beyond the issue of a representative sample) to take into account why there may be certain associations between social factors and a particular mental health outcome. For example, an algorithm used for criminal justice sentencing may make an association between race and recidivism, but not have the data to take into account the impact of existing racial biases on recidivism [66]. These kinds of issues can not only limit the benefits that people from underrepresented racial and ethnic populations may receive from AI tools, but can exacerbate discrimination against particular groups. Efforts to increase the diversity of populations in datasets used for ML mental health research are critical. Professional organizations, such as the Institute of Electrical and Electronics Engineers (IEEE), have been conducting efforts to formulate recommendations and methods for reducing bias in ML algorithms [67, 68]. One important aspect is to include input from stakeholders for stakeholders, in order to provide feedback from clinicians and mental health consumers that can inform efforts to reduce bias. In implementation of ML-informed tools and systems, some institutions have also taken an approach to create an impact assessment of the tool beforehand, so that a plan can be developed and implemented to inform relevant stakeholders of potential impacts of the tool and make efforts to minimize that impact [69]. There needs to be reflection on the ethical implications of potential AI applications in mental health throughout the stages of development. As early as the stage formulation of the question or goal of the AI application, reflection on the ethical issues may be needed, because algorithms designed to identify psychiatric genetic risk for purposes or decide on allocation of healthcare resources can potentially raise ethical challenges regarding discrimination. In domains such as insurance or employment, there is also potential for discrimination in the construction of algorithms. In the USA, because laws regarding discrimination often rely on finding discriminatory intent, it may be more difficult to address such algorithmic discrimination through the courts [70]. There may be a need to consider regulations that would make certain types of discrimination based on behavioral predictions unlawful.

8.8 Privacy/Trust

Privacy and data protection have been identified as particularly important issues when it comes to big data approaches and AI technologies. In the mental health context, an important issue is that for some AI technologies, such as digital phenotyping, data may be collected in ways that individuals may not ordinarily associate with health information or even as sensitive data (such as speed of typing or tapping patterns on digital devices). Data may be collected outside of contexts in which healthcare information is protected by existing standards, such as HIPAA. Next, the data may be highly granular, especially in combination. Some data may be de-identified, but individuals may not be aware that, in combination with other data, the risk of identifiability may increase. Currently, patients or mental health consumers may not be aware of the ways that their personal data may be shared or sold to different organizations and companies, or that those companies can generate additional behavioral or health inferences about individuals. The data protection policies of mental health applications can have repercussions for individuals in areas such as employment, insurance, litigation that people may not reasonably have expected. At the same time, privacy concerns need to be balanced against data sharing practices that advance scientific research. In Europe, the General Data Protection Regulation presents a model for stronger data protections, including stricter consent provisions for the collection of data [71]. California has enacted similar provisions in the California Consumer Privacy Act [72]. While these regulations are useful for protecting personal data, the inferences that can be drawn from the data people share may still pose concern [73]. A reliance on consent as an approach to mitigate data protection concerns can also be problematic if not giving consent means that the user will be barred from using useful services. Stronger consent rules for use of personal data are necessary, particularly in contexts outside of healthcare, and provided at appropriate reading levels. At the same time, a focus on individual consent can overlook the need to include a broader range of people for broader discussion of how data may be ethically collected and the appropriate societal goals in doing so [74].

8.9 Surveillance

Technologies utilizing AI to monitor people's behavior may also have surveillance applications that are ethically challenging [75]. There are a number of institutions and companies that have an interest in monitoring individual behavior and conducting predictive analysis of mental states for a variety of reasons. Recently, the US government had proposed monitoring data from a range of wearables and apps to identify individuals for their potential to conduct mass shootings [76]. People diagnosed with mental illness were mentioned as a particular focus of such monitoring. While there was immediate pushback to this proposal from mental health and privacy advocates, the desire to use AI technologies to monitor people with mental illness for such purposes is not surprising. The use of facial recognition and genetic technologies for surveillance purposes in China has also received attention and criticism, as these technologies have been used to conduct behavioral surveillance, particularly targeting ethnic minority populations [77, 78]. There have been some laws passed on a local level to limit the use of facial recognition technology for surveillance [79], and there is a need to consider whether more regulation is needed. The use of technologies for behavioral and mental health surveillance can undermine the trust that people have in these technologies, use of their data, and the healthcare system. In the consumer domain, the massive collection and brokering of personal data is a part of what has been termed "surveillance capitalism." Beyond the issue of access to personal data, the inferences from these data can be used in attempts to influence and manipulate individuals for marketing and political purposes that raise ethical concerns on a societal level [80].

8.10 Conclusion

As efforts move forward to formulate guidelines and identify solutions to the ethical challenges presented by AI applications in mental health, there is a need for stakeholders with expertise in a range of disciplines, as well as patients and consumers, to come together and provide input. Transparency and informed consent have been commonly identified as goals, particularly in order to address some of the data protection and privacy challenges, in order to educate users and advise them of the potential repercussions of sharing their data. With AI technologies that are used for identifying and addressing behavioral issues outside of healthcare, ensuring meaningful consent of individuals remains challenging and elusive. Even though transparency and informed consent are important components of ethical use of mental health applications of AI, there remains a need to consider regulation to protect the privacy and safety of consumers, guard against discrimination in relation to predictive technologies, and overall ensure broader discussion and action take place regarding realizing societal as well as individual benefits from behavioral and mental health applications of AI.

References

1. Yu K-H, Beam AL, Kohane IS. Artificial intelligence in healthcare. Nat Biomed Eng. 2018;2(10):719. https://doi.org/10.1038/s41551-018-0305-z.
2. Price N. Artificial intelligence in health care: applications and legal issues. The Petrie-Flom Center for Health Law Policy, Biotechnology, and Bioethics at Harvard Law School. https://petrieflom.law.harvard.edu/resources/article/artificial-intelligence-in-health-care-applications-and-legal-issues. Accessed 2 Mar 2019.
3. Libbrecht MW, Noble WS. Machine learning applications in genetics and genomics. Nat Rev Genet. 2015;16(6):321–32. https://doi.org/10.1038/nrg3920.
4. Bzdok D, Meyer-Lindenberg A. Machine learning for precision psychiatry. ArXiv:1705.10553 [Stat]. 2017. http://arxiv.org/abs/1705.10553.
5. Rose S. Machine learning for prediction in electronic health data. JAMA Netw Open. 2018;1(4):e181404. https://doi.org/10.1001/jamanetworkopen.2018.1404.
6. Scalable and accurate deep learning with electronic health records. npj Digital Medicine. n.d. https://www.nature.com/articles/s41746-018-0029-1. Accessed 29 Aug 2019.
7. Hao B, Li L, Li A, Zhu T. Predicting mental health status on social media. In: Rau PLP, editor. Cross-cultural design. Cultural differences in everyday life. Berlin Heidelberg: Springer; 2013. p. 101–10.
8. Ngiam KY, Khor IW. Big data and machine learning algorithms for health-care delivery. Lancet Oncol. 2019;20(5):e262–73. https://doi.org/10.1016/S1470-2045(19)30149-4.
9. Mols B. In black box algorithms we trust (or do we?). https://cacm.acm.org/news/214618-in-black-box-algorithms-we-trust-or-do-we/fulltext. Accessed 31 Aug 2019.
10. Price WN. Regulating black-box medicine. SSRN Scholarly Paper. Rochester, NY: Social Science Research Network; 2017. https://papers.ssrn.com/abstract=2938391.
11. Demner-Fushman D, Chapman WW, McDonald CJ. What can natural language processing do for clinical decision support? J Biomed Inform. 2009;42(5):760–72. https://doi.org/10.1016/j.jbi.2009.08.007.
12. Jackson RG, Patel R, Jayatilleke N, Kolliakou A, Ball M, Gorrell G, Roberts A, Dobson RJ, Stewart R. Natural language processing to extract symptoms of severe mental illness from clinical text: the clinical record interactive search comprehensive data extraction (CRIS-CODE) project. BMJ Open. 2017;7(1):e012012. https://doi.org/10.1136/bmjopen-2016-012012.
13. Cook BL, Progovac AM, Chen P, Mullin B, Hou S, Baca-Garcia E. Novel use of natural language processing (NLP) to predict suicidal ideation and psychiatric symptoms in a text-based mental health intervention in Madrid [Research article]. 2016. https://doi.org/10.1155/2016/8708434.
14. Althoff T, Clark K, Leskovec J. Large-scale analysis of counseling conversations: an application of natural language processing to mental health. Trans Assoc Comput Linguist. 2016;4:463–76. https://doi.org/10.1162/tacl_a_00111.
15. Denecke K, May R, Deng Y. Towards emotion-sensitive conversational user interfaces in healthcare applications. Stud Health Technol Inform. 2019;264:1164–8. https://doi.org/10.3233/SHTI190409.

16. Miner A, Chow A, Adler S, Zaitsev I, Tero P, Darcy A, Paepcke A. Conversational agents and mental health: theory-informed assessment of language and affect. In: Proceedings of the fourth international conference on human agent interaction, 123–130. HAI '16. New York, NY: ACM; 2016. https://doi.org/10.1145/2974804.2974820.
17. Luxton DD. Chapter 1—An introduction to artificial intelligence in behavioral and mental health care. In: Luxton DD, editor. Artificial intelligence in behavioral and mental health care; 2016. p. 1–26. https://doi.org/10.1016/B978-0-12-420248-1.00001-5.
18. Patel UK, Anwar A, Saleem S, Malik P, Rasul B, Patel K, et al. Artificial intelligence as an emerging technology in the current care of neurological disorders. J Neurol. 2019; https://doi.org/10.1007/s00415-019-09518-3.
19. Rothstein MA. Health privacy in the electronic age. J Leg Med. 2007;28(4):487–501. https://doi.org/10.1080/01947640701732148.
20. Martinez-Martin N. What are important ethical implications of using facial recognition technology in health care? AMA J Ethics. 2019;21(2):180–7. https://doi.org/10.1001/amajethics.2019.180.
21. Bennett CC, Doub TW. Chapter 2—Expert systems in mental health care: AI applications in decision-making and consultation. In: Luxton DD, editor. Artificial intelligence in behavioral and mental health care; 2016. p. 27–51. https://doi.org/10.1016/B978-0-12-420248-1.00002-7.
22. Masri RY, Jani HM. Employing artificial intelligence techniques in Mental Health Diagnostic Expert System. In: 2012 international conference on computer information science (ICCIS), vol. 1. 2012. p. 495–99. https://doi.org/10.1109/ICCISci.2012.6297296.
23. Singh VK, Shrivastava U, Bouayad L, Padmanabhan B, Ialynytchev A, Schultz SK. Machine learning for psychiatric patient triaging: an investigation of cascading classifiers. J Am Med Inform Assoc JAMIA. 2018;25(11):1481–7. https://doi.org/10.1093/jamia/ocy109.
24. Koh HC, Tan G. Data mining applications in healthcare. J Healthcare Inform Manag JHIM. 2005;19(2):64–72.
25. Vayena E, Blasimme A, Cohen IG. Machine learning in medicine: addressing ethical challenges. PLoS Med. 2018;15(11):e1002689. https://doi.org/10.1371/journal.pmed.1002689.
26. Char DS, Shah NH, Magnus D. Implementing machine learning in health care—addressing ethical challenges. N Engl J Med. 2018;378(11):981–3. https://doi.org/10.1056/NEJMp1714229.
27. Laranjo L, Dunn AG, Tong HL, Kocaballi AB, Chen J, Bashir R, et al. Conversational agents in healthcare: a systematic review. J Am Med Inform Assoc. 2018;25(9):1248–58. https://doi.org/10.1093/jamia/ocy072.
28. Fitzpatrick KK, Darcy A, Vierhile M. Delivering cognitive behavior therapy to young adults with symptoms of depression and anxiety using a fully automated conversational agent (Woebot): a randomized controlled trial. JMIR Mental Health. 2017;4(2):e19.
29. Riek LD. Chapter 8—Robotics technology in mental health care. In: Luxton DD, editor. Artificial intelligence in behavioral and mental health care. San Diego: Academic Press; 2016. p. 185–203. https://doi.org/10.1016/B978-0-12-420248-1.00008-8.
30. Robins B, Dautenhahn K. Tactile interactions with a humanoid robot: novel play scenario implementations with children with autism. Int J Soc Robot. 2014;6(3):397–415. https://doi.org/10.1007/s12369-014-0228-0.
31. Vanderborght B, Simut R, Saldien J, Pop C, Rusu AS, Pintea S, Lefeber D, David DO. Using the social robot Probo as a social story telling agent for children with ASD. Interact Stud. 2012;13(3):348–72. https://doi.org/10.1075/is.13.3.02van.
32. Miner AS, Milstein A, Hancock JT. Talking to machines about personal mental health problems. JAMA. 2017; https://doi.org/10.1001/jama.2017.14151.
33. Lányi CS. Virtual reality in healthcare. In: Ichalkaranje N, Ichalkaranje A, Jain LC, editors. Intelligent paradigms for assistive and preventive healthcare; 2006. p. 87–116. https://doi.org/10.1007/11418337_3.
34. Virtual reality might be the next big thing for mental health. n.d. Scientific American Blog Network website: https://blogs.scientificamerican.com/observations/virtual-reality-might-be-the-next-big-thing-for-mental-health/. Accessed 20 Aug 2019.

35. Anderson PL, Price M, Edwards SM, Obasaju MA, Schmertz SK, Zimand E, Calamaras MR. Virtual reality exposure therapy for social anxiety disorder: a randomized controlled trial. J Consult Clin Psychol. 2013;81(5):751–60. https://doi.org/10.1037/a0033559.
36. Insel TR. Digital phenotyping: technology for a new science of behavior. JAMA. 2017;318(13):1215–6. https://doi.org/10.1001/jama.2017.11295.
37. Onnela J-P, Rauch SL. Harnessing smartphone-based digital phenotyping to enhance behavioral and mental health. Neuropsychopharmacology. 2016;41(7):1691–6. https://doi.org/10.1038/npp.2016.7.
38. Torous J, Staples P, Barnett I, Sandoval LR, Keshavan M, Onnela J-P. Characterizing the clinical relevance of digital phenotyping data quality with applications to a cohort with schizophrenia. Npj Digit Med. 2018;1(1):15. https://doi.org/10.1038/s41746-018-0022-8.
39. Jain SH, Powers BW, Hawkins JB, Brownstein JS. The digital phenotype. Nat Biotechnol. 2015;33(5):462–3. https://doi.org/10.1038/nbt.3223.
40. Kantrowitz L. When Facebook and Instagram think you're depressed. 2017. Vice website: https://www.vice.com/en_us/article/pg7d59/when-facebook-and-instagram-thinks-youre-depressed. Accessed 26 Oct 2017.
41. Dans E. The rise of real-time, context-based insurance. n.d. Forbes website: https://www.forbes.com/sites/enriquedans/2017/03/12/the-rise-of-real-time-context-based-insurance/. Accessed 29 Sept 2018.
42. Martinez-Martin N, Insel TR, Dagum P, Greely HT, Cho MK. Data mining for health: staking out the ethical territory of digital phenotyping. Npj Digit Med. 2018;1(1):68. https://doi.org/10.1038/s41746-018-0075-8.
43. Cortez NG, Cohen IG, Kesselheim AS. FDA regulation of mobile health technologies. N Engl J Med. 2014;371(4):372–9. https://doi.org/10.1056/NEJMhle1403384.
44. Center for Devices and Radiological Health. Digital Health [WebContent]. n.d. FDA.gov website: https://www.fda.gov/medicaldevices/digitalhealth/. Accessed 20 Feb 2018.
45. Glenn T, Monteith S. Privacy in the digital world: medical and health data outside of HIPAA protections. Curr Psychiatry Rep. 2014;16(11):494. https://doi.org/10.1007/s11920-014-0494-4.
46. Huckvale K, Torous J, Larsen ME. Assessment of the data sharing and privacy practices of smartphone apps for depression and smoking cessation. JAMA Netw Open. 2019;2(4):e192542. https://doi.org/10.1001/jamanetworkopen.2019.2542.
47. Bloss C, Nebeker C, Bietz M, Bae D, Bigby B, Devereaux M, et al. Reimagining human research protections for 21st century science. J Med Internet Res. 2016;18(12):e329. https://doi.org/10.2196/jmir.6634.
48. Danks D, London AJ. Algorithmic bias in autonomous systems. In: Proceedings of the 26th international joint conference on artificial intelligence. 2017. p. 4691–7. http://dl.acm.org/citation.cfm?id=3171837.3171944.
49. Mittelstadt BD, Floridi L. The ethics of big data: current and foreseeable issues in biomedical contexts. Sci Eng Ethics. 2016;22(2):303–41. https://doi.org/10.1007/s11948-015-9652-2.
50. Jha S, Topol EJ. Adapting to artificial intelligence: radiologists and pathologists as information specialists. JAMA. 2016;316(22):2353–4. https://doi.org/10.1001/jama.2016.17438.
51. Luxton DD. Artificial intelligence in psychological practice: current and future applications and implications. Prof Psychol Res Pract. 2014;45(5):332–9. https://doi.org/10.1037/a0034559.
52. Sucala M, Schnur JB, Constantino MJ, Miller SJ, Brackman EH, Montgomery GH. The therapeutic relationship in e-therapy for mental health: a systematic review. Journal of Medical Internet Research. 2012;14(4). https://doi.org/10.2196/jmir.2084.
53. Torous J, Roberts LW. The ethical use of mobile health technology in clinical psychiatry. J Nerv Ment Dis. 2017;205(1):4–8. https://doi.org/10.1097/NMD.0000000000000596.
54. Rendina HJ, Mustanski B. Privacy, trust, and data sharing in web-based and mobile research: participant perspectives in a large nationwide sample of men who have sex with men in the United States. J Med Internet Res. 2018;20(7):e233. https://doi.org/10.2196/jmir.9019.

55. Nebeker C, Lagare T, Takemoto M, et al. Engaging research participants to inform the ethical conduct of mobile imaging, pervasive sensing, and location tracking research. Transl Behav Med. 2016;6(4):577–86. https://doi.org/10.1007/s13142-016-0426-4.
56. Martinez-Martin N, Kreitmair K. Ethical issues for direct-to-consumer digital psychotherapy apps: addressing accountability, data protection, and consent. JMIR Mental Health. 2018;5(2). https://doi.org/10.2196/mental.9423.
57. Chan S, Torous J, Hinton L, Yellowlees P. Towards a framework for evaluating mobile mental health apps. Telemed J E-Health: Offic J Am Telemed Assoc. 2015;21(12):1038–41. https://doi.org/10.1089/tmj.2015.0002.
58. Center for Devices and Radiological Health. Digital health—digital health software pre-certification (Pre-Cert) program [WebContent]. n.d. https://www.fda.gov/MedicalDevices/DigitalHealth/UCM567265. Accessed 2 Aug 2018.
59. Koene A. Algorithmic bias: addressing growing concerns [leading edge]. IEEE Technol Soc Mag. 2017;36(2):31–2. https://doi.org/10.1109/MTS.2017.2697080.
60. Cohen IG, Amarasingham R, Shah A, Xie B, Lo B. The legal and ethical concerns that arise from using complex predictive analytics in health care. Health Aff. 2014;33(7):1139–47. https://doi.org/10.1377/hlthaff.2014.0048.
61. Miner AS, Milstein A, Schueller S, Hegde R, Mangurian C, Linos E. Smartphone-based conversational agents and responses to questions about mental health, interpersonal violence, and physical health. JAMA Intern Med. 2016;176(5):619–25. https://doi.org/10.1001/jamainternmed.2016.0400.
62. Torous J, Onnela J-P, Keshavan M. New dimensions and new tools to realize the potential of RDoC: digital phenotyping via smartphones and connected devices. Transl Psychiatry. 2017;7(3):e1053. https://doi.org/10.1038/tp.2017.25.
63. Glymour B, Herington J. Measuring the biases that matter: the ethical and casual foundations for measures of fairness in algorithms. In: Proceedings of the conference on fairness, accountability, and transparency. FAT* '19. Atlanta, GA: Association for Computing Machinery; 2019. p. 269–78. https://doi.org/10.1145/3287560.3287573.
64. Towards trustable machine learning. Nat Biomed Eng. 2018;2(10):709. https://doi.org/10.1038/s41551-018-0315-x.
65. Tunkelang D. Ten things everyone should know about machine learning. n.d. Forbes website: https://www.forbes.com/sites/quora/2017/09/06/ten-things-everyone-should-know-about-machine-learning/. Accessed 13 Jan 2018.
66. Dressel J, Farid H. The accuracy, fairness, and limits of predicting recidivism. Sci Adv. 2018;4(1):eaao5580. https://doi.org/10.1126/sciadv.aao5580.
67. Winfield A, Halverson M. Artificial intelligence and autonomous systems: why principles matter. n.d. IEEE Future Directions website: http://sites.ieee.org/futuredirections/tech-policy-ethics/september-2017/artificial-intelligence-and-autonomous-systems-why-principles-matter/. Accessed 28 Aug 2019.
68. Policy recommendations: control and responsible innovation of artificial intelligence. 2018. The Hastings Center website: https://www.thehastingscenter.org/news/policy-recommendations-control-responsible-innovation-artificial-intelligence/. Accessed 5 Dec 2018.
69. Institute AN. Algorithmic impact assessments: toward accountable automation in public agencies. 2018. Medium website: https://medium.com/@AINowInstitute/algorithmic-impact-assessments-toward-accountable-automation-in-public-agencies-bd9856e6fdde. Accessed 31 Aug 2019.
70. Kleinberg J, Ludwig J, Mullainathan S, Sunstein CR. Discrimination in the age of algorithms. Journal of Legal Analysis. 2018;10. https://doi.org/10.1093/jla/laz001.
71. EU General Data Protection Regulation (GDPR): Regulation (EU) 2016/679 of the European Parliament and of the Council of 27 April 2016.
72. California Consumer Privacy Act of 2018.
73. Wachter S, Mittelstadt B. A right to reasonable inferences: re-thinking data protection law in the age of big data and AI. SSRN Scholarly Paper. Rochester, NY: Social Science Research Network; 2019. https://papers.ssrn.com/abstract=3248829.

74. Costanza-Chock S. Design justice: towards an intersectional feminist framework for design theory and practice. SSRN Scholarly Paper. Rochester, NY: Social Science Research Network; 2018. https://papers.ssrn.com/abstract=3189696.
75. Martinez-Martin N, Char D. Surveillance and digital health. Am J Bioeth AJOB. 2018; 18(9):67–8. https://doi.org/10.1080/15265161.2018.1498954.
76. Wachter S, Mittelstadt B. A right to reasonable inferences: re-thinking data protection law in the age of big data and AI (SSRN Scholarly Paper No. ID 3248829). 2019. Social Science Research Network website: https://papers.ssrn.com/abstract=3248829.
77. Feng E. How China is using facial recognition technology. NPR.Org. n.d. https://www.npr.org/2019/12/16/788597818/how-china-is-using-facial-recognition-technology. Accessed 11 Mar 2020.
78. China uses DNA to map faces, with help from the west. The New York Times. n.d. https://www.nytimes.com/2019/12/03/business/china-dna-uighurs-xinjiang.html. Accessed 11 Mar 2020.
79. Conger K, Fausset R, Kovaleski SF. San Francisco bans facial recognition technology. The New York Times. 2019, May 14. https://www.nytimes.com/2019/05/14/us/facial-recognition-ban-san-francisco.html.
80. Big other: surveillance capitalism and the prospects of an information civilization—Shoshana Zuboff, 2015. n.d. https://journals.sagepub.com/doi/10.1057/jit.2015.5. Accessed 11 Mar 2020.

Digital Behavioral Technology, Deep Learning, and Self-Optimization

9

Karola Kreitmair

9.1 Introduction

Digital behavioral technology (DBT), which includes wearables, mobile health technology, certain smartphone apps, and various neurodevices, is a rapidly expanding class of technology that increasingly permeates all areas of human life. Individuals use such technology to shape themselves physiologically, psychologically, behaviorally, and socially, in order to become healthier, more mindful, better rested, more creative, and more intelligent versions of themselves.

While previously DBT may have been used in an offline manner, allowing an individual to measure her own performance, e.g., a heart rate belt to be used during running, DBT now consists of massively interconnected sensor and logging technology that yields a comprehensive picture of the physiological, environmental, behavioral, neurological, genomic, and social dimensions of a given individual. Moreover, the quantity of data this technology produces is so enormous, that the only viable means of gleaning robust insights from these data is through deep learning and AI.

While AI can gain valuable inferences from data, it is also prone to some serious flaws that are ethically problematic. Rather than producing objective results, as it is largely perceived to do by the public [1], AI propagates biases that are inherent in the data. Moreover, deep learning, the learning architecture that many AIs used in DBT are built on, operates in a way that may be opaque to human observers. This means that humans cannot explain or control how an AI comes up with a particular solution or organization. Furthermore, algorithms can game the reward functions designed to force them to learn, and deliver useless or harmful results.

K. Kreitmair (✉)
Medical History and Bioethics, University of Wisconsin—Madison, Madison, WI, USA
e-mail: kreitmair@wisc.edu

F. Jotterand, M. Ienca (eds.), *Artificial Intelligence in Brain and Mental Health: Philosophical, Ethical & Policy Issues*, Advances in Neuroethics,
https://doi.org/10.1007/978-3-030-74188-4_9

Within a broader societal context of the reponsibilization of the individual with respect to her health and well-being, DBT is increasingly used for all manner of self-optimization. AI factors into this self-optimization work, and along with it so do such issues as algorithmic bias, opacity, and reward hacking. Unbeknownst to the DBT user, these flaws in AI may affect how she is pursuing self-optimization.

In what follows I will provide an overview of DBT and explain how it is involved in self-optimization. I will then address how AI is incorporated into DBT. Next I will look at some issues with AI that have ethical ramifications. Finally, I will consider how these issues impact the self-optimization work that the user is doing with the help of her AI-enabled DBT.

9.2 Digital Behavioral Technology

Digital behavioral technology (DBT) is a class of technologies that is used by individuals to attempt to alter some aspect of their physiological, neurological, psychological, or behavioral selves. It comprises wearables, mobile health technologies, smartphone apps, and a variety of neurotechnologies [2]. Individuals want to shape their bodies, minds, and lives in ways they (or in some cases others) deem desirable, such as, for example, being more productive at work, improving cognition, meditating better, becoming fitter and healthier, sleeping better, being more creative, being a better lover, kicking undesirable habits, and losing weight. They believe that by using DBT such goals become more achievable.

While there is variety amongst these technologies, most DBT tends to be *personal*, *digital*, and *mobile*. It is *personal*, because it tends to be used by only one individual at a time and monitors or modulates only that individual's functioning. It is *digital*, because it utilizes binary computing systems that enable powerful processing and online connectivity. Finally, it is often *mobile*, because it is small and lightweight enough that individuals can carry it around on their persons [2]. For example, DBT includes sensor-driven technologies such as the Fitbit, that keep track of a user's location and activity levels [3], as well as stimulation technologies such as the transcranial direct current stimulation (tDCS) device BrainStimulator [4] that promise to improve cognition and help with depression, chronic pain, and anxiety.

What makes all these technologies instances of digital *behavioral* technology, is that they are all technologies intended to alter an individual's behavior. Often such behavior is sought to be improved for instrumental reasons, e.g., for a particular person, eating less may lead to weight loss which may lead to better health and well-being. Sometimes changing the behavior is an end in itself, as for instance with technologies that seek to limit symptoms of obsessive compulsive disorder [5]. Note that this does not mean that what is being measured or directly affected is always behavior itself. With much of this technology, what is being measured or affected are physiological or neurological properties of the user. However, it is behavior that is sought to be affected.

9.2.1 Functionalities

DBT operates through a vast array of *functionalities* (see Table 9.1).

One of the most common functionalities is *tracking*. Tracking is the passive registering and recording of a particular dynamic feature of an individual. It is passive, because the user need not actively input data into a device. Tracked dimensions include a wide array of features, such as, for example, an individual's location, her heart rate, her ECG, her breathing volume, the composition of her sweat, her EEG, and her blood alcohol level (see Table 9.2 for a list of the dimensions that can be tracked). In general, tracked dimensions lend themselves to quantification, which makes them amenable to analysis.

Tracking is distinct from *logging*, in which an individual uses the technology to record various features. Logging is commonly used for features that are not readily amenable to quantification, such as keeping track of qualitative dimensions such as mood or satisfaction. For example, the mobile app eMoods allows the user to note daily mood, irritability, and anxiety levels [14], while the mobile app Semistry allows the user to log, classify, and rate sexual encounters and activities [15].

Vocal analysis is a further functionality of DBT. Smartphone apps such as Cogito's Companion [16] detect speech patterns that are associated with mental health conditions like bipolar disorder and depression. Vocal analysis, such as provided by Voicesense, can also be used as a predictor of individual behavior, from the likelihood of someone defaulting on a loan to whether an employee is suffering early signs of burnout [17].

Visual analysis is also a functionality of DBT. Programs like Google Lens permits users to gain information about objects they capture with their smartphone's camera [18]. Users can scan a flower, find out what kind it is, and where the nearest florist is located. Or they can scan someone's outfit and receive information on the brand of the item, as well as similar items available for purchase.

Table 9.1 Functionalities

Functionalities	
Type	Example
Tracking	*See* Table 9.2 *for categories*
Logging	Sex logging (e.g., *Semistry*)
Vocal analysis	Speech pattern recognition (e.g., *Companion*)
Visual analysis	Visual object recognition (e.g., *Google Lens*)
Gamification	Mental health app (e.g., *SuperBetter*)
Stimulation	TDCS (e.g., *Foc.Us Go Flow*)
Drug delivery	Nicotine delivery (e.g., *Chrono Therapeutics*)
Virtual reality	Haptic VR suit (e.g., *Teslasuit*)
Assistant	Sleep assistant (e.g., *Neurogixs Alpha AI*)

Table 9.2 Tracked dimensions

Tracked Dimensions	
Location	E.g., Strava [6]
Activity	E.g., Fitbit Versa 2 [3]
Sleep	E.g., Fitbit Versa 2 [3]
ECG	E.g., Qardio [7]
EEG	E.g., Muse [8]
HR	E.g., Fitbit PurePulse [9]
Respiration rate and volume	[10]
BAC	E.g., Bactrack Skyn [11]
Ingestion events	E.g., Proteus Ingestible Sensor [12]
Sweat composition	[13]

A further functionality of DBT is *gamification.* By introducing game-like elements such as competition (against others or oneself), badges, points, and levels, DBT seeks to capitalize on the appeal of games to compel users to adhere to their use of the technology. Apps like SuperBetter provide users with games that involve completing quests and defeating "bad guys" in an effort to improve mental health, including tackling anxiety, depression, chronic pain, and recovering from concussions [19].

DBT may also function by directly *stimulating* the body or the brain. For example, transcutaneous electrical nerve stimulation devices such as the Thync Relax Pro [20] and transcranial direct current stimulation devices such as the foc.us Go Flow [21] deliver low-intensity electric currents to particular areas of the brain in an attempt to facilitate or inhibit neuronal activity in that area [22]. This is done to modulate brain functioning and improve cognition, relieve symptoms of anxiety and depression, combat cravings, and enhance meditation [23].

Another functionality of DBT is to directly *deliver drugs* that impact the body. Chrono Therapeutics for example, delivers medication such as nicotine transdermally in specific dosages timed to coincide with detected symptoms [24].

DBT also includes *virtual reality* (VR) systems. VR systems create a computer-generated environment into which the user can immerse herself. Such systems were once clunky, expensive, and required precise positioning of computers and sensors throughout a room. Today, they have shrunken to affordable untethered headsets, like Oculus Go [25]. VR can also be extended to include "tactile" or "haptic responsiveness." For example, the Teslasuit consists of a full body suit that uses nerve and muscle stimulation to generate haptic sensations for a fully bodily immersive VR experience. This allows users to feel a virtual breeze, the warmth of a virtual sun, or the touch of an avatar [26].

Finally, a new functionality of DBT is that of AI-enabled *assistant*. For example, Google Assistant, which is integrated into Google's Smartwatch, can listen and talk to the user, integrate questions with data analyzed from various other DBT, and provide the user with information and recommendations regarding a vast array of individual-specific information in a multitude of domains [27]. Examples of domain-specific AI assistants are: (1) Baby Connect [28] [29], which helps parents keep track of the quantity and quality of soiled diapers and expressed breast milk, computes the average duration of breastfeeding sessions, recommends when to nurse and when to switch breasts, and even sends information directly to Twitter; (2) Symptomate [30] [31], which analyzes the medical symptoms a user reports and uses AI to generate a differential diagnosis; (3) Neurogixs Alpha AI [32], which analyzes sleep conditions including EEG measurements and provides customized curation and recommendations; (4) CarePod from Sensory Health Systems [33], which provides lifestyle solutions for the elderly and those with limited mobility; (5) Good Morning Routine [34] and Bedtime Routine [35], which turns on/off lights, sets alarms, opens blinds, and briefs/debriefs you on the day; (6) Controlicz [36], which allows the user to speak to and give commands to smart objects and appliances around the home; (7) WorkAssist AI [37], which analyzes information collected from health and fitness wearables, mobile apps, and online behavior, to determine if the user's behavior is "beneficial or detrimental to productivity," and accordingly generate "clear instructions/recommendations to increase productivity" [38]; and (8) Girlfriend Maya [39], a chatbot who replies to a user's utterances, like "Good night, darling," with appropriately "girlfriend"-like responses.

Having identified the functionalities of DBT, let us turn to the types of users of this technology.

9.2.2 Usage Profiles

DBT is used in different circumstances and by different actors. There are thus different *usage profiles* of DBT. Some DBT is employed within the parameters of a *clinical context*, in which a healthcare provider oversees use of the technology. For instance, the Proteus ingestible sensor, which is embedded in pills and tablets, allows healthcare providers to monitor their patients' medication adherence [40]. A further usage profile of DBT is the *research context*. This includes both research that is conducted in academic settings and research by those developing DBT intended for sale, either with or without FDA approval. A third usage profile of DBT is the *direct-to-consumer* (DTC) context. As with clinical use, DBT is used here in order to "treat" the user (in some broad sense), but this is not done within the parameters of a clinical, provider–patient relationship. In this context, users simply purchase DBT products, the vast majority of which are not FDA-regulated, and apply them as they themselves see fit [2, 41]. The final usage profile is the *third-party* context. In this arena, DBT is used by a party that is not a healthcare provider, a researcher, nor the individual herself, in order to track or affect other individuals. Employee wellness programs, schools monitoring students, military applications all fall within this usage profile.

The focus of this chapter is self-optimization and so the primary usage profile with which I am concerned is DTC use. DTC DBT has gained hugely in popularity in the past 5–10 years. Reports by market research firms show that between 2010 and 2014 there was a 500% increase in the number of non-invasive neurotechnology patents filed [42]. The global wearables industry alone was valued at 32.63 billion

US dollars in 2019, and is expected to grow to over 100 billion US dollars by 2021 [43]. This growth suggests a perception on the part of the user that DTC DBT is useful in achieving one's goals [44]. The next section will discuss how this technology is implicated in the pursuit of self-optimization.

9.3 Digital Behavioral Technology and Self-Optimization

As described above, DBT can be used in the context of different usage profiles. Self-optimization is generally done within the DTC context, although it can sometimes occur in the third-party and even the clinical relationship contexts as well. Individual consumers are increasingly using DBT in order to optimize themselves. This is occurring as patients and consumers take on more of the burden of their own health and well-being. Sometimes referred to as the "democratization of healthcare," numerous observers [45] [46] have argued that technology is contributing towards the "responsibilization" of patients and consumers. Thanks to DBT, individuals now have the capability of using technology for the purpose of monitoring their own health and well-being, and use such technology to attempt to improve these [47]. This *capability* has contributed to an *expectation* that individuals now have a responsibility of improving their own health and well-being, as we will see below.

Much of the language around DBT includes this exhortation towards self-optimization. In the media, DBT is praised as a means to take on responsibility for improving oneself. Headlines like "This New Generation of Wearables Empowers People to Take Charge" [48] and claims that "wearables empower 'busy lives' to develop a more responsible approach towards themselves" [49], illustrate how DBT is perceived as increasing consumers' agency in their quest for well-being and health.

In the bioethics literature the term "e-patient," short for "empowered patient," has been coined to describe "health consumers participating fully in their medical care" [50, p. 2]. On this view of patients, individuals have an obligation to be informed about conditions and treatment options. Access to information gives patients a responsibility to take control in medical decision-making. The acquisition of this information has been facilitated by technology, specifically DBT. Thus, the "e" in "e-patient" has come to stand for "'electronic', 'equipped', 'enabled', [...] 'engaged' or 'expert'" [50, p. 2]. As Schmietow & Marckmann [46] note, "[s]elf-empowerment turns into a self-obligation to be 'digitally engaged' and at the same time expresses a shift of priorities from externally induced healthcare to a more elusive health and self-management" (p. 627).

Sociologist Deborah Lupton has documented this embrace of self-optimization. In her extensive research on *self-quantification*, i.e., the phenomenon of individuals embracing the tracking functionality of DBT (see Table 9.1), she identifies a desire of improving the self as a central focus of self-tracking activities that are designed to radically expand self-knowledge [45, 51, 52]. As a subject from one of her interviews puts it: "Unless something can be measured, it cannot be improved" [45, p. 67]. Another states, "[y]ou want to be your best self. [...] It's studying yourself as an interesting topic in ways that you couldn't study yourself before [...] this is just giving you self-awareness into previously invisible aspects of your life" [45, p. 65]. Lupton describes this as a practice "of self-hood that conforms to cultural expectations concerning self-awareness, reflection and taking responsibility for managing, governing oneself and improving one's life chances" [45, p. 68]. "Self-tracking therefore represents the apotheosis of the neoliberal entrepreneurial citizen ideal" [45, p. 68].

This conception of the self is in line with the existential notion of the self as something to be fashioned or created. In his analysis of the Nietzschean self, Anderson [53, p. 229] describes this concept of self-hood as something normative, i.e., a *task* that one is continually setting for oneself. In this way, the self is not a static component of an individual, but is constantly fashioned through the actions an individual undertakes. As Anderson notes, there appears to be a paradox in this notion of self-creation, for surely the thing being created must already exist to do the creating. But this paradox, he points out, can be dissolved if one distinguishes between a descriptive conception of the self, that carries out the plan of self-creation [54], and a normative conception of an ideal self that is the telos of one's self-fashioning pursuit [53]. In short, self-optimization as it is embraced by users of DBT (and as it is advertised by its manufacturers) presupposes something like the existential conception of a self that is continually being created [2].

Rather than focusing on the social determinants of health and well-being, DBT shifts the onus of responsibility to the individual. Here the assumptions are that this individual ought to be both equipped and motivated to take up DBT in pursuit of optimization. As becomes clear from Lupton's research, individuals see this technology as central in the project of fashioning themselves in the existential sense discussed above. DBT is used to craft the body and mind in a way that the individual endorses, as individuals believe that using this technology enables them to create the conditions to become the kinds of persons they want to become.

Such an outlook is grounded in a belief of what matters. It matters to be the best version of oneself, to be optimal. The idea of "working on the self" is central to how users of DBT see themselves in the world. But it is also embedded in a larger culture of productivity. Individuals who are responsible for all aspects of their goings-on populate the workplace. Productivity in the workplace is enmeshed with productivity at home. The same devices that are used for productivity at work are used in the private sphere and vice versa.

Thus, the picture that emerges is one in which DBT enables self-optimization to a point where individuals are expected to take on responsibility for all dimensions of the self. I identify three avenues of self-optimization: *information*, *parameterization of behavior*, and *direct interaction*.

9.3.1 Information

Individuals can use DBT to gain *information* about various dimensions of themselves. Specifically, DBT provides information on physiological (e.g., heart rate [9]), psychological (e.g., mood logging [14]), and neurological (e.g., EEG [55]) features of the self. Often, such information is then used by individuals to adjust behavior in ways to favorably affect these dimensions. Lupton talks about individuals achieving "knowledge, awareness, problem-solving" [56].

This kind of tracked (or logged) self-knowledge is a departure from how we standardly acquire self-knowledge. It encourages the user to gather information about the self through the processing of quantitative representation, rather than gaining self-knowledge through embodied situated unconscious cognition. This places the process of gaining information about the self on par with that of gaining information about objects external to the self. As such, the individual takes a third-person approach to herself, as she encounters her body, mind, and brain as a quantifiable object that permits of manipulation [2].

9.3.2 Parameterization of Behavior

A further use of DBT is the *parameterization of behavior*. DBT can issue signals or alarms when certain tracked values fall outside of desired parameters. For example, fitness trackers can alert an individual when her heart rate drops below a certain value. Signals can be used in this way to help users refrain from behaviors they find undesirable. Alternatively, users can also be rewarded, such as with badges or points, when values are within desired parameters. These methods give users incentives to behave well and disincentivizes poor behavior.

One technology that explicitly utilizes the principles of conditioning is the Pavlok 2 [57]. The Pavlok 2 is an aversive conditioning device, that emits small electric shocks when a user engages in behavior, e.g., nail-biting, smoking, eating sweets, sleeping too late, or spending too much time on time-wasting websites, that she is seeking to curtail. In this way, it aims to reinforce desirable behavior traits.

VR is a functionality of DBT that can also be used for behavior parameterization. VR generates alternative visual and auditory phenomenological experiences and can be extended to generate embodied virtual reality experiences including "tactile" or "haptic responsiveness." For example, the Teslasuit consists of a full body suit that uses nerve and muscle stimulation to generate haptic sensations for a fully bodily immersive VR experience [26]. One use of VR is in the addiction recovery. Patients can practice saying no to drugs in triggering environments, such as crack houses or bars [58]. While such addiction recovery is usually performed within the usage profile of a clinical relationship, there are many DTC applications that are either already being used or may be used in the future. For instance, consumers can use VR in a DTC setting to attend to their nicotine cravings. Alternatively, VR can *gamify* one's fitness routine by allowing a user to immerse herself into an alternate reality where she is boxing with a virtual opponent [59].

9.3.3 Direct Interaction

An additional use of DBT that contributes to self-optimization is *direct interaction*. For instance, DTC neurostimulation devices directly stimulate (or inhibit) parts of the brain in an attempt to impact brain function. Much of this direct interaction use is based on speculative scientific claims. For instance, technology that seeks to harness the effects of non-invasive vagus nerve stimulation to dampen the sympathetic nervous system response claims to enhance focus, promote positive thinking, and curb cravings [60]. In these cases, the user attempts to self-optimization not through conscious action, but rather through direct intervention in the relevant brain region. Various other neurostimulation technologies, such as the tDCS BrainStimulator, also operate in this way [4]. A further form of direct interaction is DBT that delivers medication. For instance, wearable devices by Chrono Therapeutics assist individuals in quitting smoking by monitoring nicotine levels in the blood and administering nicotine transdermally when individuals most require it [24].

9.4 Digital Behavioral Technology and Artificial Intelligence

DBT increasingly involves artificial intelligence (AI) in order to perform the functionalities mentioned above. The AI employed in this arena is trained through deep learning. Deep learning is a class of machine learning, in which an artificial neural network extracts patterns from data with which it is supplied. Deep learning extracts these patterns at multiple layers of abstraction, ranging from the specific to the more abstract. Given enough data and a large enough network, these networks can learn very complex patterns, such as recognizing faces from visual content, meaningful elements from natural language, and medical conditions from health data. Moreover, AI learns inductively from experience. Algorithms are iteratively updated when new data are provided. Such updating occurs without being explicitly programmed by human programmers. Rather, neural networks absorb new data and adjust connection weights between nodes in a stochastic fashion. This means both that the structure of neural networks is entirely dependent on the data that are inputted, and that there are no explicit programming rules discernable by humans [61–63].

Deep learning functions thanks to the availability of *big data*. Data scientists from IBM describe big data as being made up of four dimensions: *volume, variety, velocity*, and *veracity* [64]. The *volume* of available data is staggering. 2.3 trillion gigabytes are generated every day [65], with 90% of all currently existing data having been generated in the last 2 years. Experts believe 1.7 megabytes of data are created every second for every person on earth [65].

These data come from a *variety* of sources. Five new Facebook profiles are created every second and more than 300 million photos are uploaded to Facebook every day [66]. Every minute, 16 million text messages are sent [66]. At the same time, the number of connected wearable devices worldwide has increased from 325 million in 2016 to 722 million in 2019, with forecasts predicting this number to reach one billion by 2022 [67]. Meanwhile, the internet of things (IoT) has grown from two billion connected objects in 2006 to 200 billion in 2020 [68]. Moreover, an estimated 2.3 trillion gigabytes of electronic healthcare record data were produced in 2020 [69].

Different kinds of DBT generate different kinds of data, from location information to EEG, from photographic content to blood alcohol levels. Given the diversity of sources, combining the variety of data can yield extremely well-rounded representations of individuals and populations. The only way such large quantities of data can be processed in time is through powerful transaction processing systems (TPS). This is captured in the *velocity* of the four Vs.

Finally, *veracity* refers to the quality (accuracy and applicability) of the data. Much of the data that are available are of poor quality. They are inaccurate, incomplete, and even inconsistent. Data often needs to be cleaned so that so-called "dirty data" are kept from accumulating [70].

Thanks to the availability of AI, today's DBT user need no longer be satisfied with an n-of-1 trial, where she tracks or logs her own physiological, neurological, or emotional goings-on and pores over them in an attempt to discover behaviors that are conducive to self optimization. Today's DBT is "smart" in that the wealth of data that are produced by the DBT is combined with enormous amounts of data produced by other devices, including data integrated across different devices and platforms, in order to be fed into powerful machine learning programs that use deep learning to glean insights from the combined data. Thus, the data from an individual's wrist-worn fitness tracker, or an individual's EEG device, or an individual's location are just a small fraction of an ocean of information that reveals much more powerful insights about aspects of people's selves (their bodies, minds, environment), than any individual could glean on her own.

There are multiple ways in which AI can be integrated with DBT. First and foremost, DBT uses AI for *data analytics*. As noted above, the vast quantity of data requires deep learning in order to glean usable insights. In turn, deep learning makes such data incredibly valuable. Companies learn a considerable amount about individuals' health, behavior, activities, and beliefs from such AI-aided data analysis [71]. These insights not only reveal patterns at the population level, but also the individual level [72]. In addition to companies using this information in their own product and service development, it is standardly sold to other companies, such as marketing and health insurance companies. Moreover, with the increasing involvement of "Big Tech" in the healthcare arena, companies like Google are mining medical records [73], to expand their reach.

A further way AI may be integrated with DBT is for *restorative purposes*. Deep learning can be harnessed to allow individuals who are blind to regain some of the capabilities of sighted people. For example, AIServe [74] combines computer vision and AI to provide users with a wearable device that analyzes the environment, detects different elements in the surroundings, such as bikes, cars, and people, and then gives voice navigation instructions to the user. Deep learning is also employed in hearing aids. Thanks to AI, devices can learn what kind of environments a user tends to be in and learn to filter out desirable sounds from non-desirable ones, e.g., an interlocutor's voice from background noise [75]. AI-enhanced hearing aids can also directly translate language, transcribe what is being heard and said by the user, monitor brain functioning, and interface with a smart assistant [75].

Beyond restoration, AI-enabled DBT may also be used for *enhancement* purposes. Neurostimulation devices, such as the BRAINtellect 2, may be worn during sleep and employ AI to translate the user's brain waves into engineered music-like sound waves that are believed to enhance memory, learning, and wellness [76].

AI is increasingly employed to improve *VR* and augmented reality *experiences*. Thanks to deep learning, user preferences and virtual environment layouts can be updated in real time. For example, the Hololens2 is a deep learning enabled mixed reality system that generates constantly updated realities for the user which allow users to visually and tactilely interact with objects in their environment [77]. Hololens2 can also create avatars that deliver the words an individual is speaking in a different language [78].

AI is also used to enhance DBT to include *assistants*. In a previous section, I outline the various assistant functionalities. Given the self-optimization purpose of DBT, AI-enabled assistants are employed for health and wellness recommendations. AI is used to integrate vast quantities of data from DBT and other sources to provide recommendations to the user. Companies like LifeQ, for instance, use deep learning to develop models and algorithms that translate data from wearables and mobile apps into usable information. LifeQ provides individualized information for consumers on how to modify their behavior in real time to achieve their health and well-being goals, for insurance companies and corporate wellness programs on the health risks of individuals, and for clinicians on the treatment options, progress, and compliance of their patients [79]. Other assistants go beyond health and wellness, integrating advice from both "life and work." Galahad AI has introduced a virtual personal assistant, VYou, that helps individuals forgo short-term temptations in favor of long-term goals. Explicitly set up to allow users to engage in self-optimization, VYou "empower[s] people to better manage the most important and challenging aspects of their personal lives including their time, health, relationships, and money" [80].

Clearly, there are many ways in which AI is integrated into DBT. Particularly interesting from the perspective of self-optimization are smart assistants that advise and guide users on their quest to better themselves. Such assistants may even be enhanced via "affective computing"—a form of computing that allows assistants to deliberately involve and influence emotions and other affective phenomena [81].

Users gain beliefs about themselves through AI-enabled DBT that go beyond data captured by the technology. The technology takes on the role of guide and advisor, and can impact a user's reflective beliefs of self-worth. It can prompt a user to feel good about herself, for instance when a DBT assistant explicitly commends her for her behavior, or implicitly when the user recognizes that her behavior is within desirable parameters. An individual can also be made to feel poorly by DBT, either by being explicitly admonished by an assistant, or by finding herself failing at remaining within measurable parameters. Affective computing may increase this effect.

Of note, deep learning not only affects the insights that are gleaned from this technology, but in turn also affects the very way DBT functions. The AI-driven insights from data are fed into the functioning of the devices themselves. For instance, as a technology learns patterns from the behavior of a particular user along with the behavior of all the other users, it updates the parameters of what counts as "normal" and even "desirable" behavior.

9.5 Problems with Artificial Intelligence

As described, AI is a driving force for this technology. But there are well-documented flaws with deep learning. Learning algorithms are not static, yielding results through mathematical models. Rather they are iterative, constantly updating themselves as a result of the inputted data [1]. As already noted, inputted data are increasingly emerging from more and more sources, with the internet of things contributing to the well-spring of big data. This iterative process is vulnerable to systemic bias. If the data on which an algorithm is trained and which it uses to update itself contain bias, then the insights that such algorithms yield will be systematically biased. This can shift parameters of what may count as "normal," "correct," or "desirable."

Such algorithmic bias has been seen in healthcare decision-making—where black patients' needs are underestimated compared to those of white patients [82], and job hiring—where women's resumes were systematically scored lower than those of comparable men [83].

A further concern about AI is algorithmic opacity. Deep learning is not built on explicit theoretical rules inputted by human users. Rather, the internal workings of such an AI system are a "black box," that operates in whatever way yields the appropriate results. This means that users have no way of knowing how or why a certain system generates a particular result [84]. It also means that beyond pointing to an algorithm that has yielded correct results in other instances, no justification can be provided for a particular result in a particular instance [85].

A widely cited example of problematic algorithmic opacity is the Correctional Offender Management Profiling for Alternative Sanctions (COMPAS) tool. This is a decision support tool intended to help US judges to predict the likelihood of a defendant's risk of recidivism [86]. Judges can use a defendant's risk score to determine an appropriate length and severity of sentence. However, an investigation by ProPublica showed that COMPAS systemically overestimates the risk of recidivism for black offenders and underestimates the risk of recidivism for white offenders, even though race is not a data point that is available to the deep learning algorithm [87, 88]. But since COMPAS's inner workings are hidden from users, defendants have no recourse for arguing that in their case, the algorithm's results are unjustified. Moreover, since programs such as COMPAS are proprietary and thus subject to trade secrecy, defendants cannot even find out what inputted data were used in determining their recidivism score [89].

A further concern regarding AI is reward hacking [1]. The principle way in which so-called AI "agents" (i.e., algorithms) learn is through reinforcement learning. As in Skinnerian behaviorist psychology, correct behavior is sought to be reinforced. In reinforcement learning, an agent performs an action and is provided with feedback for that action from the environment. The feedback after each action is provided as a "reward" or a "punishment," where the size of the reward or punishment is a measure of how close the agent's action came to the correct action [90]. The goal of the AI agent is to maximize its rewards and minimize punishment. This way, through trial-and-error, the agent iteratively hones in on the correct result. However, this behaviorist approach can be gamed. An often-cited example is a cleaning robot who is rewarded for minimizing the amount of dirt it sees. Such a robot can simply close its eyes and thus receive a reward, even though this obviously does not fulfill the purpose intended by the human designer [91].

The problems of algorithmic bias and reward hacking are tenacious. Even when a given instance of bias or hacking is rectified, this provides no guarantee that a further instance will not arise. The opacity with which deep learning operates makes any attempt to correct structural biases or failures difficult.

9.6 Possible Effects of Problems with AI on DBT's Role in Self-Optimization

Given AI's prevalence in DBT, the aforementioned problems with AI may well manifest themselves in the self-optimizing function of DBT. I will look at how algorithmic bias, algorithmic opacity, and reward hacking affect the three avenues of self-optimization identified previously, i.e., direct interaction, information, and parameterization of behavior.

As mentioned above, neurostimulation devices use *direct interaction* for enhancement purposes. Tracked EEG sleep data is processed through deep learning algorithms in order to generate and emit music-like waves intended to improve wellness and cognition. Meanwhile devices like the Neuvana stimulate the Vagus nerve in order to enhance mental acuity, promote positive thinking, or curb cravings [60]. As AI is increasingly used by such devices, the risk of algorithmic bias becomes more significant. What the AI recognizes as a desirable EEG state will depend on a big data set sourced from many users over many individual instances. But any biases that inhere in the incoming data will be reflected in the parameters that the algorithm identifies as "normal" or "desirable." Perhaps users of this technology are more likely than the general population to aim for mental acuity or heightened concentration—a supposition that is plausible given that such individuals are more likely to employ such technology in the first place. As a result, parameters "endorsed" by the AI may be skewed. This is problematic, because an EEG state that is in fact representative of the general population would be identified as inadequate or undesirable. With the technologies discussed here, this may result in individuals receiving neurostimulation at times or in ways that are not in fact beneficial. Egregious instances of unwarranted neurostimulation may be rather noticeable and thus may quickly be weeded out. But slight shifts in desirable parameters may be less obvious and thus more insidious as they may contribute to a skewing of what is considered normal. Beyond neurostimulation, such an effect can be problematic within the self-optimization modes of *information* and *parameterization of behavior*.

Self-optimization through *information* relies heavily on the data analytics action of the AI. Algorithmic bias that corrupts such analytics may lead to a number of problematic consequences. As described previously, vocal analysis is used to identify a range of conditions, including which users suffer from depression or are likely to default on a loan. Algorithmic bias in such functionalities can lead to individuals being categorized as suffering from a condition when in fact they are not, or being categorized as not suffering from a condition when in fact they are. This could result both in users being discriminated against, and in users not receiving the assistance that the technology is intended to provide. If we consider the use of vocal analysis in determining something like an individual's likelihood to default on a loan, it is easy to see how algorithmic bias could yield unfair results on the basis of non-praiseworthy and non-blameworthy properties, much like was the case with the COMPAS tool in criminal contexts. Vocal analysis might indirectly discriminate against individuals with certain properties. This can occur even if these properties are ommitted from the data, because they are considered too sensitive. For instance, an individual's socioeconomic status (SES) may not be a data point that is available to an algorithm, because it might be discriminatory. However, individuals of low SES reliably have higher rates of smoking, of being exposed to secondhand smoke, and of being sick from smoking-related diseases than individuals of higher SES [92]. Whether an individual smokes or is exposed to secondhand smoke is reflected in that individual's vocal qualities, and is thus useable data for analytic algorithms. Consequently, even if the data available to an AI is free of any mention of SES, the algorithm may still carve up the population along SES dimensions, thanks to correlated proxy features. Moreover, because of algorithmic opacity, affected parties may have no way of knowing how an algorithm arrives at a particular categorization, and thus cannot assess whether such results are justified in a given case.

As noted previously, vocal analysis is also used for restorative purposes, for instance in AI-enabled hearing aids that filter out voices from background noise. Certain vocal features are reliably correlated with racial groups [93]. Algorithmic bias may lead to certain kinds of voices, possibly the voices of certain racial groups, being less audible to users of smart hearing aids than others. If we assume that in general people prefer talking to people whom they can understand well, then such bias might subtly affect the kinds of people with whom users choose to spend time. What's more, this effect may happen unbeknownst to users, who may not be able to point to why they are engaging in conversation with one group of people rather than another.

Alternatively, with AI-enabled translation performed by such smart hearing aids, thanks to algorithmic bias, translated utterances may contain words that inadequately express the speaker's intended semantic content. While this may merely be frustrating if it occurs in a morally neutral way, it is worrisome if utterances exhibit unintended discriminatory or offensive language. The latter has been shown by Microsoft's chatbot Tay to be a real risk [94]. Not only might this portray users as unfairly racist, sexist, or prejudiced in some other way, with prolonged use it may even erode a user's beliefs about her own views, as I have argued elsewhere [95].

The clearest example of how the flaws in AI may clash with the self-optimization function of DBT is within the mode of *parameterization of behavior*. As already mentioned, algorithmic bias can cause parameters of what counts as "normal" or "desirable" to shift. This is particularly the case with AI-enabled tracking technology, which is designed to identify appropriate values or value-ranges of trackable dimensions (see Table 9.2) and alert, praise, punish, or reward the user on the basis of her adherence to those values or value-ranges. We see this, for instance, in the Pavlok 2 that administers a small electric shock to the user if she falls outside of certain parameters, for instance if she exercises for too short of a duration. An AI-enabled version of such behavior-parameterizing tracking technology will obviate the need for human input on what constitute acceptable parameters, e.g., an appropriate duration for exercising, by arriving at such parameters on the basis of big data, including from users of the technology and a wide array of other sources. Then, this DBT can simply punish a user if she deviates from what the algorithm has deemed as desirable.

Thanks to algorithmic bias, however, the parameters towards which the DBT is steering the user may not actually correspond to parameters that should be sought after. Moreover, because of algorithmic opacity, a user may not be able to see why parameters are what they are, and thus why she falls short. This might result in users feeling unjustifiably discouraged about their performance, or conversely unwarrantedly accomplished. The ramifications of such effects will vary. For some users falsely appearing to fall outside of parameters or erroneously appearing to fall within them will have little consequence. For others who perhaps place great importance on being within certain bounds, wrongful categorization may be unduly burdensome. Users who exhibit compulsive behaviors, such as individuals with eating or exercising disorders, already employ DBT to contribute to their disorders [96, 97]. Skewing parameters of what counts as "normal" or "desirable" could exacerbate this harmful phenomenon.

Beyond the DTC context, such skewing of parameters may be problematic in the third-person context. When tracking technology is used by corporate wellness programs to dole out rewards and punishments [98], or by insurance companies to determine premiums [99], miscategorizations of users may lead to injustices in the same vein that have occurred with the COMPAS tool. Moreover, just as with COMPAS, the presence of algorithmic opacity makes redressing any such injustices hard if not impossible.

Finally, smart assistants are becoming widespread amongst AI-enabled DBT. As noted earlier, there are life assistants designed to help users with a huge array of tasks, including determining whether their behavior is beneficial or detrimental to productivity [37], whether their behavior will help them achieve their health and well-being goals [79], and whether they are successfully forgoing short-term temptations in favor of long-term goals [80]. Such DBT also assists insurance companies and corporate wellness programs in determining the health risks of individuals [79].

However, such smart assistants may be vulnerable to reward hacking. Assistants are "rewarded" for giving "good" "instructions and recommendations" [38] to users, where "good" is determined by running algorithms that mine big data from a plethora of sources. But honing in on what exactly constitutes "good" instructions and recommendations may be algorithmically burdensome. It may, for some algorithmically opaque reason, be easier to generate instructions and recommendations that can appear as "good," but are actually not. This is an instance of reward hacking. When the AI manages to find such "imposter" results, it receives its reward and has thus fulfilled its purpose. Of course, while this may be beneficial from the perspective of the AI, it is not beneficial at all for the human user. But again, weeding out such reward hacking issues is aggravated by algorithmic opacity. As humans rely more on their smart assistants to guide them in their pursuits of healthier, happier, more productive lives, issues of reward hacking will become more and more serious, as they threaten to undermine individuals' autonomy in fashioning themselves. Similarly, when corporate wellness reward and punishment structures, as well as insurance premiums, depend on a user's adherence to assistant-provided instructions and recommendations, reward hacking can contribute to an unjust distribution of burdens and benefits.

9.7 Conclusion

AI-enabled DBT has enormous potential to affect the way users engage in activities of self-optimization. Individuals are increasingly taking on the burden of engineering their own health, wellness, and productivity in explicitly engaged and active roles. This surging involvement is predicated on a belief that technology now exists to facilitate this effective self-fashioning. Algorithmic bias, algorithmic opacity, and reward hacking undermine this pursuit, often in ways that are unknown to the user. If individuals truly are to be empowered, these issues with deep learning must be addressed.

References

1. Osoba OA, IV WW. An Intelligence in our image: the risks of bias and errors in artificial intelligence. Rand Corporation; 2017.
2. Kreitmair KV. Dimensions of ethical direct-to-consumer neurotechnologies. AJOB Neurosci. 2019;10(4):152–66. https://doi.org/10.1080/21507740.2019.1665120.
3. Fitbit official site for activity trackers and more. n.d. https://www.fitbit.com/home. Accessed 15 Dec 2019.
4. The Brain Stimulator tDCS Devices—Shop and stimulate your life today! n.d. https://thebrain-stimulator.net/. Accessed 15 Dec 2019.
5. Brunelin J, Mondino M, Bation R, Palm U, Saoud M, Poulet E. Transcranial direct current stimulation for obsessive-compulsive disorder: a systematic review. Brain Sci. 2018;8(2):37. https://doi.org/10.3390/brainsci8020037.
6. Strava. Run and cycling tracking on the social network for athletes. n.d. https://www.strava.com/. Accessed 15 Feb 2020.

7. Qardio. Qardio official store. Qardio. n.d. https://store.getqardio.com/. Accessed 15 Feb 2020.
8. Muse—Meditation made easy. Muse. n.d. https://choosemuse.com/. Accessed 15 Feb 2020.
9. Fitbit PurePulse® continuous wrist-based heart rate. n.d. https://www.fitbit.com/purepulse. Accessed 9 Jan 2020.
10. Chu M, Nguyen T, Pandey V, Zhou Y, Pham HN, Bar-Yoseph R, Radom-Aizik S, Jain R, Cooper DM, Khine M. Respiration rate and volume measurements using wearable strain sensors. Npj Digit Med. 2019;2(1):1–9. https://doi.org/10.1038/s41746-019-0083-3.
11. BACtrack Skyn™—The World's 1st wearable alcohol monitor. n.d. BACtrack Skyn. https://skyn.bactrack.com/. Accessed 15 Feb 2020.
12. Many drugs could come equipped with ingestible sensors. Pharmacy Times. n.d. https://www.pharmacytimes.com/contributor/timothy-aungst-pharmd/2018/05/many-drugs-could-come-equipped-with-ingestible-sensors. Accessed 15 Feb 2020.
13. Bariya M, Nyein HYY, Javey A. Wearable sweat sensors. Nat Electron. 2018;1(3):160–71. https://doi.org/10.1038/s41928-018-0043-y.
14. LLC Y. Mood tracker by eMoods—Advanced Journal & Reporting Insights for Bipolar. EMoods - easy mood charting for bipolar/manic depression and other mood disorders. n.d. https://emoodtracker.com. Accessed 15 Dec 2019.
15. Semistry. Sexual statistics and measurements. n.d. https://www.semistry.com/. Accessed 15 Dec 2019.
16. Cogito spins out Companion app to detect depression in voice—Business Insider. n.d. Retrieved December 15, 2019, from https://www.businessinsider.com/goldman-salesforce-cogito-companion-detects-depression-in-voice-2018-12. Accessed 15 Dec 2019.
17. Voicesense. n.d. https://www.voicesense.com/. Accessed 15 Dec 2019.
18. How to use google lens to identify objects using your smartphone. Digital Trends. n.d. https://www.digitaltrends.com/mobile/how-to-use-google-lens/. Accessed 15 Dec 2019.
19. Roepke AM, Jaffee SR, Riffle OM, McGonigal J, Broome R, Maxwell B. Randomized controlled trial of superbetter, a smartphone-based/internet-based self-help tool to reduce depressive symptoms. Games Health J. 2015;4(3):235–46. https://doi.org/10.1089/g4h.2014.0046.
20. Thync Relax Pro review. Wareable. 2017. https://www.wareable.com/wearable-tech/thync-relax-pro-review.
21. Go flow 4mA tDCS and so-tDCS stimulator by Foc.us. Focus - take charge. n.d. https://foc.us/go-flow/. Accessed 9 Feb 2020.
22. Brunoni AR, Nitsche MA, Bolognini N, Bikson M, Wagner T, Merabet L, Edwards DJ, Valero-Cabre A, Rotenberg A, Pascual-Leone A, Ferrucci R, Priori A, Boggio P, Fregni F. Clinical research with transcranial direct current stimulation (tDCS): challenges and future directions. Brain Stimul. 2012;5(3):175–95. https://doi.org/10.1016/j.brs.2011.03.002.
23. Landhuis E. Do D.I.Y. brain-booster devices work? Scientific American. n.d. https://www.scientificamerican.com/article/do-diy-brain-booster-devices-work/. Accessed 18 Dec 2019.
24. Chrono Therapeutics. Chrono therapeutics. n.d. https://www.chronothera.com. Accessed 29 Nov 2019.
25. Oculus Go: standalone VR headset. Oculus. n.d. https://www.oculus.com/go/. Accessed 18 Dec 2019.
26. Teslasuit does full-body haptic feedback for VR. Engadget. n.d. https://www.engadget.com/2016/01/06/teslasuit-haptic-vr/. Accessed 29 Nov 2019.
27. Google assistant. n.d. https://assistant.google.com/explore. Accessed 18 Dec 2019.
28. Software S. Baby connect: baby tracker and log for android, iPhone, iPad, Kindle and for the web. n.d. https://www.babyconnect.com/. Accessed 18 Dec 2019.
29. Baby connect. Google assistant. n.d. https://assistant.google.com/services/a/uid/0000006d500622ec?hl=en-US. Accessed 18 Dec 2019.
30. Symptomate – Check your symptoms online. n.d. https://symptomate.com. Accessed 18 Dec 2019.
31. Symptomate. Google assistant. n.d. https://assistant.google.com/services/a/uid/000000a04396478b?hl=en-US. Accessed 18 Dec 2019.

32. NEUROGIXS. n.d. http://neurobeat-a.com/product/product.php?ptype=view&prdcode=1811280001&catcode=10000000&page=1&searchopt=&searchkey=. Accessed 18 Dec 2019.
33. Care pod. Google assistant. n.d. https://assistant.google.com/services/a/uid/000000e0fa0c2605?hl=en-US. Accessed 8 Jan 2020.
34. Good morning routine. Google assistant. n.d. https://assistant.google.com/services/a/uid/0000004f0e04d1da?hl=en-US. Accessed 8 Jan 2020.
35. Bedtime routine. Google assistant. n.d. https://assistant.google.com/services/a/uid/0000001807f0256f?hl=en-US. Accessed 8 Jan 2020.
36. Controlicz—The voice of your Domoticz Home Automation System. n.d. https://www.controlicz.com/. Accessed 8 Jan 2020.
37. Flow Harmonics. n.d. https://flowharmonics.com/works.html. Accessed 8 Jan 2020.
38. Flow Harmonics. n.d. https://flowharmonics.com/people.html. Accessed 8 Jan 2020.
39. Girlfriend Maya. Google assistant. n.d. https://assistant.google.com/services/a/uid/000000efa718ac50?hl=en-US. Accessed 8 Jan 2020.
40. Belknap R, Weis S, Brookens A, Au-Yeung KY, Moon G, DiCarlo L, Reves R. Feasibility of an ingestible sensor-based system for monitoring adherence to tuberculosis therapy. PLoS One. 2013;8(1):e53373. https://doi.org/10.1371/journal.pone.0053373.
41. Illes J. Neuroethics: anticipating the future. Oxford University Press; 2017.
42. SharpBrains. Market report on pervasive neurotechnology: a groundbreaking analysis of 10,000+ patent filings transforming medicine, health, entertainment and business. 2018. https://sharpbrains.com/pervasive-neurotechnology/. Accessed 24 May 2021.
43. Grand View Research. Market analysis report: wearable technology market size, share & trends analysis report by product (wrist-wear, eye-wear & head-wear, foot-wear, neck-wear, body-wear), by application, by region, and segment forecasts, 2020–2027. 2020. https://www.grandviewresearch.com/industryanalysis/wearable-technology-market. Accessed 26 May 2021.
44. Pfeiffer J. Quantify-me: consumer acceptance of wearable self-tracking devices. 16. 2016.
45. Lupton D. The quantified self. Wiley; 2016.
46. Schmietow B, Marckmann G. Mobile health ethics and the expanding role of autonomy. Med Health Care Philos. 2019;22(4):623–30. https://doi.org/10.1007/s11019-019-09900-y.
47. Sharon T. Self-tracking for health and the quantified self: re-articulating autonomy, solidarity, and authenticity in an age of personalized healthcare. Philos Technol. 2017;30(1):93–121. https://doi.org/10.1007/s13347-016-0215-5.
48. McGillin F. This new generation of wearables empowers people to take charge. The Doctor Weighs In. 2017. https://thedoctorweighsin.com/this-new-generation-of-wearables-empowers-people-to-take-charge/.
49. Day1Tech. How Wearables are transforming the Healthcare Industry. Medium. 2019. https://medium.com/@Day1Tech/how-wearables-are-transforming-the-healthcare-industry-e103f7985c8.
50. Meskó B, Radó N, Győrffy Z. Opinion leader empowered patients about the era of digital health: a qualitative study. BMJ Open. 2019;9(3):e025267. https://doi.org/10.1136/bmjopen-2018-025267.
51. Sumartojo S, Pink S, Lupton D, LaBond CH. The affective intensities of datafied space. Emot Space Soc. 2016;21:33–40. https://doi.org/10.1016/j.emospa.2016.10.004.
52. Lupton D. How do data come to matter? Living and becoming with personal data. Big Data Soc. 2018;5(2):2053951718786314. https://doi.org/10.1177/2053951718786314.
53. Anderson RL. What is a Nietzschean self? 1. Oxford University Press. 2012. https://www.oxfordscholarship.com/view/10.1093/acprof:oso/9780199583676.001.0001/acprof-9780199583676-chapter-9. Accessed 15 Apr 2020.
54. Sartre J.-P. Being and nothingness (Barnes HE, Trans.; Original ed. edition). Washington Square Press; 1993.
55. Advanced EEG technology—backed by science. EMOTIV. n.d. https://www.emotiv.com/our-technology/. Accessed 9 Jan 2020.
56. Lupton D. 'It's made me a lot more aware': a new materialist analysis of health self-tracking. Media Int Aust. 2019;171(1):66–79. https://doi.org/10.1177/1329878X19844042.

57. The Science. n.d. Pavlok. https://pavlok.com/science/. Accessed 29 Nov 2019.
58. Strauss G. Virtual reality: a powerful new tool for addiction treatment centers. Medium. 2019. https://blog.limbix.com/virtual-reality-a-powerful-new-tool-for-addiction-treatment-centers-a9dc7437fdd9.
59. Top 15 best VR fitness games for a total body workout. n.d. https://www.vrfitnessinsider.com/top-15-best-vr-fitness-games-total-body-workout/. Accessed 29 Nov 2019.
60. Science: Vagus nerve stimulation - Neuvana. n.d. https://Neuvanalife.Com/. https://neuvanalife.com/science/. Accessed 5 Feb 2020.
61. LeCun Y, Bengio Y, Hinton G. Deep learning. Nature. 2015;521(7553):436–44. https://doi.org/10.1038/nature14539.
62. Helbing D. Societal, economic, ethical and legal challenges of the digital revolution: from big data to deep learning, artificial intelligence, and manipulative technologies. ArXiv:1504.03751 [Physics]. 2015. http://arxiv.org/abs/1504.03751.
63. Luxton DD, Anderson SL, Anderson M. Chapter 11—Ethical issues and artificial intelligence technologies in behavioral and mental health care. In: Luxton DD, editor. Artificial intelligence in behavioral and mental health care. Academic Press; 2016. p. 255–76. https://doi.org/10.1016/B978-0-12-420248-1.00011-8.
64. Infographic: the four V's of big data. IBM Big data and analytics hub. n.d. https://www.ibmbigdatahub.com/infographic/four-vs-big-data. Accessed 29 Nov 2019.
65. Becoming a data-driven CEO. Domo. n.d. https://www.domo.com/solution/data-never-sleeps-6. Accessed 29 Nov 2019.
66. Marr B. How much data do we create every day? the mind-blowing stats everyone should Read. Forbes. n.d. https://www.forbes.com/sites/bernardmarr/2018/05/21/how-much-data-do-we-create-every-day-the-mind-blowing-stats-everyone-should-read/. Accessed 29 Nov 2019.
67. Global connected wearable devices 2016–2022. Statista. n.d. https://www.statista.com/statistics/487291/global-connected-wearable-devices/. Accessed 29 Nov 2019.
68. The growth in connected IoT devices is expected to generate 79.4ZB of data in 2025, according to a new IDC forecast. IDC: The Premier Global Market Intelligence Company. n.d. https://www.idc.com/getdoc.jsp?containerId=prUS45213219. Accessed 29 Nov 2019.
69. The digital universe of opportunities: rich data and the increasing value of the internet of things sponsored by EMC. n.d. https://www.emc.com/leadership/digital-universe/2014iview/index.htm. Accessed 7 Feb 2019.
70. Big data volume, variety, velocity and veracity. InsideBIGDATA. 2013. https://insidebigdata.com/2013/09/12/beyond-volume-variety-velocity-issue-big-data-veracity/.
71. Emerging Issues Task Force, International Neuroethics Society. Neuroethics at 15: the current and future environment for neuroethics. AJOB Neurosci. 2019;10(3):104–10. https://doi.org/10.1080/21507740.2019.1632958.
72. Martinez-Martin N, Insel TR, Dagum P, Greely HT, Cho MK. Data mining for health: staking out the ethical territory of digital phenotyping. Npj Digit Med. 2018;1(1):68. https://doi.org/10.1038/s41746-018-0075-8.
73. Copeland R. WSJ news exclusive. Google's 'project nightingale' gathers personal health data on millions of Americans. Wall Street J. 2019. https://www.wsj.com/articles/google-s-secret-project-nightingale-gathers-personal-health-data-on-millions-of-americans-11573496790.
74. AiServe. Computer vision navigation. n.d. https://www.aiserve.co/. Accessed 19 Nov 2019.
75. Livio AI. StarkeyPro. n.d. https://starkeypro.com/products/wireless-hearing-aids/livio-ai. Accessed 19 Nov 2019.
76. Braintellect—B2v2. Sleep deeply and wake refreshed. n.d. https://braintellect.com/. Accessed 20 Nov 2019.
77. Making the HoloLens 2: Advanced AI built Microsoft's vision for ubiquitous computing. Innovation Stories. 2019. https://news.microsoft.com/innovation-stories/hololens-2-shipping-to-customers/.
78. Demo: The magic of AI neural TTS and holograms at Microsoft Inspire 2019. n.d. https://www.youtube.com/watch?time_continue=32&v=auJJrHgG9Mc. Accessed 22 Nov 2019.
79. Solutions – LifeQ. n.d. https://www.lifeq.com/solutions/. Accessed 22 Nov 2019.

80. Vyou. Galahad. n.d. https://galahadai.com/vyou/. Accessed 29 Nov 2019.
81. MIT Media Lab: Affective Computing Group. n.d. https://affect.media.mit.edu/. Accessed 22 Nov 2019.
82. Obermeyer Z, Powers B, Vogeli C, Mullainathan S. Dissecting racial bias in an algorithm used to manage the health of populations. Science. 2019;366(6464):447–53. https://doi.org/10.1126/science.aax2342.
83. Amazon scraps secret AI recruiting tool that showed bias against women. Reuters. 2018. https://www.reuters.com/article/us-amazon-com-jobs-automation-insight-idUSKCN1MK08G.
84. Burrell J. How the machine 'thinks': understanding opacity in machine learning algorithms. Big Data Soc. 2016;3(1):205395171562251. https://doi.org/10.1177/2053951715622512
85. Paudyal P, William Wong BL. Algorithmic opacity: making algorithmic processes transparent through abstraction hierarchy. Proc Hum Fact Ergon Soc Annu Meet. 2018;62(1):192–6. https://doi.org/10.1177/1541931218621046.
86. Corbett-Davies S, Pierson E, Feller A, Goel S. A computer program used for bail and sentencing decisions was labeled biased against blacks. It's actually not that clear. Washington Post. n.d. https://www.washingtonpost.com/news/monkey-cage/wp/2016/10/17/can-an-algorithm-be-racist-our-analysis-is-more-cautious-than-propublicas/. Accessed 10 Feb 2020.
87. Julia Angwin JL. Machine bias [Text/html]. ProPublica. 2016. https://www.propublica.org/article/machine-bias-risk-assessments-in-criminal-sentencing.
88. JeffLarsonJA.HowweanalyzedtheCOMPASrecidivismalgorithm[Text/html].ProPublica.2016. https://www.propublica.org/article/how-we-analyzed-the-compas-recidivism-algorithm.
89. Wexler, R. Opinion. When a computer program keeps you in jail. The New York Times. 2017. https://www.nytimes.com/2017/06/13/opinion/how-computers-are-harming-criminal-justice.html.
90. Arulkumaran K, Deisenroth MP, Brundage M, Bharath AA. A brief survey of deep reinforcement learning. IEEE Signal Process Mag. 2017;34(6):26–38. https://doi.org/10.1109/MSP.2017.2743240.
91. Amodei D, Olah C, Steinhardt J, Christiano P, Schulman J, Mané D. Concrete problems in AI safety. ArXiv:1606.06565 [Cs]. 2016. http://arxiv.org/abs/1606.06565.
92. CDCTobaccoFree. Cigarette and tobacco use among people of low socioeconomic status. Centers for Disease Control and Prevention. 2019. https://www.cdc.gov/tobacco/disparities/low-ses/index.htm.
93. BaughJ.Linguisticprofiling.BlackLinguist.2005;https://doi.org/10.4324/9780203986615-17.
94. Garcia M. Racist in the machine: the disturbing implications of algorithmic bias. World Polic J. 2016;33(4):111–7. https://doi.org/10.1215/07402775-3813015.
95. Kreitmair KV. Commentary: neuroprosthetic speech: pragmatics, norms, and self-fashioning. Camb Q Healthc Ethics. 2019;28(4):671–6. https://doi.org/10.1017/S0963180119000616.
96. Levinson CA, Fewell L, Brosof LC. My Fitness Pal calorie tracker usage in the eating disorders. Eat Behav. 2017;27:14–6. https://doi.org/10.1016/j.eatbeh.2017.08.003.
97. Plateau CR, Bone S, Lanning E, Meyer C. Monitoring eating and activity: links with disordered eating, compulsive exercise, and general wellbeing among young adults. Int J Eat Disord. 2018;51(11):1270–6. https://doi.org/10.1002/eat.22966.
98. Ajunwa I, Crawford K, Ford JS. Health and big data: an ethical framework for health information collection by corporate wellness programs. J Law Med Ethics. 2016;44(3):474–80. https://doi.org/10.1177/1073110516667943.
99. Strap on the Fitbit: John Hancock to sell only interactive life insurance. Reuters. 2018. https://www.reuters.com/article/us-manulife-financi-john-hancock-lifeins-idUSKCN1LZ1WL.

10 Mental Health Chatbots, Moral Bio-Enhancement, and the Paradox of Weak Moral AI

Jie Yin

10.1 Major Concerns Regarding Moral Bio-Enhancement

When artificial intelligence (AI) is used to enhance cognitive skills, especially for moral deliberation, the ethical concerns that have been discussed in most literature center around two major issues are: privacy and autonomy [1]. The use of advanced neurotechnology, such as transcranial magnetic stimulation, leads to concerns about privacy. The downside of moral bio-enhancement, such as negative effects on autonomy, has been frequently discussed. For example, John Harris (2011) argues that the use of biochemicals such as SSRIs to enhance morality poses a threat to moral agency because it deprives the recipients of the freedom to fail [2]. Huang (2019) recently argued for the same view from a different perspective, claiming that moral bio-enhancement is objectionable in that it undermines human moral agency by obstructing accurate self-reflection and choice-making on that basis [3].

The abovementioned objections to moral bio-enhancement address its desirability yet place less focus on its feasibility. The trend of talking about what should be done and needs to be done is mostly shaped by philosophers concentrating on the normative questions instead of scientific ones because science has not yet revealed a detailed guide to implementing moral bio-enhancement. Despite his belief in a possible future when moral enhancement can be achieved, Wiseman (2016) does not consider mainstream biological and neuroscientific methodology work, such as taking blood samples or using fMRIs. He contends that "the present brain-reductive

This chapter revises and updates portions of a Chinese paper by Jie Yin. (尹洁，"弱道德人工智能可行吗——从精神医学用途到道德增强?"，医学与哲学, July 2020, Vol. 41. No. 13: 1–7).

J. Yin (✉)
School of Philosophy, Fudan University, Shanghai, China

Center for Biomedical Ethics, Fudan University, Shanghai, China
e-mail: jieyin@fudan.edu.cn

F. Jotterand, M. Ienca (eds.), *Artificial Intelligence in Brain and Mental Health: Philosophical, Ethical & Policy Issues*, Advances in Neuroethics,
https://doi.org/10.1007/978-3-030-74188-4_10

approach to studying moral functioning might be intrinsically incapable of handling such a complex set of phenomena as moral functioning encompasses and thus they make for a very poor foundation if moral enhancement enthusiasts want to present a credible scientific basis for the neuromodulation of moral behavior and judgment through pharmaceutical or other means" ([4], pp. 134–135]).

10.2 Weak Moral AI as a Socratic Assistant

On the other hand, the idea of moral AI, which added an extra layer to the already complex problem of moral bio-enhancement, encounters doubts as well as applause. Despite the similar concerns briefly discussed above, the development and progress of computer science seem to bring us new hope of making moral machines by advancing moral info-enhancement rather than moral bio-enhancement.

Savulescu and Maslen (2015) propose a model of moral AI, arguing that the common objections against moral AI will not be a problem for their model, as moral AI helps agents to overcome their natural psychological and physiological limitations [5]. Owing to the limitations of human psychology and evolutionary mechanisms, humans are disposed to develop an in-group affinity, which is simply a natural tendency that cannot be denied, so moral development will not extend its application to out-group people. Persson and Savulescu (2017) argue that because of "moral hard-wiring," sociocultural intervention cannot be the only way of achieving moral progress, and biomedical enhancement, albeit not being the only one, must be one of the most important ways to enhance morality [6]. Savulescu believes that human morality is hard-wired in some ways but also agrees that a weak moral AI can be useful even if it merely assists moral deliberation without having much effect on motivation.

Savulescu and Maslen (2015) argue that what they have in mind resembles what Lara and Deckers (2019) propose as "weak moral AI" [7]. They hold that for moral enhancement, motivation should not be the only important thing to consider. If we can bypass the motivation and use "exhaustive enhancement," we leave all moral decisions to machines. For instance, a computer chip can be implanted into the brain and tell the body to follow the instructions, an AI system can be set up to give us the order to follow, or a political system that sanctions such actions can be guided by the AI system. The difference lies in the fact that the direct model can force humans to act according to the AI system output, but the latter is external and needs the compliance of humans. Both models make humans redundant, albeit saving the hard work of moral deliberation and possible errors and avoiding the issue of undermining privacy and autonomy.

Lara and Deckers (2019) argue for a comparatively "modest" model of an AI assistant, and such weak moral AI merely asks questions in a Socratic way in which the assumption of the agent might be questioned, the concept they understand might be clarified, useful information that is necessary for the best inference and decision-making is provided, or the coherency of the argument is examined [7]. In other words, weak moral AI does not make the decision for the agent and merely helps to strengthen the argumentation itself in the form of dialogue. In this way, the ethics machine as a robot does not aim to internalize any moral doctrine within a moral agent. It is different from an autonomous moral machine that can function by itself, without relying on humans. For this reason, the usual concern over trust will not be an issue because humans ultimately make the call, not machines. If our trust is about whether computers can offer reliable guidance on strengthening the argument, as Lara and Deckers would argue, such concern is not necessary, as machines are much more reliable in accomplishing those tasks, as long as the algorithm is technically feasible.

By interacting with the ethics machine, one learns how to conduct their thinking in a conceptualized, well-informed, and logical manner. This is a cognitive "purification" or "improvement" process. Like a logic teacher, the machine can detect illogical reasoning and conceptual fallacy while conversing and interacting with a user, asking questions such as the following: "do you think you hold consistent moral views regarding justice in the previous two cases?," "do you think you are impartial in treating the two?," "what makes you think these two cases are different?" These questions seem to head in the right direction because they resemble the pedagogy adopted by Socrates in Plato's dialogues. The concern, however, is not whether such interaction is genuinely Socratic but whether such pedagogy can function well for the purpose of being able to provide advice based on fixed rules set by the algorithm, as this might not render the artificial intelligence sufficient for making morally better people. However, the good side of the story is that Lara and Deckers remove the burden of accountability from the assistant to the agent who uses and interacts with the ethics machine. It is indeed in this way that the agent themselves can retain autonomy and authenticity.

10.3 The Paradox of a Weak Moral AI: A Philosophical Argument

10.3.1 The Impotence of Moral Judgment: Humeanism Vs. Anti-Humeanism

A defense of moral AI enhancement seems to be plausible in some respects, according to Lara and Deckers (2019) [7]. They argue that such a Socratic assistant has quite a few advantages over "traditional" enhancement via motivation. It avoids the problem of undermining autonomy and privacy. It is certainly safer because no computer chip or device needs to be implanted in the human brain or body. Further, it leaves the final moral judgment open to the agent, allowing sufficient space for respecting moral pluralism. Moreover, it functions not only as the midwife for moral deliberation but also as a method for education toward the goal of moral progress. However, it does not seem to fit moral psychology if a moral AI model merely bypasses motivation. Humans simply do not bypass motivation and act in accordance with moral judgment. A reasonable inquiry would be whether a moral AI assistant would be useful without having any influence on motivation. I suggest framing the question in another way, asking whether any moral judgment can function by itself without effecting any motivation change.

To answer the question of how to understand the relationship between moral judgment and motive, philosophers come up with contrasting views and explanations. In meta-ethics, there is a more general and commonly discussed theme of moral motivation, in which Humeanism and Anti-Humeanism contribute significantly to the literature [8]. Philosophers aim to explore whether there is a necessary connection between moral judgment and motivation. Another way to frame the question would be to ask whether the mere presence of moral judgment enables a motivation to act. Humeanism holds that moral judgment cannot by itself motivate the agent to act but requires an additional act that can generate desire, or there is preexisting desire or "conative" state, which is sometimes called "pro-attitudes." A strong and even radical version of anti-Humeanism comes from Plato, holding that the apprehension of the forms of moral ideas should be able to provide overriding motivation for the agent to act. A similar view holds that a moral judgment will not count as genuine if it does not provide sufficient motivation to act in accordance with the judgment that recommends certain action. Thus, when one speaks of moral judgment, he/she has to understand that attributing moral judgment to someone already means the one to whom the very moral judgment is attributed has a corresponding moral motivation.

However, we do not have to take such a view for granted. The central debate on whether moral judgment can by itself render sufficient motivation to act illuminates the algorithm design of moral AI. This is because if Humeanism is true and additional resources are necessary for generating motivation, a weak moral AI as a Socratic assistant does not provide much help, as it can at most provide sound moral judgment. Moral AI acting as a Socratic assistant cannot handle anything like motivational or noncognitive elements, the point of which resembles what Persson and Savulescu discuss. I do not have to enumerate all Humean arguments to argue against the weak moral AI model. However, if there is still a question as to whether moral judgment alone can provide motivation, Socratic assistant moral AI may not be able to motivate the agent to act according to what the moral judgment provides.

The ethics machine seems extremely similar to the online tests in which we participate. Think of the classic MIT media lab research on the moral decision-making of autonomous vehicles. The online survey comprises all types of ethical dilemmas, from which the tester is asked to choose between different options. The typical question is whether the participant would prefer to continue with the original route of a car and kill person A on the pedestrian line or whether they would select the option of steering the wheel and killing person B on the other side of the road. The test confers various combinations of values to A and B, such as that of an old man and a baby, or that of one person with high socioeconomic status (SES) and the other with low SES. Most people would find it difficult to come up with a satisfying answer, as it is always bad to kill anyone. However, the test is designed to force the participant to rank valuable things (including human beings and animals). Let us assume that we have a detailed picture about the composition of choices made by different groups of people and that the data allows us to, based on the ranking of valuable things in life, design an algorithm and build an interactive AI as moral assistant that imitates the response and reaction of actual human beings. Thus, we might possibly "communicate" with the machine such that through getting the input of our beliefs and preferences, we could engage in a kind of dialogue that will result in a recommendation. However, it is not clear whether such dialogue will influence motivation. More importantly, it is not even clear what kind of answer should be recommended. Empirical studies do not, in principle, entail normative answers as regards what ought to be adopted for solving ethical dilemmas, although the result of such studies might help analyze human moral psychology. However, moral reasoning is far more complicated than current empirical studies have informed us. Psychologists sometimes perceive moral judgment as an ad hoc rationalization of one's intuitive responses to conflicting cases, and this view also endorses the idea that cognitive and affective processing work separately. For example, Greene et al. (2004) hold that our intuitive response to moral dilemmas depends on which subsystem has more force at any given time [9], but quite a few philosophers think Greene et al.'s argument is based on a problematic hypothesis regarding morally irrelevant factors and that such a hypothesis rests on a questionable view of the neurobiological underpinnings of moral reasoning [10].

10.3.2 Moral Hard-Wiring and the Limits of Weak Moral AI

A weaker moral AI might not work because those "hard" parts in our morality are simply impossible to remove through valid and even sound moral arguments. In their paper "Moral Hard-wiring and Moral Enhancement," Persson and Savulescu propose that because of the characteristics of human moral psychology, biomedical moral enhancement might not be the only way of making moral progress but should be one of the most important ways. They aimed Powell and Buchanan's view that sociocultural means are more efficient and safer than biomedical means making the latter unnecessary. Persson and Savulescu (2017) hold that, while some attitudes can be changed by sociocultural intervention through understanding and accepting better rational arguments such as what we see in the case of racism and sexism, they face difficulty in explaining motivational or noncognitive elements [6].

In other words, what Persson and Savulescu (2017) hold to be "moral hard-wiring" refers to the part in humans that is hardly changed in a moral developmental environment and not sensitive to rational persuasion [6]. Unlike racism and sexism, nepotism and cronyism cannot be eradicated by sociocultural intervention. The existence of the hard-wiring part shows that some stubborn prejudices are not easily corrected. Under such circumstances, the application of a weak moral AI is extremely limited and seems to be impotent, if it can only raise some questions that aim to make the agent aware of the hidden logical fallacy or conceptual ambiguity. For some people, pointing out the inconsistency in their logical reasoning does work in that they actually admit such a mistake when they are cognitively capable of doing so, which further affects their motivation and might trigger moral acts. However, this does not work for all. Some people are entirely cognitively capable of realizing their fault in logical reasoning but will not change, or will merely stick to their assumption and not put themselves into others' shoes, which makes any moral persuasion, not to mention a moral AI assistant, useless. Amoralists also fall into the latter category by not acting on what they understand as right or good.

10.3.3 Motivation Ethics: The Evaluation of Moral Agents and Actions

Motivation matters not only because intuitively we do value the act conducted for the sake of morality. The answer to the question above hinges on how much we *ought to* value motivation compared to logical and reasonable moral deliberation. A coherent moral theory, including both moral acts and agents, might be helpful in understanding how important motivation could be and what it means to bypass motivation. Moral theories, as Matthew Coakley (2017) writes, either tackle the agent or the behavior. For example, when morally evaluating objects, either we evaluate what people do, asking questions such as whether what the agent does is right or wrong, or we evaluate the moral character of the agent, asking questions such as whether they are virtuous or have a specific moral character [11]. The major ethical theories, as one sees in the history of ethics or moral theory, fall into the category of discussing the rightness either of the action or of the agent. Coakley raises an interesting and meaningful question, namely, if under the framework of one of the moral theories some action can be judged as right, can this be compatible with the conclusion drawn from another moral theory that judges the moral character of the very agent? Intuitively, there seems to be an easy answer for this question because we tend to assume coherency among different types of moral evaluations, but Coakley displays how difficult it is by examining major theories such as consequentialism and deontology and concludes by proposing motivation ethics as a possible alternative. By motivation ethics, he means "agents are morally better the more motivated they are to promote the [overall] good" ([12], p. 59), thereby combining the evaluation of motive and act. Stonestreet points out that Coakley's project hinges on whether readers accept his conception of the [overall] good as "*the* moral good" (italics from quote) [12], which seems to be a very fair objection that Coakley needs to address. However, setting that assumption aside, Coakley does propose something important that is easily missed, namely, the moral agent problem, which says that if an ethical theory places too much emphasis on the evaluation of the action, it falls short of evaluating the agent, thus causing the problem of incoherency among different routes of moral evaluation. We do not need to accept Coakley's version of motivation ethics, but he reminds us of an important question, namely, the need to consider both motivation and act. The theoretical reflection on the importance of motivation shows how weak a Socratic assistant moral AI might be because motivation is bypassed. Although the model proposed by Savulescu and Maslen and Lara and Deckers seems to avoid the problem that is mostly discussed in the name of privacy and autonomy because motivation is bypassed, the reason that they might be self-defeating is that they drop this dimension.

Emphasizing an act without taking care of motivation is certainly against Kantian ethics doctrine, which holds that moral worth merely comes from the act done from the motive of duty [13]. Motives other than those of duty cannot be morally right; even compassion and sympathy do not count. Albeit intuitively unappealing, Kantian doctrine tells us something important about the essence of morality. As Savulescu and Maslen envisage, a weak moral AI can function as a monitor of the moral environment, moral organizer, moral prompter, moral adviser, protection from immorality, and, finally, as preserver of moral autonomy and group-level moral AI. However, if none of these functions can somehow indirectly generate a motive of duty in the agent who uses or interacts with such AI, then one who takes Kantian doctrine seriously will think that a Socratic assistant as moral AI is useless.

If a weak moral AI could successfully work as envisaged, then at most it could offer assistance to generate a fairly reasonable moral decision, rather than making the actual decision for the agent herself or himself, not to mention acting morally. The agent, no matter how much information they acquire from the moral AI machine, has to make the final call on what to do and execute the act that they choose. This leaves a gap in which moral AI might be impotent, namely, if we adopt a view of "motivation ethics" and evaluate the agent rather than merely focusing on the act, then we do not merely care about the output of a moral act done by the agent. Under such circumstances, although it is not necessarily the case that the agent will not act in accordance with what is recommended by the moral machine, it is very likely that the motives of the agent are unknown and thus suspicious. From a strong Kantian viewpoint, there is no moral value in any act that is not purely motivated by duty. This probably will not be a problem for anyone who holds a view, say, merely evaluates the consequence of the action. For example, in cases in which the group-level AI helps distribute resources, usually the collective decision instead of an individual effect will be the subject of evaluation. In such cases, criteria based on consequentialism or utilitarianism are usually adopted. It is unlikely that the view from motivation ethics has to be applied in such a scenario.

10.4 Chatbots Used in Mental Health and How the Case Sheds Light on the Feasibility of a Weak Moral AI

In order to get close to achieving feasibility, it is better to have a concrete case at hand. Unfortunately, there seems to be no concrete case for a weak moral AI as a Socratic assistant until now, but AI applications, such as chatbots, have been targeted at alleviating mental health disorders, which might reveal something different from what I just discussed.

Chatbots, as systems that can converse and interact with humans by using text messages, audio messages, or even videos, have been recently used for the purpose of providing first-line support for patients [14], mostly afflicted by depression and autism, but there is little evidence for their effectiveness, and there are no consistent outcome measures [15]. The general conclusion on the use of chatbots for psychiatric purposes is that they have not been shown to be extensively effective but have therapeutic potential [16]. Currently, most innovations in chatbots for mental health and therapy have been implemented at the industry or entrepreneur level, and very few hospitals or psychiatric clinics have adopted such practices. The rationale behind this might be that clinics and hospitals are still waiting for further improvement and refinement of chatbots, or it is reasonable for the professionals to hold the view that chatbots might not be able to replace the real-life diagnosis and treatment in psychiatric clinics or hospitals.

For now, whichever form they take (virtual or non-virtual), chatbots used for mental health can merely provide people who need initial self-checks of their mental states and mentoring with mood and behavior adjustment in an accessible and convenient way. In most cultural contexts, psychiatric diseases are still stigmatized. AI applications can perfectly match the needs of those who prefer not to go public with their records, but such convenience also brings about limitations in effectiveness. Compared to traditional on-site psychiatric diagnosis and therapy, there is invariably missing information on such encounters between users and chatbots. However, on the positive side, users report that chatbots are helpful in monitoring moods and changing behavior. Researchers are also optimistic, holding that AI technology should be able to detect the irregularity in human response and, based on the well-designed embedded algorithm, try to "nudge" the user in the direction of positive feelings and thoughts. Vaidyam et al. (2019) reviewed conversational agents or chatbots in the field of psychiatry and concluded that the mental health field has the opportunity to gain more from chatbots than any other field of medicine [16].

While it is one thing to say that, implemented correctly and ethically, chatbots can be a useful psychiatric tool, it is another thing to say that they can nudge human beings at a more ethical level. Although both kinds of practice seem to endorse certain character traits such as being prosocial, moral enhancement, even in a weak sense as a Socratic assistant, is much more complex. Borenstein et al. (2016) considered moral enhancement practiceas "robotic nudges" that could withstand the challenge of moral paternalism [17]. Doubt is mainly cast on why we have to create "better people." The very idea of morally enhancing human beings, no matter what form it takes, through biomedical means or chatbots, can be objectionable because of the very nature of intrusion on humans. Criticism from such a perspective, as readers might be aware, is very commonly directed at moral enhancement in general. Besides, Borenstein et al. (2016) also raise concerns around issues of infeasibility as well as undesirableness, questioning how we would be able to confirm what kind of framework is suitable for defining what "ethical" means [17].

Despite these doubts, by proposing and analyzing the case of "nudging toward social justice," they conclude that robots can be useful in nurturing certain character traits in human beings. Although reasonable doubts around which prosocial behaviors are worthy of pursuit remain to be discussed, they hold that strategies can always be found to get us to know which prosocial behavior we aim at promoting as long as we confirm first what robots can do. It is worth noting that confirming what robots can do requires us to constantly acquaint ourselves with the innovation of AI technology before jumping hastily to any conclusion merely based on moral challenges that are hard to meet at first. What this means is that speculation on what AI technology might do could be limited by our imagination. Philosophy can evolve together with the development of technological innovation, especially regarding the terminology in the field of applied ethics. If that is true, then we might update our views on what can be done through moral enhancement by AI or, specifically, weak moral AI as a Socratic assistant. We do not know whether such weak moral AI can indirectly influence motivation, and maybe it can. However, someone might raise a doubt by saying that this is an empirical question, not a philosophical one. Then the question turns into something like this: if weak moral AI does influence motivation in an indirect manner, can the Humeanism regarding moral judgment be accommodated? I think the answer can be affirmative, since Humeanism, while insisting on a pro-attitude component, is compatible with the alternative that a moral judgment is never a purely epistemic one but already integrated with the "conative" part. This can happen when, as a matter of fact, weak moral AI elicits or changes moral motivation in the agent through Socratic dialogue around things such as the consistency of the argument or the clarity of the definition of concept, although it might not aim to achieve such a bold goal or have a precise account of how such mechanism was at first when the project was launched. In a nutshell, it is known to all that the psychological mechanism is unclear at the moment, so we cannot say with certainty whether a weak moral AI is feasible, but given the chatbot case, if mood and behavior can be changed, then moral attitude or character is probably not a very long way to go unless further empirical evidence shows an unbridgeable barrier to nudging.

10.5 Can Moral Info-enhancement Interventions Be "Medically Indicated"?

Setting aside the goal of ultimately affecting moral motivation, moral info-enhancement as a "weak" version might also be therapeutically useful. Casal (2015) raises the interesting point that moral bio-enhancement can be useful in certain circumstances, as "moral therapy" rather than for the general purpose of enhancement [18]. Carter (2017) specifically argues that moral enhancement can be "medically indicated" [19]. The contention is based on the view that empathy plays a core role in moral decision-making and behavior. According to Persson and Savulescu (2012), empathy is "a capacity to imagine vividly what it would be like to be another, to think, perceive, and feel as they do" [20]. Empathy can further contribute to sympathy, resulting in altruism necessary for moral decision and action. Those who cannot imagine the experiential state of others and/or cannot respond to the state are (cognitively) deficit in moral reasoning.

To claim for a therapeutic use presupposes considering a lack of empathy as a case of mental disorder. It remains to be discussed whether a lack of empathy can be classified as mental disease, since there is no unanimous definition on mental disorder. However, as Carter (2017) notes, if a deficiency of empathy fits the criteria for mental disorder, then we would at the very least be able to consider such deficiency as candidate for treatment through moral enhancement techniques [19]. But treating a deficit of empathy or excess of aggression can face serious challenges. Moral enhancement might not be agreeable or acceptable among those who are diagnosed with a lack of empathy. Although on the whole societal level, moral compliance definitely contributes to social collaboration; for individuals, moral deficiency can be a kind of advantage, because in most circumstances empathy makes life easier. This does not exclude the indirect benefit brought by saving one from being put in jail or being the victim of violence due to those actions resulting from a lack of empathy.

Meanwhile, since considering empathy as the core of moral deficit is open to further discussion, advocating for providing moral enhancement as treatment is not without its problems. It raises further ethical and regulatory concerns when moral enhancement technology is considered to be akin to therapeutics, such as, when the enhancement (as treatment) should be implemented, to whom it might be applicable, and where to draw the line between therapy and enhancement [21], the same old question that enhancement technology-related ethical concern always centers around.

However, if we accept a normative definition of mental disease and consider a lack of empathy as mental disorder, we might be inclined to consider it as a candidate for moral therapy through moral bio-enhancement. The worry is, given so many concerns mentioned above, moral bio-enhancement as therapy might has a long way to go. But if empathy is a genuinely cognitive deficit, in principle, it can be cultivated through clarifying concepts, providing and emphasizing mostly relevant information for the judgment, correcting logical fallacy, etc. These are exactly what moral info-enhancement could accomplish, and so a weak moral AI may work as a middle ground between traditional moral education and innovative but controversial moral bio-enhancement. As compared to traditional moral education, moral info-enhancement has better efficacy, convenience, and possible economic utility. As compared to moral bio-enhancement, it is safer and can eliminate worries about autonomy and privacy.

Weak moral AI, as is envisaged, aims at enhancing moral agent's ability of information processing, and thus mainly works on cognitive level, so it does not expect mood or emotion change as the effect. It is "weak" in that it merely purifies moral reasoning by facilitating the processing of morally relevant information through dialogue. The very nature of weak moral AI as such avoids abovementioned problems for moral bio-enhancement, such as infringement on privacy and autonomy. But even if ignoring those concerns and considering moral bio-enhancement as promising instrument for better moral compliance, weak moral AI can always work as a safer first step, that is, a trial implemented before moral bio-enhancement technology is ready to step in.

Instead of facing the situation that most people are disinclined to accept the bio-enhancement therapy, info-enhancement offers a solution as it merely tries to help the agent achieve better moral reasoning through a "soft" method. In this way, the medicalization worry is gone, too. As Conrad (1992) defines, medicalization consists of "defining a problem in medical terms, using medical language to describe a problem, adopting a medical framework to understand a problem, or using a medical intervention to 'treat' it" ([22], p. 211). As I see it, the definition of medicalization as such is not value-laden, and its being good or not depends on how it is used. Medicalization has one use that we should be aware, and as Carter (2017) points out, it is the "medicalization of deviance to provide medical social control" ([19], p. 350). If moral bio-enhancement is considered to be applicable in serious cases such as psychopathy and "moral deficiency disorder" thereby unavoidably faces the charge on medical social control through medicalization, then moral info-enhancement as therapy can play a moderate role in a much accessible way, for a much larger group of people that simply believe they could achieve better moral compliance and better social life through cognitive training on moral reasoning.

10.6 Concluding Remarks

Now we can get back to asking whether there is a paradox inherent in the model of weak AI and what we can learn from such inquiry. From a philosophical point of view, the paradox of a weak moral AI as a Socratic assistant seems to lie in the fact that, by bypassing motivation, moral info-enhancement avoids the problem of infringement of privacy and undermining autonomy. However, setting feasibility problems aside, without having such an effect on the agent, mere moral info-enhancement is external and impotent, and based on motivation ethics and Kantian ethics, it is undesirable by diminishing the essence of being moral.

Nevertheless, as the recent chatbot case shows, mood and behavior can be altered by conversing and interacting with AI applications, implying the possibility that mental states can be changed through similar weak AI methods, and moral enhancement in this sense might as well head in a feasible direction. Of course, this is not to assert that weak moral AI or any other kind of moral enhancement can be successful. Admittedly, the hope of augmenting moral functioning requires tremendous work on biochemical and neurobiological mechanisms as well as on the psychosocial environmental context that has shaped moral behavior and character. The discussion on moral info-enhancement as an alternative or middle-ground therapy in brain and mental health also shows promising application of a weak moral AI model. We could be even more optimistic when finding mental health chatbots feasible, especially regarding its possibility of changing mental states. Besides, the whole discussion also shows that the philosophical argument that initially goes against the feasibility of such weak moral AI can be referred back and reconsidered once empirical evidence is available, which reminds us of the dubiousness of purely theoretical arguments in dealing with practical issues, especially around future science and technology, for which human imagination can never be wild enough.
FundingThis work was supported by NSFC (National Natural Science Foundation of China). Grant Number 71843009.

References

1. O'Brolcháin F, Gordijn B. Ethics of brain–computer interfaces for enhancement purposes. In: Clausen J, Levy N, editors. Handbook of neuroethics. Dordrecht: Springer; 2015. p. 1207–26.
2. Harris J. Moral enhancement and freedom. Bioethics. 2011;25(2):102–11.

3. Huang PH. Moral enhancement, self-governance, and resistance. J Med Philos. 2018;43(5):547–67. https://doi.org/10.1093/jmp/jhy023.
4. Wiseman H. The myth of the moral brain: the limits of moral enhancement. Cambridge: The MIT Press; 2016.
5. Savulescu J; Maslen H. Moral enhancement and artificial intelligence: moral AI? In: Romportl J, Zackova E, Kelemen J, editors. Beyond artificial intelligence. The disappearing human-machine divide. Dordrecht: Springer; 2015. p. 79–95.
6. Persson I, Savulescu J. Moral hard-wiring and moral enhancement. Bioethics. 2017;31(4):286–95. https://doi.org/10.1111/bioe.12314.
7. Lara F, Deckers J. Artificial intelligence as a socratic assistant for moral enhancement. Neuroethics. 2020;13:275–87. https://doi.org/10.1007/s12152-019-09401-y.
8. Rosati CS. Moral motivation. In: Zalta EN, editor. The Stanford Encyclopedia of Philosophy. Stanford, CA: Stanford University; 2016. https://plato.stanford.edu/archives/win2016/entries/moral-motivation/. Accessed 30 Aug 2019.
9. Grenne JD, et al. The neural bases of cognitive conflict and control in moral judgment. Neuron. 2004;44:389–400.
10. Glannon W. Brain, body and mind: neuroethics with a human face. New York: Oxford University Press; 2011.
11. Coakley M. Motivation ethics. London: Bloomsbury; 2017.
12. Stonestreet EL. Notre dame philosophical reviews. 2017. https://ndpr.nd.edu/news/motivation-ethics/. Accessed 30 Aug 2019.
13. Kant I. Practical philosophy (trans. Gregor M). New York: Cambridge University Press; 1999.
14. De Jesus A. Chatbots for mental health and therapy—comparing 5 current apps and use cases. 2019. https://emerj.com/ai-application-comparisons/chatbots-mental-health-therapy-comparing-5-current-apps-use-cases/. Accessed 28 May 2020.
15. Abd-alrazaq AA, et al. An overview of the features of chatbots in mental health: a scoping review. Int J Med Inform. 2019;132:103978. https://doi.org/10.1016/j.ijmedinf.2019.103978.
16. Vaidyam AN, et al. Chatbots and conversational agents in mental health: a review of the psychiatric landscape. Can J Psychiatry. 2019;64(7):456–64.
17. Borenstein J, Arkin R. Robotic nudges: the ethics of engineering a more socially just human being. Sci Eng Ethics. 2016;22:31–46. https://doi.org/10.1007/s11948-015-9636-2.
18. Casal P. On not taking men as they are: reflections on moral bioenhancement. J Med Ethics. 2015;41(4):340–2.
19. Carter S. Could moral enhancement interventions be medically indicated? Health Care Anal. 2017;25:338–53.
20. Persson I, Savulescu J. Unfit for the future: the need for moral enhancement. Oxford: Oxford University Press; 2012.
21. Daniels N. Normal functioning and the treatment-enhancement distinction. Camb Q Healthc Ethics. 2000;9(3):309–22.
22. Conrad P. Medicalisation and social control. Annu Rev Sociol. 1992;18:209–32.

The AI-Powered Digital Health Sector: Ethical and Regulatory Considerations When Developing Digital Mental Health Tools for the Older Adult Demographic

11

Camille Nebeker, Emma M. Parrish, and Sarah Graham

11.1 Introduction

Globally, the population of older adults is rapidly growing with estimates of those over 60 years of age doubling by 2050—an increase from representing 12% to 22% of the population—and it is predicted that 80% of these older adults will live in low-to-moderate–income countries [1]. Mental illness affects nearly 20% of the older adult population (60 years above) in the United States and approximately 15% globally with 6.6% of all disabilities facing older adults attributed to mental and neurological disorders [1]. Moreover, older adults are less likely to perceive the need for or use mental healthcare [2]. The majority of these individuals have limited or no access to mental healthcare and those who potentially have access may not be able to afford care [3]. These data support careful consideration of how information and

C. Nebeker (✉)
Herbert Wertheim School of Public Health, University of California, San Diego, La Jolla, CA, USA

Research Center for Optimal Digital Ethics in Health (ReCODE.Health), University of California, San Diego, La Jolla, CA, USA

Sam and Rose Stein Institute for Research on Aging, University of California, San Diego, La Jolla, CA, USA

Design Lab, University of California, San Diego, La Jolla, CA, USA
e-mail: nebeker@eng.ucsd.edu

E. M. Parrish
San Diego State University/University of California, San Diego Joint Doctoral Program in Clinical Psychology, La Jolla, CA, USA

S. Graham
Sam and Rose Stein Institute for Research on Aging, University of California, San Diego, La Jolla, CA, USA

Department of Psychiatry, University of California, San Diego, La Jolla, CA, USA

F. Jotterand, M. Ienca (eds.), *Artificial Intelligence in Brain and Mental Health: Philosophical, Ethical & Policy Issues*, Advances in Neuroethics,
https://doi.org/10.1007/978-3-030-74188-4_11

communication technologies (ICTs) that support digital health research, including the use of AI-powered tools, can be used to support older adults generally and, particularly, those with mental illness.

The use of digital strategies to advance health promotion, disease prevention, and treatment research has exploded over the past decade. A recent study that assessed funding for digital health research by the National Institute of Health (NIH) revealed that the National Institute of Mental Health was among the top investors across all NIH institutes, along with the National Cancer Institute and the National Institute of Drug Abuse [4]. Based on NIH funding and the scientific literature, digital tools employed in health research are used to support observational and intervention studies via mobile apps, remote and wearable sensors, and social media platforms. As examples, one study used a SenseCam device, which is an outwardly facing wearable camera that records a first-person point of view and can help the wearer with recalling events of the day to assist older adults with memory loss [5]; smartphones are ubiquitous tools that can be used for brief assessments like real-time ecological momentary assessment (EMA), an alternative to retrospective self-report that is used for repeated sampling of users' current behaviors and experiences [6]; and social media platforms are being leveraged to identify and mitigate health risk factors like social isolation and loneliness; for example, a study conducted by Quinn (2018) enrolled residents of a retirement community to evaluate the effects of social media use on cognitive decline and found positive effects on information processing and cognitive function [7]. There is evidence to suggest that higher social technology use is associated with better self-rated health, fewer chronic illnesses, higher subjective well-being, and fewer depressive symptoms [8] and that older adults have similar levels of online social connectedness to younger adults [9]. Yet, the literature also reports that high use of social media among teens and young adults may be detrimental to mental health and well-being [10–12]. Clearly, more research is needed to better understand both potential benefits and risks to health.

Digital health research has increased among the older adult demographic due, in some part, to the increased adoption of smartphone technologies by those over 60 [13]. Smart phones can host apps designed to improve mood, avert loneliness, and promote self-reliance, the latter which is critical for aging independently [14]. However, while there is increasing interest and growth in the use of technology to support independent and healthy living, it is important to consider both the potential benefits in the use of assistive technologies along with risks of potential harm—especially when involving older adults combined with mental illness. Factors that influence willingness to adopt technology among older adults include trustworthiness of the vendor, practices that align with privacy expectations, and usability of the product [15, 16]. For example, a study conducted by Andrews et al. (2019) found that mobile app graphics and jargon familiar to digital natives negatively impacted adoption for the older adult demographic; whereas Wang et al. (2019) identified privacy preferences and control of data to be important factors in technology adoption [14, 17].

Smart homes offer another potential digital solution for the ongoing mental and physical healthcare of aging adults. Smart home technology is designed to gather data about the dweller's health, location, and environment. Smart homes first emerged in the late 1990s and can serve to monitor various aspects of a person's activity, behavior, and health through digital technology, particularly the use of video, audio, wearable, and environmental sensors combined with artificial intelligence analytic techniques. Due to the complexity of collecting these data, most research is in the pilot or planning stage and sample sizes are limited [18]. Many smart home applications are focused primarily on monitoring the daily activities of an older adult within their home environment to aid in the clinical detection and/or diagnosis of aging-related impairments like dementia in a real-world (as opposed to laboratory) environment, such as the Dem@Home platform [19, 20].

Key features of a smart home are: (a) automation (ability to accommodate automatic devices or perform automatic functions), (b) multifunctionality (ability to perform various duties), (c) adaptability (ability to adjust to meet the needs of users), (d) interactivity (ability to interact with or allow for interaction among users), and (e) efficiency (ability to perform functions in a time-saving, cost-saving, and convenient manner) [21]. In order to provide support for an individual with dementia, and promote aging in place, a smart home could advance beyond monitoring and assist with routine things like self-care, medication adherence, meal preparation, and safety support (e.g., prevent falls, wandering behavior, or dangerous situations like a fire) and provide socialization [22]. Problems with the detection and diagnosis approach include lack of evidence that sensor signals, activities, behaviors, etc., and are indeed causal to diagnosis of dementia or other mental health disorders, lack of standardization across algorithms that evaluate these data, and the fact that these platforms do not necessarily interact with the individual with dementia in a meaningful way that could promote aging in place [18, 23–25].

In this chapter, we describe regulatory and ethical frameworks used in the US and challenges with conducting ethical digital health research including special considerations when developing tools to meet the needs of older adults, particularly those who suffer from a mental illness. We narrow the scope of mental illness to those with dementia, as dementia is one of the most common mental illnesses affecting older adults [26, 27] and is a priority target of the digital therapeutic sector [28]. Dementia is defined as a major neurocognitive disorder by the DSM-5 that affects six cognitive domains: complex attention, executive function, learning and memory, language, perceptual-motor function, and social cognition [29]. People living with dementia are a particularly vulnerable group who may warrant additional protections from harms associated with biomedical and behavioral research studies due to reduced cognitive ability impacting decisional capacity [30]. Furthermore, cognitive impairments may impact their ability to provide true informed consent, highlighting the need for special protections [31].

What may be unique about big data and digital technologies that are powered by AI-tools is the extent to which our existing regulations apply to, and are able to be carried out by, ethics review boards. First, we briefly reflect upon regulations that guide current human research protections and speak to gaps exposed when not all involved in the digital health research sector are bound by regulations. We then describe three commonly accepted ethical principles used in the review of biomedical and behavioral research (Belmont Report) [32] and introduce a fourth principle advanced by authors of the Menlo Report in response to increased availability of information and communication technologies (ICT) (e.g., smartphones and wearables) [33]. To contextualize challenges introduced by digital health research designed for use by older adults, we provide a use case based on smart home technologies. We then apply a decision-making checklist developed to assist behavioral scientists that includes five intersecting domains including ethical principles, risk/benefit assessment, access and usability, privacy, and data management [34].

11.2 Regulatory Gaps

As health research is an international endeavor, it is important to acknowledge global efforts to elevate ethical practices in research involving humans. The Declaration of Helsinki developed by the World Medical Association [35] governs much human research globally. The Common Rule is the rule of ethics that governs research supported by the US Department of Health and Human Services [36]. Globally, there are standards and procedures for operationalizing ethical health research involving humans that are country and/or organization specific [37], each with a goal to communicate expectations that speak to the ethical and responsible conduct of biomedical and behavioral research. In addition to regulations, professional societies (e.g., American Psychiatric Association, World Health Organization) have established codes of ethics that address professional expectations and in addition speak to research participant protections specific to the discipline to foster norms among affiliated members [38, 39].

An exception to these regulatory requirements has emerged over the past decade. New forms of research have emerged that are un- or underregulated as organizations began leveraging big data from a plethora of sources to conduct predictive analytics concurrent with the emergence of direct to consumer mobile apps and passive sensor technologies [40]. As noted, US federal regulations are in place to guide research supported by the federal government and that which falls under the US Food and Drug Administration's oversight (e.g., developing drugs or devices, including some digital therapeutics). This means that regulations apply to those in the more traditional research settings like universities yet do not necessarily apply to research being planned or carried out by citizens or digital therapeutic (DTX) start-ups and technology giants that have entered the digital health sector. In fact, much of the research taking place in the digital health sector is unregulated because the products fall under the "wellness" domain (e.g., Fitbit), which the FDA may not evaluate [41]. The FDA considers a product to be a medical device when the intended use refers to a specific disease or condition [42]. For regulated researchers who include wellness devices or apps as tools for research, they do receive review by an ethics review committee. This regulatory gap is problematic for many reasons including inconsistency between regulated and unregulated researchers specific to: (a) formal training in research design and methods and (b) acculturation with respect to awareness of ethical norms and practices. The potential impact on society is important to consider, especially to vulnerable populations, as consumers are not likely familiar with these regulatory gaps nor the potential risks of harm introduced by AI and sensor technologies [43]. Furthermore, the FDA requires patient engagement in device development for those products that will be used in digital medicine [44]. The involvement of patients in the development of digital health devices is a critical step forward, as historically patients have not had a voice at the table, particularly older adults [45].

11.3 Ethical Principles

The ethical principles that undergird much of biomedical and behavioral research described in the Belmont Report include Respect for Persons, Beneficence, and Justice [32]. These principles, published in 1979, later inspired the US federal regulations for human research protections [46], which were adopted by 18 federal agencies and are now referred to as the Common Rule. These principles were deemed relevant for guiding ethical research practices and have, for the most part, stood the test of time. Several years ago, and in response to the increase in information and communication technologies (ICTs) and related ICT research (ICTR), the Menlo Report was developed, which applied the three Belmont principles to ICTR and added a fourth principle of Respect for Law and Public Interest [33].

The important contribution of the Menlo Report is its attention to how existing regulations and practices are not sufficient to address the current challenges introduced by interactions between people and communication technologies. Those developing the report included cybersecurity experts working with the federal department of homeland security and who were familiar with the potential impact of ICTs. Digital research is built upon complex and ubiquitous computing communication technologies, and our discordant regulatory structures, law, and social norms create ethical gaps. The Menlo Report speaks to our limited understanding of the scale and speed with which risks can manifest and begins to elevate awareness of these gaps and potential harms as well as provide solutions. A key factor in the Menlo Report is recognizing that ICTs create distance between the researchers and people who participate in the research, elevating the potential risks of harm beyond an individual human research participant to include a range of stakeholders who may be affected. As such, the report encourages contemplation of harms that extend beyond the direct research subject and suggests that researchers and ethics review boards carefully evaluate the impact of technologies and information communications across various stakeholders, including bystanders. This includes becoming familiar with laws (e.g., information privacy, trespass statues) and regulations and committing to accountable practices. The overlay of ICTR consideration to the existing principles is to both enhance awareness and increase understanding as we use these tools to support health research while at a crossroads of policy and governance gaps. A revised Ethical Impact Assessment has been created to reflect new principles added in the Menlo Report [47].

Collectively, the principles, each briefly described below, are useful in making the research process more transparent and for engaging in dialogue about the ethical dimensions of research.

Respect for Persons. The principle of "Respect for Persons" speaks to study participation being voluntary and that people are recognized as autonomous agents who are able to determine what is in their best interest. It is demonstrated through the informed consent process whereby a person who is eligible to participate in a research study, and has the decisional capacity, is given information deemed necessary to make a choice about volunteering as a study participant. Even in more conventional biomedical research, there is much debate about the effectiveness of the consent document and communication process as well as efforts to make improvements [48]. The concerns primarily focus on how complex study information is delivered, who delivers the information, and how information is influenced by culture, religion, and literacy [49]. In digital health research where thousands of people can be enrolled via a mobile phone, the consent delivery may be occurring via an e-consent process which introduces a number of new challenges, primarily how users proccss information on a screen and their tendency to click and agree without reviewing the content [50]. While e-consent is feasible for older adults, additional challenges are introduced specific to technology-enabled research that may compromise their ability to provide informed consent, including unfamiliarity with terminology used and lack of technological literacy [51].

Beneficence. The goal of the principle of "Beneficence" is to minimize possible harms and maximize possible benefits. This occurs when an ethics review board systematically evaluates the risk of harms to the individual participant against the possible benefits of knowledge gained from the study to those represented by participants and society [32]. Evaluating the probability and magnitude of potential harms is challenging, yet ethics review boards typically have the expertise necessary to make risk assessment and management decisions that allow the research to move ahead. If not, they can outsource to obtain the necessary expertise. If the risk to benefit determination identifies risks that are unacceptable in relation to possible benefits, reviewers may decline approval such that a study will not be conducted. When ICTs support digital health research, it is often difficult to identify possible risks in advance and subsequently understand how best to manage those risks, and moreover, the appropriate expertise may not be readily available. This is in part due to the scale and speed at which a risk can develop and our limited understanding of the dynamics between the physical and connected world. The Cambridge Analytica fiasco is one example of ICTR with unknown downstream risks of harm. In this case, an academic researcher deployed a personality survey via the Facebook platform. Responses were then used to profile participants, including those who were contacts of the initial participant and ultimately were believed to have influenced how citizens voted in the US 2016 presidential election [52]. In the connected world of today, there are information-centric harms that need to be considered with respect to data confidentiality and sensitivity of information, recognizing that the potential risk of harm will vary by individual and also extend beyond the individual. When it comes to older adults, there are privacy considerations to incorporate when assessing risk. In a recent study of older adult privacy preferences, researchers noted a significant difference in privacy attitudes when compared to younger adults and adolescents, with older adults being significantly more likely to identify as fundamentalists (40%) compared to younger adults (6.7%) [53]. What this means is that older adults: have a high value for privacy, believe they own and have control of their information, support laws and regulations to secure privacy rights, and may be willing to share personal data with a trusted entity [53]. While technologies can add value to aging in place and may be acceptable due to the potential benefits, privacy, and the risk of privacy violations, is an important factor to consider [54]. Clearly, applying the principle of beneficence will require input from diverse stakeholders to better understand the potential risks of harm and how best to mitigate those in digital health research targeting older adults.

Justice. The principle of "Justice" is to encourage the fair selection of research participants and equitable distribution of risks and possibility of benefit [32]. Those who are included in the research should represent people who may benefit from the knowledge gained. In conventional regulated research, it is possible to review the research protocol and evaluate the study inclusion and exclusion criteria to determine alignment with the principle of justice. However, in unregulated digital health, the idea of justice is difficult to evaluate in that those who have access to a product or app are those who become the data source upon which algorithms are derived [55]. Issues of bias in training data used to inform algorithm development are well documented [56, 57]. That being said, managing the bias is dependent on organizational standards for accountability and transparency that drive fair and ethical decision processes—these standards are only now being developed (see Ethically Aligned Design, IEEE) [58]. Recently the National Science Foundation in the US allocated funding to examine Fair, Ethical, Accountable and Transparent AI [59], and there are a number of initiatives globally working to advance ethical AI [60]. There are studies underway to assess the acceptability of in-home monitoring systems that can communicate cognitive changes and other health problems to caregivers and clinicians [25, 61, 62]. The promise of tech-enabled health research, particularly, digital geriatric mental health research, is the potential benefit of creating greater access to services needed by a growing older adult demographic [63].

Respect for Law and Public Interest. The Menlo Report, published by the Department of Homeland Security in 2012, applied the Belmont Report principles (above) to ICT and cybersecurity research and added this fourth principle of "Respect for Law and Public Interest." The goal of this report was to encourage those involved in ICT research to engage in legal due diligence and be transparent and accountable in methods and results. This principle, if and when applied, may bridge the gap in our current regulatory environment where wellness products are unregulated and not bound by existing regulations for human research protections. That being said, there is no requirement for any unregulated research to adhere to either the Belmont or Menlo Reports. This is concerning, especially when considering digital geriatric mental health research and the vulnerability of this demographic. The future of this research is of critical public health relevance, and plans to advance this work need to be supported [63].

Drawing on ethical principles along with factors relevant to digital research, Table 11.1 presents questions that researchers, ethics boards, research participants, and technologists may consider across the research cycle of development, implementation, and reporting. This list is not intended to be comprehensive, but more practical ideas for how to think about ethics when involving older adults in digital mental health research.

With this background on our current regulatory environment and ethical principles developed to guide responsible research practices, we present a use case to contextualize how digital health research, including AI-powered tools, are deployed. Our use case describes the development of a smart home platform intended to be interactive with the user—it simulates a type of assisted-living facility in the home environment that may enable aging in place for older adults with dementia.

11.4 Smart Home Use Case

Amiribesheli and Bouchachia in 2019 introduced a smart home platform specifically tailored for three end users: persons with mild dementia symptoms, their caregivers, and geriatric psychiatrists. These researchers first identified problem scenarios specific to this population including repetitive speech, dehydration, loneliness, learning to use new devices, nighttime wandering, forgetfulness, and challenges with vision. They developed smart home technology that can intervene on five different levels to assist with these problems: (1) inviting awareness, (2) suggesting, (3) prompting, (4) urging, and (5) performing. For example, in the case of monitoring dehydration, the system would be preprogrammed by a caregiver regarding the number of times an individual should drink per day; environmental sensors would detect movements; software would recognize the activity of drinking; the system would log the number of times this activity occurred throughout the day; if the individual was not drinking enough, the system would prompt the individual to drink more through inviting awareness and suggesting; if the individual ignored these prompts, the system would send an alarm to the caregiver; and finally, the system would maintain logs over time that could be viewed by the individual's physician to assess trends.

11.5 Discussion and Case Analysis: Ethical Dimensions of Smart Home Technologies

With this use case as an example of technology and AI at the intersection of geriatric mental health, we evaluate responsible practices across the key domains of consent, access and usability, risks and benefits, privacy, and data management.

Table 11.1 Prompts to guide application of ethical principles to geriatric digital mental health research

Key factors	Access and usability	Risks and benefits	Privacy	Data management
Ethical principles				
Respect for persons	Consent provides: Relevant information within Terms of Service/Privacy policy in plain language Access to definitions Access to visual and audio versions of information Possibility of bystander involvement	Consent conveys: Risks and risk management strategies Evidence unknown risks possible benefits to the person, people like them, and to society	Consent conveys: Nature of personal information collected Data sharing plan Privacy policy risks	Consent conveys: Data collection process Data storage and security who will have data access protocols for data sharing
Beneficence	Includes a plan for return of group and individual study information Study design includes features to increase access and usability for older adults Short and long-term use has been or will be tested with older adults Rights of all stakeholders are considered	Study design is responsive to privacy preferences Evidence to support tech reliability/validity Evidence is peer-reviewed Risks are known and mitigated Risks are unknown Potential benefits outweigh possible risks of harm	Privacy expectations are respected Participant data are not shared or sold to a third party Participant contact information is not exploited	Data collection by party external to the research team Potential of data collected on or about a bystander Data are accessible to the participant Data are transferrable to the EHR Data ownership is clear
Justice	Device or App tested with older adults Requires internet Requires smartphone AI trained on data inclusive of older adults	Legal harms are known Potential risk of discrimination is transparent Risks of harm no greater for 65+ demographic	Bias is managed to reduce: Economic harm Social harm Discrimination Profiling	System vulnerabilities are publicly disclosed Data are not used to target groups or people

(continued)

Table 11.1 (continued)

Key factors	Access and usability	Risks and benefits	Privacy	Data management
Respect for Law and Public Interest	AI is accountable Algorithms are documented and transparent	Data and privacy protections are compliant	Increase trust Protect privacy	Data encryption meets expected standards Storage is HIPAA compliant Data are deidentified

Source: This work is published with permission and reflects an adaptation of the Digital Health Checklist developed for Researchers (DHC-R). It was developed by Camille Nebeker, EdD, MS, licensed under a Creative Commons Attribution-Non-Commercial 4.0 International License 2018 [64]

11.5.1 Informed Consent and Agency

Informed consent becomes especially important when such a large volume of personal data is collected in digital research and needs to account for both the content of the information to deliver and the process for delivery and cognitive capacity. When it comes to enrolling in a smart home study, it may be that our norms and practices for obtaining informed consent need to change—especially in the era of big data [31]. To be effective and to mitigate low technology literacy, informed consent needs to be adapted to the unique needs of older adults. A qualitative study of older adults found that adults are concerned about the potential intrusiveness of smart home technologies, but may not be aware of the extent of security risks, highlighting the need of informed consent that is adapted to technological literacy [65]. This suggests that presenting information so that the text is visually accessible and incorporates options for accessing unfamiliar terms (i.e., cloud storage) is needed. Moreover, how data is collected, transferred, stored, and shared must be clearly stated in the consent delivery and in a manner that conveys the granularity, volume, and personal nature of the data collected. How to do this well is a topic for further research.

11.5.1.1 Content and Delivery

An example of how informed consent can be improved was the subject of an exercise conducted with residents of a retirement community in southern California who were enrolled in a longitudinal study designed to assess the extent to which AI could detect cognitive and physical decline. The study received approval from the local ethics review board and used a traditional method for obtaining informed consent—which being several pages of paper with a 12-point font and little white space. Several residents were asked to take that consent form and imagine what it would look like if it was designed for them. Comments included less information, adjustable text size, clearly defined sections, progress indicator, video explanations, definitions, and electronic receipts. This human-centered design process

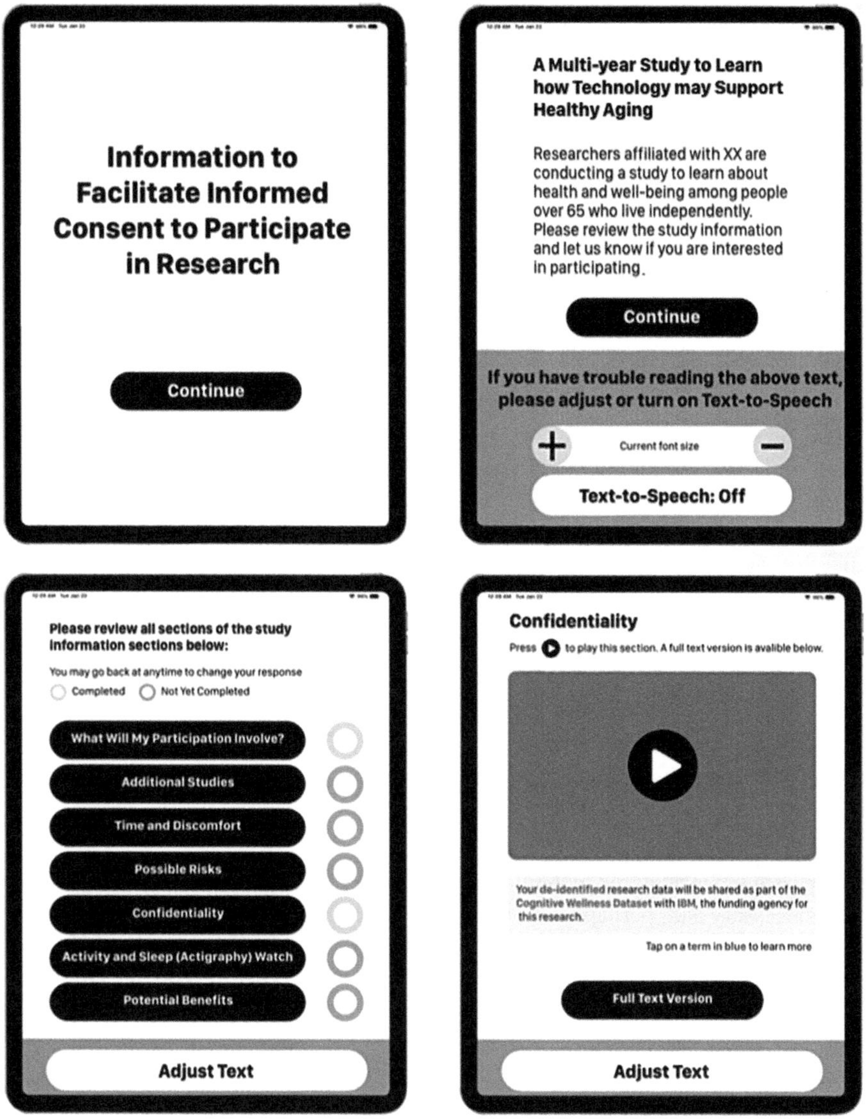

Fig. 11.1 Informed Consent Prototype. Informed by older adults, the panels show how information can be presented to increase accessibility to content that influences decision-making

resulted in a prototype (see Fig. 11.1) that is now being tested through design workshops [66].

11.5.1.2 Cognitive Capacity

Another consideration when obtaining consent to involve older adults is the potential for cognitive decline, either at the time of consent due to cognitive impairment or over the course of a longitudinal study. Ienca et al. [31] suggest that the inclusion of advanced directives may be important at the time of consent. Perhaps there could even be a plan to assess decisional capacity regularly and consult with doctors to ensure that the individual is still capable of consenting. Tools developed to assist with assessing decisional capacity exist [67, 68] and could be adapted to detect barriers introduced by low technology and data literacy. For example, when an individual is no longer able to give ongoing consent, or when the initial decision to enroll in a smart home study is made by a caregiver or physician [69], the ethical principle of Respect for Persons entitles those with diminished capacity to added protections [32].

11.5.1.3 Bystanders

With smart home sensor technologies, the risk for bystanders to be captured and subsequent rights and agency comes into play [70]. A guest in the home of an older adult with an AI listening device may not explicitly consent to having their voice recorded and analyzed. If an older adult points out this device to see if the guest would consent to having their voice recorded, they then may need to self-disclose that they have smart home technology to monitor their physical and mental health. This could be an added burden, but also potentially stigmatizing to the individual who may not wish to disclose their personal health information. Perhaps one avenue around this would be to have the AI device turn off automatically when detecting the voice of an unfamiliar individual.

11.5.2 Usability and Accessibility

11.5.2.1 Usability

In the case of smart home technology, an individual may view the technology aides as unnecessary and, perhaps, intrusive. Assessing need and perceived usability is an important first step when designing technologies for older adults [17, 71]. Customizability may be important to help increase accessibility as an individual could tailor their smart home tools and functionality based on their desires and comfort. The research protocol and subsequent deployment of a smart home surveillance and intervention system must allow the user to make decisions about what actions are monitored and what actions remain private. The user could also have a dialogue with their healthcare provider to determine what interventions would be most helpful considering their comfort level.

11.5.2.2 Accessibility

Smart home technologies are costly to install and maintain. This presents a problem for the researcher, who will need to acquire large amounts of funding to acquire enough data to prove efficacy. Perhaps more importantly is that the users who may ultimately benefit from these technologies may not be able to afford them, further widening a healthcare gap across different socioeconomic statuses. Additionally, the smart home case begs the question of whether these technologies would be appropriate for the non-technologically savvy individuals who may be overwhelmed by this technology in their homes.

11.5.3 Risks and Benefits

The first step is to identify potential risks of harm and determine if those harms are manageable and, if so, whether the benefits of knowledge to be gained outweigh these risks.

Risks, including physical, psychological, legal, and social risks, have their conceptual roots in traditional behavioral and biomedical research, yet remain relevant when using new technologies whether it be in research or in clinical care [47]. Specific to clinical applications, a user may be exposed to a physical risk if the equipment malfunctions or fails to perform as intended (e.g., fails to report critical data), resulting in harm to the user. A psychological risk might be the perceive threat of the technology invading their personal space. Another psychological (and physical) risk may be management of expectations that someone is "watching" in real time and will intervene if there is a need for assistance, which could be something agreed upon in advance or not [72]. Whether and the extent to which a clinical care team may intervene (or not) should be made clear during the informed consent process so that a user understands what to expect. With respect to the law, there is a lack of current legal guidance regarding smart home use, and thus, no solutions in place to address conflicts between smart home service providers and users [73]. Socially, users may feel as if the presence of the technology in their home is stigmatizing or fear it may reduce opportunities for face-to-face contact; however, more often than not, the benefits of increased social inclusion are thought to outweigh these risks [73].

With respect to possible benefits, smart home technologies can facilitate access to quality healthcare, enhance comfort, monitor health conditions, provide support to users, and foster social inclusion [73]. It is important to evaluate the probability and magnitude of potential harms against the potential benefits prior to making an informed decision. In a research context, the evaluation of risks and benefits associated with studies involving health technologies falls to the ethics review board (i.e., IRB in the US, REC in EU) and the research team. This risk to benefit evaluation includes consideration of the study design and the potential of the research to contribute new knowledge to the scientific enterprise. Within the healthcare application, the decision to adopt new technologies must put benefits of use ahead risks by selecting technologies, including smart home technology, that are properly vetted through rigorous research. Lastly, the informed consent process is critical to decision-making and information conveyed to the users, whether they be patients or research participants, must include a description of the possible risks of health technologies, how those risks are managed, and ultimately be acceptable by those choosing to accept. Moreover, it is important that not only consumers consider the direct and indirect effects of health technologies but that developers, clinicians, and researchers be aware of known and unknown downstream effects.

11.5.4 Privacy

The inclusion of new technologies expands the scope to include a broader conceptualization of risk specific to privacy [33]. Notably, new digital health technologies invite a risk of data security and privacy breaches [74, 75]. One of the primary categories of risk associated with ICTs are those of confidentiality, and the possibility that one's personal data may be stolen and misused, [47] which may become especially important when dealing with a technology that is in someone's home. There is also risk of third parties intercepting and subsequently modifying or falsifying personal data (e.g., cyberattacks) [76]. It is important that these risks are explicitly discussed with older adults when considering the use of new smart home technologies, along with risk mitigation solutions and related limitations.

One important consideration associated with pervasive sensor technologies in smart homes is the potential to violate privacy preferences for those who live in the home. It is necessary to understand the privacy attitudes of older adults in order to prevent privacy-related harm. Privacy may be a barrier to adopting technology for older adults, but that the practical utility of the technology outweighs this [69]. A 2008 participatory evaluation of smart home interventions found that older adults were not concerned about privacy in regard to smart homes [77]. However, the technology studied in this chapter does not include newer interventions such as AI, which should be examined. One study also notes that older adults found monitoring acceptable if they were able to decide who could view their data and the circumstances under which their data could be accessed [65]. However, it does not provide a clear guideline for how to establish this customized data sharing in practice. To best meet the needs of older adults, privacy settings should be transparent, adaptable, and customizable.

11.5.5 Data Management

Specific to data management and confidentiality, knowing how data are collected, transmitted, stored, and shared are factors that can influence the risk to benefit evaluation in deciding whether a smart home is suitable for a potential research participant. A careful assessment of who has access to these data (caregivers, physicians, social workers, family/friends, associations) as well as how data are transmitted and stored is important to convey during the consent process. Moreover, in the era of rapidly changing ICTs, data management must be dynamic and monitored as risks may arise due to instability within the platform supporting or other intermediaries supporting the research [47]. Dynamic data management can be achieved through continuous updating and monitoring of database management systems to ensure it is up to date, usable, and that the participant continues to be safe while using the technology. This will require development of new software and methods capable of managing and processing big data obtained from users (e.g., Health-cyber-physical systems) [78].

11.6 Conclusion

There is much promise in the use of AI and technological innovations to promote healthy aging [25, 62, 77]. However, the field must have a better gauge on the associated perils, including understanding what are known, as well as unknown risks of potential harms to move forward. All stakeholders must be mindful of barriers to obtaining meaningful informed consent for the direct, as well as indirect participants of research that intersects with information and communication technologies. Moreover, when considering whether potential research benefits outweigh the probability and magnitude of potential harms, evaluating the complex systems that are undergirding data collection, transmission, storage, and sharing of personal needs to be informed by experts through a dynamic "living" review process. Moving forward, considerations of diverse stakeholders, laws, and public interests are essential to informing ethical principles that will guide responsible digital health research across all demographics [31].

References

1. World Health Organization. Mental health of older adults. 2017. https://www.who.int/en/news-room/fact-sheets/detail/mental-health-of-older-adults.
2. Klap R, Unroe KT, Unutzer J. Caring for mental illness in the United States: a focus on older adults. Am J Geriatr Psychiatry. 2003;11(5):517–24.
3. National Alliance on Mental Illness. Out-of-Network, Out-of-Pocket, Out-of-Options: The Unfulfilled Promise of Parity Arlington, VA. 2016. https://www.nami.org/About-NAMI/Publications-Reports/Public-Policy-Reports/Mental-Health-Parity-Network-Adequacy-Findings-/Mental_Health_Parity2016.pdf.
4. Dunseath S, Weibel N, Bloss CS, Nebeker C. NIH support of mobile, imaging, pervasive sensing, social media and location tracking (MISST) research: laying the foundation to examine research ethics in the digital age. NPJ Digit Med. 2018;1(1):20171.
5. Hodges S, Williams L, Berry E, Izadi S, Srinivasan J, Butler A, et al., editors. SenseCam: a retrospective memory aid. International Conference on Ubiquitous Computing. Berlin: Springer; 2006.
6. Cain AE, Depp CA, Jeste DV. Ecological momentary assessment in aging research: a critical review. J Psychiatr Res. 2009;43(11):987–96.
7. Quinn K. Cognitive effects of social media use: a case of older adults. Social Media Society. 2018;4(3):2056305118787203.
8. Chopik WJ. The benefits of social technology use among older adults are mediated by reduced loneliness. Cyberpsychol Behav Soc Netw. 2016;19(9):551–6.
9. Sinclair TJ, Grieve R. Facebook as a source of social connectedness in older adults. Comput Hum Behav. 2017;66:363–9.
10. Hunt MG, Marx R, Lipson C, Young J. No more FOMO: limiting social media decreases loneliness and depression. J Soc Clin Psychol. 2018;37(10):751–68.
11. Hogue JV, Mills JS. The effects of active social media engagement with peers on body image in young women. Body Image. 2019;28:1–5.
12. Steers M-LN, Wickham RE, Acitelli LK. Seeing everyone else's highlight reels: how Facebook usage is linked to depressive symptoms. J Soc Clin Psychol. 2014;33(8):701–31.
13. Anderson M, Perrin A. Tech adoption climbs among older adults. Washington, DC: Pew Research Center; 2017.

14. Andrews JA, Brown LJ, Hawley MS, Astell AJ. Older adults' perspectives on using digital technology to maintain good mental health: interactive group study. J Med Internet Res. 2019;21(2):e11694.
15. Fox G, Connolly R. Mobile health technology adoption across generations: narrowing the digital divide. Inf Syst J. 2018;28(6):995–1019.
16. Hanson VL. Influencing technology adoption by older adults. Interact Comput. 2010;22(6):502–9.
17. Wang S, Bolling K, Mao W, Reichstadt J, Jeste D, Kim H-C, et al. Technology to support aging in place: older adults' perspectives. Healthcare. 2019;7(2):60.
18. Liu L, Stroulia E, Nikolaidis I, Miguel-Cruz A, Rios RA. Smart homes and home health monitoring technologies for older adults: a systematic review. Int J Med Inform. 2016;91:44–59.
19. Majumder S, Aghayi E, Noferesti M, Memarzadeh-Tehran H, Mondal T, Pang Z, et al. Smart homes for elderly healthcare-recent advances and research challenges. Sensors (Basel). 2017;17(11):2496.
20. Andreadis S, Stavropoulos TG, Meditskos G, Kompatsiaris I, editors. Dem@ home: Ambient intelligence for clinical support of people living with dementia. European Semantic Web Conference. Berlin: Springer; 2016.
21. Lê Q, Nguyen HB, Barnett T. Smart homes for older people: positive aging in a digital world. Fut Internet. 2012;4(2):607–17.
22. van Kasteren Y, Bradford D, Zhang Q, Karunanithi M, Ding H. Understanding smart home sensor data for ageing in place through everyday household routines: a mixed method case study. JMIR Mhealth Uhealth. 2017;5(6):e52.
23. Marzano L, Bardill A, Fields B, Herd K, Veale D, Grey N, et al. The application of mHealth to mental health: opportunities and challenges. Lancet Psychiatry. 2015;2(10):942–8.
24. Cook DJ, Schmitter-Edgecombe M, Jönsson L, Morant AV. Technology-enabled assessment of functional health. IEEE Rev Biomed Eng. 2018;12:319–32.
25. Demiris G, Hensel BK. Technologies for an aging society: a systematic review of "smart home" applications. Yearb Med Inform. 2008;17(01):33–40.
26. Seitz D, Purandare N, Conn D. Prevalence of psychiatric disorders among older adults in long-term care homes: a systematic review. Int Psychogeriatr. 2010;22(7):1025–39.
27. Ferri CP, Prince M, Brayne C, Brodaty H, Fratiglioni L, Ganguli M, et al. Global prevalence of dementia: a Delphi consensus study. Lancet (London, England). 2005;366(9503):2112–7.
28. Brown EL, Ruggiano N, Li J, Clarke PJ, Kay ES, Hristidis V. Smartphone-based health technologies for dementia care: opportunities, challenges, and current practices. J Appl Gerontol. 2019;38(1):73–91.
29. American Psychiatric Association. Diagnostic and statistical manual of mental disorders. 5th ed. Washington, DC: American Psychiatric Publishing; 2013.
30. Gurrera RJ, Moye J, Karel MJ, Azar AR, Armesto JC. Cognitive performance predicts treatment decisional abilities in mild to moderate dementia. Neurology. 2006;66(9):1367–72.
31. Ienca M, Vayena E, Blasimme A. Big data and dementia: charting the route ahead for research, ethics, and policy. Front Med (Lausanne). 2018;5(13):13.
32. Department of Health E, and Welfare. The Belmont report: ethical principles and guidelines for the protection of human subjects of research. Washington, DC; 1979.
33. Dittrich D, Kenneally E. The Menlo report: ethical principles guiding information and communication technology research. US Department of Homeland Security; 2012.
34. Nebeker C, Bartlett Ellis RJ, Torous J. Development of a decision-making checklist tool to support technology selection in digital health research. Transl Behav Med. 2019;10(4):1004–15.
35. World Medical Association. World medical association declaration of Helsinki: ethical principles for medical research involving human subjects. JAMA. 2013;310(20):2191–4.
36. U.S. Department of Health and Human Services Office for Human Research Protections. Code of Federal Regulations—Title 45 Public Welfare CFR 46. 2016. https://www.hhs.gov/ohrp/regulations-and-policy/regulations/45-cfr-46/index.html.
37. Protections of HR. International compilation of human research standards. In: Services USDoHaH, editor. 2019.

38. American Psychological Association. Ethical principles of psychologists and code of conduct. 2017.
39. World Health Organization. Code of conduct for responsible research. 2017.
40. Rothstein MA, Wilbanks JT, Brothers KB. Citizen science on your smartphone: an ELSI research agenda. J Law Med Ethics. 2015;43(4):897–903.
41. Guidance for Industry and Food and Drug Administration Staff. General wellness: policy for low risk devices. Food and Drug Administration; 2019. Contract No.: FDA-2014-N-1039.
42. Coravos A, Goldsack JC, Karlin DR, Nebeker C, Perakslis E, Zimmerman N, et al. Digital medicine: a primer on measurement. Digit Biomarkers. 2019;3(2):31–71.
43. Wexler A, Reiner PB. Oversight of direct-to-consumer neurotechnologies. Science. 2019;363(6424):234–5.
44. U.S. Food and Drug Administration. Patient engagement in the design and conduct of medical device clinical investigations: draft guidance for industry, food and drug administration staff, and other stakeholders. 2019.
45. James TA. The future of patient engagement in the digital age. Lean Forward [Internet]. 2018. https://leanforward.hms.harvard.edu/2018/10/10/the-future-of-patient-engagement-in-the-digital-age/.
46. Porter J, Koski G. Regulations for the protection of humans in research in the United States. In: The Oxford Textbook of Clinical Research Ethics, vol. 156; 2008.
47. Bailey M, Kenneally E, Dittrich D, editors. A refined ethical impact assessment tool and a case study of its application. International Conference on Financial Cryptography and Data Security. Berlin: Springer; 2012.
48. Kadam RA. Informed consent process: A step further towards making it meaningful! Perspect Clin Res. 2017;8(3):107.
49. Grady C. Enduring and emerging challenges of informed consent. N Engl J Med. 2015;372(9):855–62.
50. Wilbanks J. Design issues in E-consent. J Law Med Ethics. 2018;46(1):110–8.
51. Jayasinghe N, Moallem BI, Kakoullis M, Ojie MJ, Sar-Graycar L, Wyka K, et al. Establishing the feasibility of a tablet-based consent process with older adults: a mixed-methods study. Gerontologist. 2019;59(1):124–34.
52. Gibney E. The scant science behind Cambridge Analytica's controversial marketing techniques. Nature. 2018.
53. Mao W, Vysyaraju AR, Nebeker C. Aging in place, AI, and privacy preferences. In: IBM Conference on AI in healthy aging. Cambridge, MA; 2018.
54. Boise L, Wild K, Mattek N, Ruhl M, Dodge HH, Kaye J. Willingness of older adults to share data and privacy concerns after exposure to unobtrusive in-home monitoring. Geron. 2013;11(3):428–35.
55. Francis I. Using classical ethical principles to guide mHealth design. Online J Nurs Inform. 2017;21(3).
56. Gianfrancesco MA, Tamang S, Yazdany J, Schmajuk G. Potential biases in machine learning algorithms using electronic health record data. JAMA Intern Med. 2018;178(11):1544–7.
57. Nordling L. A fairer way forward for AI in health care. Nature. 2019;573:S103–S5.
58. Institute of Electrical and Electronics Engineers. Ethics in action—the IEEE Global Initiative on Ethics of Autonomous and Intelligent Systems. 2019. https://ethicsinaction.ieee.org/.
59. Foundation NS. Artificial intelligence (AI) at NSF. 2019. https://nsf.gov/cise/ai.jsp.
60. Russell S, Dewey D, Tegmark M. Research priorities for robust and beneficial artificial intelligence. AI Mag. 2015;36(4):105–14.
61. Chan M, Campo E, Esteve D, Fourniols JY. Smart homes—current features and future perspectives. Maturitas. 2009;64(2):90–7.
62. Mihailidis A, Cockburn A, Longley C, Boger J. The acceptability of home monitoring technology among community-dwelling older adults and baby boomers. Assist Technol. 2008;20(1):1–12.
63. Fortuna KL, Torous J, Depp CA, Jimenez DE, Areán PA, Walker R, et al. A future research agenda for digital geriatric mental health care. Amsterdam: Elsevier; 2019.

64. Nebeker C. Decision-making checklist to support ethical geriatric digital mental health research: ReCODE Health. 2019. https://recode.health/tools]
65. Melenhorst A-S, Fisk AD, Mynatt ED, Rogers WA. Potential intrusiveness of aware home technology: perceptions of older adults. In: Proceedings of the human factors and ergonomics society annual meeting. Los Angeles, CA: Sage; 2004.
66. Wang S, Nebeker C. Co-designing tech to support aging in place: prototype of digital informed consent. In: Poster presented at the 2019 IBM/UC San Diego Quarterly Artificial Intelligence Health Aging Meeting. La Jolla, CA: UC San Diego; 2019.
67. Jeste DV, Palmer BW, Golshan S, Eyler LT, Dunn LB, Meeks T, et al. Multimedia consent for research in people with schizophrenia and normal subjects: a randomized controlled trial. Schizophr Bull. 2008;35(4):719–29.
68. Jeste DV, Palmer BW, Appelbaum PS, Golshan S, Glorioso D, Dunn LB, et al. A new brief instrument for assessing decisional capacity for clinical research. Arch Gen Psychiatry. 2007;64(8):966–74.
69. Courtney KL, Demiris G, Rantz M, Skubic M. Needing smart home technologies: the perspectives of older adults in continuing care retirement communities. Inform Prim Care. 2008;16(3):195–201.
70. Nebeker C, Harlow J, Espinoza Giacinto R, Orozco-Linares R, Bloss CS, Weibel N. Ethical and regulatory challenges of research using pervasive sensing and other emerging technologies: IRB perspectives. AJOB Empirical Bioethics. 2017;8(4):266–76.
71. Peek STM, Wouters EJ, Luijkx KG, Vrijhoef HJ. What it takes to successfully implement technology for aging in place: focus groups with stakeholders. J Med Internet Res. 2016;18(5):e98.
72. Michie S, Yardley L, West R, Patrick K, Greaves F. Developing and evaluating digital interventions to promote behavior change in health and health care: recommendations resulting from an international workshop. J Med Internet Res. 2017;19(6):e232.
73. Marikyan D, Papagiannidis S, Alamanos E. A systematic review of the smart home literature: a user perspective. Technol Forecast Soc Chang. 2019;138:139–54.
74. Filkins BL, Kim JY, Roberts B, Armstrong W, Miller MA, Hultner ML, et al. Privacy and security in the era of digital health: what should translational researchers know and do about it? Am J Transl Res. 2016;8(3):1560.
75. Khan S, Hoque A. Digital health data: a comprehensive review of privacy and security risks and some recommendations. Comput Sci J Moldova. 2016;71(2):273–92.
76. Talal M, Zaidan A, Zaidan B, Albahri A, Alamoodi A, Albahri O, et al. Smart home-based IoT for real-time and secure remote health monitoring of triage and priority system using body sensors: multi-driven systematic review. J Med Syst. 2019;43(3):42.
77. Demiris G, Oliver DP, Dickey G, Skubic M, Rantz M. Findings from a participatory evaluation of a smart home application for older adults. Technol Health Care. 2008;16(2):111–8.
78. Zhang Y, Qiu M, Tsai C-W, Hassan MM, Alamri A. Health-CPS: healthcare cyber-physical system assisted by cloud and big data. IEEE Syst J. 2015;11(1):88–95.

AI Extenders and the Ethics of Mental Health

12

Karina Vold and José Hernández-Orallo

12.1 Introduction

Consider two scenarios. Helen is an 80-year-old woman who was diagnosed with Alzheimer's a few years ago. She used to use physical labels and other cues at home to help support her memory. But she now wears augmented reality glasses that label objects in her vision range, detect hazards when she is manipulating objects, indicate where things are, keep her agenda, recognize the faces of family and friends, and help her while navigating beyond her home and in many other everyday scenarios. Helen's grandson, Lewis, is a 10-year-old boy with ADHD (attention deficit hyperactivity disorder). He uses a special device that monitors his activity (movements, speech, gaze, etc.), issues recommendations through a gamified scoring system, sends indicators to teachers and family, and performs other well-thought smooth interventions. The device has boosted Lewis's self-confidence and focus, and his academic results are improving. Lewis is also allowed to use his device for exams, becoming the envy of his classmates.

While scenarios such as those above are not yet possible with current clinical technologies, with current trends moving toward digital health applications, they may become commonplace in the future. The distinctive features of both these cases are: (1) the person is *tightly coupled* with a tool (a device, such as a smartphone, or a wearable, such as an augmented reality headset) and (2) the tool integrates

K. Vold (✉)
Institute for the History and Philosophy of Science and Technology, University of Toronto, Toronto, ON, Canada
e-mail: karina.vold@utoronto.ca

J. Hernández-Orallo
Valencian Research Institute for Artificial Intelligence, Universitat Politecnica de València, València, Spain
e-mail: jorallo@upv.es

F. Jotterand, M. Ienca (eds.), *Artificial Intelligence in Brain and Mental Health: Philosophical, Ethical & Policy Issues*, Advances in Neuroethics,
https://doi.org/10.1007/978-3-030-74188-4_12

artificial intelligence (AI) capabilities (image recognition, face detection, navigation, event recognition, natural language processing, speech recognition, prediction, etc.). Together these features put tension on an "internalist" view of the mind, according to which any process outside the brain is considered as subsidiary to human cognition. In contrast, these two characteristics are in perfect alignment with the extended mind thesis (EMT), the view that the human mind and cognition are sometimes constituted by more than just the brain [1].

In Hernández-Orallo and Vold, we introduced the notion of "AI extenders," as any external tool that uses AI capabilities and is sufficiently tightly coupled with a person's cognitive system that it should be considered a cognitive extender more broadly [2]. While in contemporary philosophy, there has been quite a lot of discussion of the theory and potential of EMT; these discussions focus on relatively simple technologies. There has been almost no consideration of what more sophisticated emerging AI and data-enabled technologies can do qua cognitive extenders. We argue that with the use of artificial intelligence, there is a strong case that some of our cognition is taking place outside our brains and having deeper effects for replacing, enhancing, or regulating parts of our cognitive activity. In the examples above, we have two cases of AI extenders being used to help people with very different mental conditions. How are "AI extenders" going to serve and affect mental health, as well as our philosophical and ethical interpretation of treatments and interventions? Answering this question is the goal of this chapter.

The range of mental conditions is structured in well-known classifications, such as the ICD-10 Classification of Mental and Behavioral Disorders published by the World Health Organization (WHO) [3]. We will analyze some of these conditions under the perspective of the EMT and will investigate how current and future AI extenders, according to the capabilities they provide, may lead us to new therapeutic possibilities, new ethical challenges and a philosophical re-understanding of some aspects of mental health.

The rest of the chapter is organized as follows. Section 12.2 reviews the EMT and its evolution toward more flexible interpretations. Section 12.3 gives a definition of AI extenders, as a particular case of cognitive extenders, making it distinctive from other uses of technology for mental health. Section 12.4 selects a few mental conditions and interprets them under the EMT. Section 12.5 is more explicit about the capabilities that AI tools can extend, how these tools can be applied to a diversity of mental conditions, and how this can change in the future, especially if the EMT is accepted by clinicians and patients. Section 12.6 explores some of the positive and negative impacts of AI extenders, when used either with or without a therapeutic motivation. Finally, Sect. 12.7 gives a series of recommendations to AI designers and clinicians, and open questions for future research.

12.2 What Is the Extended Mind Thesis?

For a long time now, most scientific investigation into the mind, e.g., in neuroscience and cognitive science, has considered the brain to be the sole physical locus of the mind. According to these brain sciences, the mind is an information processing system that sits in between sensory inputs and motor outputs, and which functions by performing computations on inner representations of the world [4]. This "internalist" view is "neuro-centric" in the sense that all of the relevant inner representations and computations are thought to be instantiated in neural networks in the brain, while everything beyond the brain is considered an input source or an arena for outputs [5]. A result of this demarcation of the mind is that mental disorders have also tended to be demarcated based on this assumed boundary of the mind, that is, mental disorders are thought to be brain disorders.

Over the past two decades, however, a new picture of the mind has gained popularity, which, if true, would challenge this orthodox view. The EMT maintains that human thought and reason are not entirely "in the head." Instead, the effective circuits of human thought and reason sometimes crucially involve the technologies we use and even our social networks and institutional structures, such that the physical locus of the mind is "extended" beyond the brain [1]. The technologies that are often cited as examples of "extenders" range from humble writing utensils, such as pens and pencils, and the external symbols they create [1], to more sophisticated technologies, such as smartphones, as well as many things in between, including Scrabble tiles and Venn diagrams. We can state the EMT as follows:

> Extended Mind Thesis (EMT) = Representational vehicles (or information-bearing structures) located beyond the brain can be partly constitutive of an agent's mental states and processes.

The EMT accepts the claim that the mind is an information processing system—a core commitment of cognitive science—but maintains that the relevant information-bearing structures, that is, the vehicles of mental representations can sometimes be instantiated by non-biological elements, beyond the brain. To put it simply, if the EMT is correct, then there is more to the mind than the brain. And, accordingly, at least in some cases, in order to explain and treat mental disorders, we may need to look beyond the brain. Indeed, a number of defenders of the EMT have argued that certain disorders, such as Alzheimer's, borderline personality disorder, and autism, can be better understood, assessed, and treated by taking a wider lens on physical locus of the mind—we will discuss these examples in Sect. 12.4. But first we will discuss some of the arguments that have been used to support the EMT.

In their seminal paper titled, "The Extended Mind," Clark and Chalmers motivate the EMT by considerations of parity between external representational vehicles and internal cognitive parts [1]. They describe a scenario intended to motivate their view which involves two people—Otto and Inga. Inga has a well-functioning biological memory that allows her to recall the location of a museum she wishes to visit and to successfully navigate her way there. Despite having Alzheimer's, Otto also performs this task quite well, but he does so by relying on a notebook, which he uses as an external memory tool—recording important information and consulting his notes whenever needed. Clark and Chalmers argue that "in all the relevant respects," Otto's notebook plays the same functional role in guiding his behavior as Inga's biological memory does for her, and so the information stored in Otto's notebook should count as a part of the constitutive machinery of Otto's mind just as the information stored in Inga's brain does for her. Hence, their argument is based on the idea that external resources can make equivalent functional contributions to one's cognitive processes as internal resources can. They write:

> "If, as we confront some task, a part of the world functions as a process which, *were it done in the head*, we would have no hesitation in recognizing as part of the cognitive process, then that part of the world *is* (so we claim) part of the cognitive process." [1]
> (p. 8; emphasis in the original)

This idea has come to be known as the "parity principle." Arguments based on considerations of functional parity, like Clark and Chalmers, represent the first wave of arguments for the EMT. They have come under criticism on several fronts, which eventually led to the second-wave of arguments for the EMT. While we will not rehearse all of these criticisms here, there are a few worth mentioning because they are particularly relevant to the discussion at hand.

One long-standing debate around the appeal to functional parity has been over how to characterize the relevant functional role that Clark and Chalmers appeal to (this amounts to a commitment to the view that philosophers call "role functionalism"). This functional role determines in what ways external components must be similar to internal ones, in order to count as a genuine part of the mind. Clark and Chalmers, for example, mention three features they think are important for capturing the ways in which the information in Otto's notebook is on par with the information in Inga's brain: (1) both are a constant in the agent's life, (2) both are readily accessible when the agent needs them, and (3) both are relied upon, such that the agent trusts and endorses the information without hesitation. Among other objections, these features have come under criticism as being too coarse, for at least two reasons. Some argue that certain important features are missing and that if the particular nuances of human biological cognitive functions were included, then it would be unlikely that Otto's notebook, or indeed any external resources could really count as extensions [6]. While others have argued that these three conditions alone are too coarse as they would allow for all sorts of external objects to count as extenders, leading to absurd scenarios where everything that one reads on the Internet, for example, or the entire library that one visits, is a part of their mind [7].

The parity argument has further been criticized for relying on an overly normative picture of the mind: by letting some notion of a "healthy" biological mind set, the baseline for what could count as a possible extension of the mind. For example, imagine a "healthy" biological mind, like Inga's, started to rely on external objects to enhance her memory. Extended mind theorists are generally quite supportive of the idea that cognitive extenders enhance healthy minds—they allow us to go beyond what the naked brain can do. But in order for the parity principle to support these cases, one would have to imagine a case that involved someone (e.g., a Martian with superhuman cognitive capacities) who (purely internally) had mental capacities that go beyond those of a "normal" human. It is a roundabout way of arguing for what should be a straightforward commitment of the EMT.

Ultimately these limitations, along with other philosophical challenges facing the parity principle, were in part what motivated "second-wave" extended mind theorists to instead appeal to a criterion of functional *contributions* rather than functional *parity*. Second-wave arguments focus on the different but complementary contributions that external and internal resources make to bring about cognitive functions [8, 9]. This style of argument can straightforwardly defend the possibility of enhancing "normal" or "healthy" minds, and hence is able to overcome the first-wave focus on compensation for biological deficits. For this reason, it is likely that AI extenders, as we will define them in the next section, will need to appeal to complementarity arguments. Part of what makes AI applications so useful for humans is how they can go beyond our own cognitive capacities: processing more information, at faster speeds, and in new ways.

Complementarity arguments tend to focus on *how* external resources can be appropriately integrated with internal resources such that they can jointly govern cognitive activities and behavior, even though their functional contributions are not strictly analogous (as parity had demanded). One of the central issues for these views is articulating exactly how inner and outer resources need to be integrated. Some version and selection of the features that Clark and Chalmers defend—constancy, accessibility, and reliability—are often endorsed, though rarely the exact set [9–11]. Heersmink, for example, has recently defended a multidimensional framework, including the dimensions of information flow, reliability, durability, trust, procedural transparency, informational transparency, individualization, and transformation [11]. We will not engage this debate here, but for the purposes of this chapter, we broadly endorse Heersmink's definition of "coupling" as well as his view that each of these dimensions is a matter of degree. Perhaps it is most important to note that all of these dimensions are relational. That is, they depend on how a particular agent stands in relation to the tool—Does the agent rely on the tool in order to complete a cognitive task? Does the agent trust the information provided by the tool (rarely questioning its veracity)? How individualized is the tool for that particular agent (i.e., how difficult would it be for another agent to use it)? This would have to be assessed on a case-by-case basis.

The EMT also includes cases beyond non-biological artifacts; sometimes people rely on the minds of others as cognitive extenders. Such cases are known as "socially extension." Clark and Chalmers, for example, discuss the possibility of Otto relying on Inga's mind, rather than on his notebook [1]. So long as Inga is constantly in his life, the information in her mind (about where the museum is located) is accessible to him when he needs it, and he relies on the information (trusting its veracity), thereby we could say that Otto's mind extends into Inga's. Supporting this idea is a growing body of research on the social distribution of cognition, which tends to focus on the psychology of memory and group decision-making. The theory of transactive memory developed by Daniel Wegner, for example, maintains that memory processes (including encoding, storage, and retrieval) are sometimes shared across stable dyads (with a particular focus on highly interdependent couples) and groups [12, 13]. Wegner explains that transactive memory systems involve a "set of individual memory systems in combination with the communication that takes place between individuals" ([12], p. 186). Memory processes are, thus, not reducible to the internal processes of any particular individual in the system, as the communication between individuals is an essential part of the process. Extended mind theorists have appealed to this paradigm as one example of how a single agent's cognitive process can be distributed across multiple agents [14].

One reason that AI extenders present a paradigm shift, as we will argue in the next section, is that in some important respects they are more like cases of social extension than extension into static artifacts. For one, AI extenders have some sophisticated cognitive capacities that have previously only existed in humans, e.g., speech recognition, facial recognition, pattern recognition, and so forth. This means that while traditionally there may have been some cognitive tasks that could have only been completed through socially distributed systems, in the future there may be opportunities for individuals to rely on AI extenders rather than other people. This could have plenty of upside for people who are deficient in certain capacities, like executive control functions, and who have traditionally had to rely on members of their social network for help—we will discuss the condition of borderline personality disorder as an example of this in the next section.

There are several advantages to relying on an AI extender over another person. For one, in the case of highly interdependent couples it is often found that each individual relies on the other in various ways (e.g. Inga remembers where the museum is located, while Otto remembers the best place to park), so there is typically some degree of reciprocity, or symmetry [13]. But asymmetric socially extended beliefs might also be possible. For example, Clark and Chalmers discuss the possibility that someone relies on his regular waitress at the restaurant he frequents to determine what food he should order, thereby offloading his decision-making [1]. These asymmetric cases seem to imply, however, that one person is always being paid or "used" for cognitive labor. AI extenders present the opportunity to asymmetrically rely on a cognitively sophisticated device. Furthermore, other agents may be less stable than an AI extender device, which one can own and carry with them everywhere they go.

12.3 What Is an "AI Extender"? How This Differs from Standard Kinds of Cognitive Extension

As mentioned, the tools that are often cited as examples of cognitive extenders include both simple technologies, such as Otto and his notebook [1], as well as more advanced tools, such as smartphones [15]. Both of these technologies had transformative effects on human cognition. The use of writing tools to create external symbols and write words down represented a major shift in human intellectual history: a move from the oral tradition to literacy. The smartphone has likewise been transformative—a democratized and powerful personal computer that travels with users wherever they go. While extended mind theorists have discussed smartphones [15], it rarely gets mentioned how the computational power of smartphones has grown dramatically over the last few years, not only because of their processors, but also because of the connected use of cloud services, with many apps running or refining pre-trained deep neural networks and other technologies. We argue that this increased use of machine learning, and other functionalities brought by artificial intelligence, is qualitatively different from the kinds of cognitive extension that preceded it in several ways: these systems can perceive, navigate, make complex decisions, recognize and produce language, plan, identify emotions, etc., all in complex and changing situations.

To more precisely characterize what an AI extender is, we start from a definition of *cognitive extender* as given by Hernández-Orallo and Vold, which was adapted from Hutchins [2, 16]:

> A cognitive extender is an external physical or virtual element that is coupled to enable, aid, enhance, or improve cognition, such that all – or more than – its positive effect is lost when the element is not present.

Again, crucially, the external physical or virtual element must be *appropriately* "coupled" to be considered a cognitive extender. That is, the right conditions must be met in order for an artifact to count as a literal part of an agent's mind (i.e., on a par with the brain); and only when these conditions are met is the mind extended.

In contrast, AI extenders are an increasingly important and distinctive subkind of cognitive extender that are distinguished by their use of a particular kind of technology—AI—and their distinct implications (to be discussed in later sections). Here is a more precise definition:

> An AI extender is a cognitive extender that is "fueled" by AI. This means that some AI technology is directly responsible for the cognitive capability that the extender is able to deploy, in conjunction with its user.

With the above two definitions, what counts as an AI extender is precise, as far as what count as AI is precise. Today, we associate AI with a range of possibilities such as image and speech recognition, machine translation and planning, many of them realized through machine learning. However, AI will cover more and more *cognitive* capabilities, as we further characterize in Sect. 12.5.

Those working in psychiatry and clinical neuroscience may be familiar with related terminologies such as "cognitive assistants" or "cognitive prosthetics." It is therefore important to be clear about what distinguishes AI extenders from these well-entrenched applications and what justifies the introduction of this new category. The crucial distinction here is that, as a subspecies of cognitive extenders, AI extenders challenge the internalist view of the mind: AI extenders are a part of an agent's wide existing cognitive system, they perform cognitive functions for a human agent—just as the brain does. It is this rather strong metaphysical claim that distinguishes them from mere "cognitive assistants" or "cognitive prosthetics." This metaphysical claim is important for many ethical and policy implications that we will discuss in the next sections, and how these systems have to be regulated and built. For instance, following the examples we gave in the introduction, the definition above includes Helen's augmented reality glasses, which means that they have to be designed, in the first place, by considering how Helen's cognitive processing is going to change with their use. This may be an important, but secondary design principle, when considering some other devices that the above definition excludes, such as autonomous vehicles or cognitive robots that interact with humans occasionally (but are not tightly coupled), or an exoskeleton or a "smart" shoe that use ML algorithms to stabilize the body, but are not really providing cognitive functions to the user.

To be more precise, we might say that AI extenders (and cognitive extenders more broadly) are themselves a subset of cognitive assistants, while not all cognitive assistants are extenders. Certainly, AI extenders provide cognitive assistance or "cognitive services" [17]. For example, Helen's augmented reality glasses and Lewis's monitor system perform sophisticated cognitive processes: from perception to planning, and as such they both serve as cognitive assistants. But AI extenders are distinguished from the broader category of cognitive assistants by the degree of *coupling* (following, e.g., the dimensions from Heersmink, as mentioned above) [11]. It is this tight coupling that, so the argument for the EMT maintains, warrants us calling them a part of the agent's mind and thereby challenges the commitment to internalism. A cognitive extender is not merely processing inputs and producing outputs (such as an online machine translator); it is the locus of states that are created and accessed at any time by the human agent.

AI extenders should similarly be distinguished from cognitive prosthetics (sometimes also referred to as "orthosis"), a term that comprises many systems that can be attached to (and detached from) humans and can help or completely replace some lost or nonexistent cognitive human function. Originally, a cognitive prosthesis was defined as "a compensatory strategy that changes the environment and focuses on functional activities […] designed specifically for rehabilitation purposes" ([18], p. 41). However, many so-called cognitive prosthetics are simple software or hardware devices that are not tightly coupled and that are not "fueled" by AI; in some cases, there is no information processing or otherwise intelligent processing happening on them, like a stick compared to a leg. In other cases, no attachment (or coupling) takes place. In learning environments, for example, any device in a classroom is said to be cognitive prosthetics [19]. In our view, even if the trend today is to use the notion of cognitive prosthetics for interventions that involve some computing technology [20], many of these cannot be considered AI extenders due to a lack of appropriate coupling. As a result, many cognitive prosthetics do not carry with them any interesting metaphysical claim. They do not challenge the internalist picture of the mind, and as a result they come with a distinct (though perhaps overlapping) set of risks and opportunities from those we will discuss around AI extenders.

The categories of AI extenders and cognitive prostheses may be overlapping at times (i.e., they are not mutually exclusive), but they are also not identical. Some cognitive prostheses really are appropriately coupled to an agent and do make use of AI technologies. For instance, one of the early AI extenders was COACH (Cognitive Orthosis for Assisting aCtivities in the Home), a device that uses AI to observe, supervise, and assist people with dementia, "learn from his or her actions, and issue prerecorded cues of varying detail" [21] or Solo, another prototype that used planning and other AI techniques to help "cognitively impaired clients and their caregivers in managing their daily activities" [22]. These intelligent assistance devices for people with dementia are perhaps the best current clinical examples of AI extenders. Many of these research prototypes are now superseded by commercial products, and in cases targeting the general public, such as Ellie, Woebot, and Tess, with some of them known as virtual cognitive behavior therapists [23] a term borrowed from the early days of ELIZA, the famous computer therapist [24]. Because the relevant dimensions that characterize the appropriate coupling necessary for cognitive extension are relational in nature, these AI-driven cognitive prosthetics may in some cases also count as AI extenders.

Our definition of AI extenders also suggests why second-wave arguments for the extended mind thesis are better able to support the possibility of AI extenders. This is because the way that machine learning systems process information is likely to be relevantly dissimilar from the ways that humans do. Furthermore, as we have seen, the performance capacities of AI extenders far exceed what a notebook or a calculator can do, or even (in some respects) what a human mind can do. Indeed, AI can do much more than analogous functions (as a parity-driven argument for the EMT would require). As covered by this and other volumes, AI can lead to better diagnosis, prognosis, and treatments in mental health [25], and robotic and virtual systems are treating people with dementia, autism, and other conditions, educating children with developmental disabilities, on top of a range of possibilities for training, consultation, and healthcare management. Meanwhile, the area of affective computing is making machines able to detect and react to emotional states, where machine learning can create high-level representations from sensors on the body and brain–computer interfaces, detecting normal and abnormal situations.

Under the scope of AI extenders, we consider all these possibilities, with the condition that the system must be appropriately coupled with the person such that the effect is lost without the extender. An occasional or detached use of a robotic therapist is not an AI extender (not coupled). The use of augmented reality to treat a phobia (so that the patient is "cured") is not either (the effect is permanent) though both of these technologies might be considered as cognitive assistants [26].

Finally, this range of examples of AI extenders is indicative of just how broad and inclusive the category is intended to be. It can include a rather heterogenous set of technological applications. A companion robot could be an AI extender for the same reasons that we consider social extenders to be, for example. Some kinds of ambient intelligence (beyond smart homes and buildings), such as the persuasive mirror, could also count [27]. Probably the most obvious cases will involve software tools, such as decision-making support systems, or tools that are designed in ways that easily satisfy the relevant dimensions of coupling: applications on one's smartphone, for example, are well-suited to fit these criteria, because of how portable our phones are, how much personal information they track, how readily accessible their applications are to us, how likely we are to trust and rely on the information they provide us, and so forth. Even though AI extenders can be heterogenous in terms of their physical properties and instantiations, what makes them a cohesive category (distinct from cognitive assistances, cognitive prosthetics, and even other kinds of cognitive extenders), worthy of discussion is the role they play in the cognitive lives of humans and the ethical and design considerations that emerge from this context. This is true even though we are still in early stages of developing AI for use in clinical settings (especially systems that interact directly with the user). For this reason, our chapter focuses more on future possibilities around how AI could be used to extend cognition, in the context of mental health, exploring the risks, and the design and policy implications around how we might deal with these future scenarios.

12.4 How the Extended Mind Can Change Our Understanding, Assessment, and Treatment of Cognitive Disorders

By now, numerous authors have described how the EMT can improve either how we understand, assess, or treat mental and behavioral disorders though few have focused specifically on the kinds of intelligent assistive technologies that we have termed "AI extenders" [28–33]. In this section, we will review some of these works.

When it comes to *understanding* disorders, the central point that extended mind theorists tend to make is that there can be constitutive factors that lie outside the brain, and hence to fully understand a disorder, one cannot look to the brain alone. Simply put, there are cases of cognitive impairment that do not involve impairments of the brain. The issue of *assessment* is related to this point. Some of the standardized tests for cognitive function assume an internalist picture, focusing only on what the brain of a patient is capable of (e.g., by testing them without tools or assistive technologies). In doing so, these tests often disregard the real-life circumstances of the individual, which may involve the use of tools that make essential contributions to their cognitive functioning [31–33]. As a result, test scores can skew the picture of how "well" a patient is really doing and what they are really capable of. Hence, even if a patient has a cognitive impairment with a neuro-explanation, this might not impair their functioning in everyday life.

Finally, the matter of *treatment* is about how to view the different techniques available for rehabilitation. Researchers working on cognitive impairment in various domains have drawn distinctions between different kinds of rehabilitative strategies [34, 35], which several extended mind theorists have employed to help draw out the difference between the internalist and externalist views on treatment [31, 33]. "Restorative" strategies aim to directly address an individual's cognitive impairment by restoring their ability to perform tasks *in just the same way* that a non-impaired individual would. "Compensatory" strategies, on the other hand, attempt to circumvent impairment by helping the individual perform the same tasks but in different ways, namely by using assistive technologies [31]. Cognitive prosthetics and cognitive orthosis tools, like COACH and SOLO discussed above, were built as compensatory strategies—ways of substituting for biological deficits that could not be directly addressed [18, 21]. Because the internalist picture says that all cognition is a function of the brain alone, on this view restorative strategies must involve repairing one's internal neuro-capacities, as this is the only "true" way to improve a person's cognition. King explains that an internalist would view compensatory strategies as a second-best option; while assistive technologies might help an individual *compensate* for impairment, they do not actually fix the problem [31]. In contrast, because the EMT sees cognition as constitutively involving more than just the brain, it can view both rehabilitative strategies as genuinely restoring cognitive capabilities.

In what follows we will discuss five illustrative examples of cognitive disorders that extended mind theorists have argued can be understood in light of the EMT.

12.4.1 Alzheimer's Disease

In their now much-discussed example, Clark and Chalmers describe Otto as suffering from Alzheimer's disease, a degenerative cerebral condition characterized by a slow deterioration of multiple higher cognitive functions [1, 3]. They describe Otto as being able to function normally, despite his deteriorating biological memory, by relying on his "extended" memory, namely the information that he records in his notebook. The example suggests that by taking this wider view of the mind, we might develop new ways of assessing and treating Alzheimer's, as well as other kinds of dementia.

Drayson and Clark discuss a compelling real-life case that brings this point to life [33]. An inner-city group of Alzheimer's sufferers had scored so dismally on standard tests for Alzheimer's, such as the Consortium to Establish a Registry for Alzheimer's Disease (CERAD) protocol that doctors had anticipated the patients would need to be relocated to full-care hospitals. Yet, the patients perplexed doctors as they continued to be able to cope with the demands of daily life and to successfully live alone in a complex urban environment. Upon making home visits, doctors found that these patients had each transformed their home environments, creating ingenious personalized cognitive tools, props, and aids that supported their memory: from open storage cabinets, to notes and labels indicating what to do and when, as well as who each person was in their family photos.

Because the CERAD protocol only tests biological memory, it could not explain how these patients continued to effectively function in the world. What is more, if doctors had taken the tests of internal memory as the only standard, these patients might have been forcefully relocated to controlled hospital settings much sooner than necessary. Drayson and Clark note that the relocation of Alzheimer's patients is often a fateful turning point in which their conditions become more severe [33]. From the EMT perspective, this is not surprising; one would predict that removing patients from their supportive environments would have a detrimental effect on their functioning. The same could be said about the example of Helen, who we discussed in our introduction. Helen relied on augmented reality glasses that had been designed to support her memory and perception. Her case is an evolution where most of the physical cognitive tools, props, and aids in her home are replaced by a single tightly coupled device, an AI extender, going much beyond what Drayson and Clark found in the real case. Hence, by taking the wider "extended" view of memory and other cognitive functions not only are we able to explain how these patients could continue to function, despite low test scores, we might also rethink how we assess and treat sufferers of Alzheimer's.

12.4.2 Learning Disabilities and Disorders

A learning disability affects the way a person is able to problem-solve, plan, and acquire new knowledge and skills [3]. King argues that adopting the neurocentric internalist view of the mind commits one to problematic views about the cognitive capabilities of learning-disabled individuals, whereas the EMT allows us to more accurately assess and treat them [31]. King describes the fictional case of learning-disabled woman named Dana. Without any tools, Dana struggles to compare and evaluate various relevant factors for making decisions. She is, however, perfectly capable of making good complex decisions when she uses a graphic organizer such as a Venn diagram. This visuospatial way of representing information allows her to evaluate relevant factors and to reach a good conclusion. Indeed, when using this assistive technology, Dana's decision-making skills are just as good as anyone else's ([31], p. 49). So is Dana capable of making decisions or not? How should we assess her capacities?

Notice that Dana's use of an assistive technology counts as a compensatory rehabilitative strategy—much like the Alzheimer's patients who had relied on external resources in their environments. Hence, internalists would have to maintain that Dana has not really been "cured" because her capacities have not been restored, and so, even with the assistive technology, she cannot really make good complex decisions. Indeed, King nicely explains that the internalist is committed to an inverse relation between the extent to which an individual relies on external tools and the extent to which *she,* as an agent, is really engaging in some cognitive process. This means that Dana is only "doing" as much, cognitively speaking, as her neurons are doing ([31], p. 56). And therefore, she only merits "cognitive credit" for what her neurons do, not the cognitive work that is done by whatever assistive devices she employs. The internalist is, thus, forced to say that Dana has less cognitive capacities and deserves less cognitive credit than someone who could perform the same task "intracranially." This carries practical implications, as by adopting the stricter restorative conception of "cure," we may give preference for alternative internalist treatments, even if they are less effective (or have undesirable side effects) for the well-being of the patient.

King argues that we should resist internalism in favor of the EMT, which instead allows us to say that while Dana may need to rely on graphic organizers, she is quite capable of making complex decisions. On this view, teaching Dana how to effectively use assistive technologies (that will be readily available in her everyday life) is as good as any restorative strategy. Furthermore, Dana should get the same "cognitive credit" as someone who is able to complete a similar task without the assistive technology (i.e., internally) [31]. Heersmink and Knight have similarly argued that education and assessment should take into account how agents are able to assemble and use tools in their environment as extended cognitive systems, discussing in particular students' use of the Internet during exams [32]. We can draw a similar lesson for the 10-year-old Lewis, who (as we described in our introduction) relies on an AI extender to help aid some of the learning related symptoms of ADHD. If one takes the extended approach, Lewis should get recognition and credit for his improved academic results, even though he could not achieve these results without his assistive device.

12.4.3 Addiction

The ICD-10 clinical definition of dependence syndrome, henceforth "addiction," includes a cluster of physiological, behavioral, and cognitive phenomena in which the use of a substance takes on a much higher priority or value in one's life than is usual ([3], p. 69). Diagnostic symptoms include a compulsion to take the substance, difficulties in controlling substance taking behavior, and neglect of alternative pleasures or interests. Levy argues that adopting the EMT is useful in treating addiction because of how it can help support self-control interests [28].[1]

Internalism promotes the idea that the only way to recover from addiction is to change one's mind: addiction is entirely a matter of "will-power" and the addict just needs to "say no" to their cravings ([28], pp. 219–220). On this view, addicts tend to be held more responsible for not overcoming their addictions. But Levy cites research on "ego-depletion"—the idea that self-control draws on a limited pool of (internal) mental resources—which suggests that addicts have depleted self-control, and thus, they experience more difficulty in resisting their cravings than one who craves but is not addicted. If this is true, then it may be "literally impossible" for an addict to resist, taking the substance they crave when it is available to them and their will-power is depleted ([28], p. 219). The EMT is useful here as it points us to the agent's wider environment to look for new methods of treatment. Levy suggests, for example, that environmental modifications, including the use of technology and social support systems, can be quite effective at helping overcome addiction (other work in "positive computing" also supports this) [36]. Social support is perhaps closer to the use of AI extenders insofar as they can detect when a person is feeling weak in controlling their impulses and needs coaching or nudging.

As just one example, a system could learn what kind of situations make self-control more difficult for a particular individual. For instance, smokers usually associate tobacco with some situations (e.g., pubs, coffee) and less with others (e.g., going for a walk). So for a particular person, a machine learning system could detect that the person is likely to have depleted self-control when meeting with certain friends or going to certain places where they used to smoke. Suggesting walking routes that avoid smoking zones, or even reminding the agent that these situations may be challenging (and perhaps directing them to resources) might help them in controlling their impulses.

12.4.4 Borderline Personality Disorder

Bray argues that the EMT can offer a better understanding of certain personality disorders, such as borderline personality disorder (BPD) [29]. Personality disorders usually involve lasting and inflexible adverse patterns of thinking and feeling about oneself and others that impair how an individual functions in many aspects of life [3]. Because they can include affective dysregulation, cognitive and perceptual distortions, and impulsive behavior, they tend to be thought of as a subclass of mental and behavioral disorders [3].[2] BPD is characterized by emotional instability and disorders of mood that affect how a person relates with others. Its symptoms include a deficit in one's ability to perform certain high-level cognitive tasks, such as emotional regulation and impulse control, disturbed patterns of thinking or

[1] This is not an exhaustive list, but rather a selection from the ICD-10 Diagnostic guidelines for dependence syndrome.

[2] They are categorized as mental disorders in the ICD-10.

self-perception, and "a liability to become involved in intense but unstable relationships" ([3], p. 160).

Bray suggests that because people with BPD have a meta-cognitive deficit, it is possible that they are more likely to rely on those around them as a way to help supplement their internal, biological deficit [29]. Recall the possibility of social cognitive extension—where one agent relies on the information and abilities of another agent as an "extension" of her cognitive system. Bray argues that people with BPD may form particularly close dyads with others, especially romantic partners, friends, or family members, in an attempt to make use of their executive functions, as this is the only kind of coupling that can fill the deficit in their own biological cognitive system.

This could explain why people with BPD tend to form "unusually intense" relationships, why they suffer from a fear of abandonment, and why they are typically "devastated" when these relationships end [29]. As we say above, in order to cognitively extend, one needs a tight coupling with the external element, e.g., a stable, reliable, and high-bandwidth connection with the "extender" (among, perhaps, other features) [11]. But reliance on others has certain inherent drawbacks, Bray explains: "Imagine what it would be like if important parts of your own brain were able to detach themselves at will and wander away for indeterminate periods, perhaps never to return." [29] For the BPD sufferer, this is how he views the people in his life, with whom he has formed close "couplings." If these people up and leave, he would be left without the ability to self-regulate, to control his emotions or impulses. This also points to one way in which AI extenders could be used for treatment—AI extenders could replicate some of these metacognitive skills while at the same time being more reliable than social extenders. We will pick up on this below where we discuss the benefits of AI extenders.

12.4.5 Autistic Disorders

Autistic disorders are "pervasive developmental disorders" characterized by deficits in social interactions and social communication, and restricted and repetitive patterns of behaviors, interests, or activities ([3], p. 198). Krueger and Maiese maintain that while there is no current consensus on the cause of autism, the most popular explanations over the last several decades appeal to a theory of mind deficit [37]. The result of this has been to think of the disorder as a disturbance confined to the head of the individual—that is, to assume an internalist perspective. This in turn has shaped the typical treatment and intervention strategies, which are generally aimed at helping individuals develop their mind-modeling capacities. While this may be one helpful technique, Krueger and Maiese argue this perspective overlooks the fundamentally embodied and relational factors which contribute to autism and in doing so also overlooks potential treatment strategies [37].

Krueger and Maiese argue, for example, that people with autism typically suffer from "style blindness"—they have "a perceptual inability to extract socially salient information from the qualitative kinematics of others' actions" ([37], p. 24). This explains why they often cannot pick up on subtle social cues (e.g., non-verbal communicative behaviors, such as gestures and facial expressions) and why they struggle to understand figurative language.

But there is evidence that this lack of access to social norms is not "in principle," as people with autism can access and abide by social norms of their own group (i.e., other autistics) and when norms are made explicit ([37], p. 23). Hence, instead of only employing restorative strategies with the expectation that people with autism need to develop their internal mind modeling capacities, we might also use AI extenders for compensatory strategies aimed at making social cues (such as categorizing the tone of voice or body language for the user) and the meaning of figurative language more "visible".

Another example is how we think about and treat the characteristic movements and behaviors of people with autism, which can consist of "hand-flapping, finger-snapping, tapping objects, repetitive vocalizations, or rocking back and forth" ([37], p. 27). According to Krueger and Maiese, these are typically viewed as meaningless reflexes or nervous tics, but in fact these behaviors (sometimes called "self-stimulations" or "self-stims") may be strategic deployments used to organize incoming sensory information—for example, to occlude signal noise when incoming information threatens to be overwhelming or to heighten arousal in order to better access salient information (ibid). But standard "internalist" treatment programs have traditionally tried to eliminate or suppress self-stims, whereas a wider approach could recognize their important role as embodied cognitive coping strategies, and even try to foster these strategies. For instance, an AI extender could be designed to find and produce appropriate stims (the most effective and least visible for other people).

12.5 The Specific Effects of AI Extenders on Mental Health

An AI extender can come in different forms: a device (e.g., a tablet), a wearable (e.g., a watch), or an app or service that is available across different platforms (physical personal assistant and computer). These tools can be generic (e.g., a navigation system or an agenda), can be addressed to a range of mental health issues (e.g., a monitoring system), or can be devised for a particular condition (e.g., an anti-stress app).

If we start with generic AI extenders, they are usually devised to improve or compensate for one or more cognitive abilities. In Hernández-Orallo and Vold, we identified 14 cognitive abilities in which AI can extend cognition (the full definition can be found in that paper) [2]. These capabilities are reproduced in Table 12.1. The table also includes examples of AI extenders for each ability, either in general (non-clinical applications, second column) or for mental health (clinical application, third column). General applications may be motivated by comfort or efficiency; e.g., most people do not use GPS navigation devices to compensate for any limitation, but as an enhancement. Clinical applications of AI extenders aim at—although not exclusively—compensatory uses. The capabilities in Table 12.1 have effects on

Table 12.1 The left-hand column indicates 14 cognitive capabilities that can be extended by AI, the middle column provides examples of the kind of AI application that can achieve this (full account in Hernández-Orallo and Vold 2019). The right-hand column shows particular clinical examples in mental health

Capability	General examples	Clinical examples
MP: Memory processes	Automated reminders or prompts; new customized mnemonics to improve long-term memory, or tag our experiences with related people, concepts and other situations to improve episodic memory	Apps telling an Alzheimer's patient whether something has already been done, said or visited before
SI: Sensorimotor interaction	Pattern-recognition systems; mixing representations through generative models; intelligent sensors and actuators	Haptic/robotic clothing aides for people with Parkinson's disease
VP: Visual processing	Object and scene recognition or color-recognition tools for visually impaired; facial recognition; augmented reality; intelligent filters; and lens	Scene sketchers to contrast visual hallucinations in patients with schizophrenia
AP: Auditory processing	Voice-to-text applications for hearing-impaired; highlighting parts of speech that might be missed; following multiple conversations and prompting the user based on modeled interests; music apps	Ambient and speech recognition systems to contrast auditory hallucinations in patients with schizophrenia
AS: Attention and search	Modeling user interests and goals to focus our attention; e.g., through text search or summaries, web search engines, or with object recognition	Attention focusing devices for patients with attention deficit disorders
Pl: Planning	Automated agendas, daily task planners, and prompts based on modeling of user's goals and interests	Daily task organizers for a patient with moderate mental retardation
CE: Comprehension and expression	Digital writing assistance tools using natural language processing (e.g., Grammarly); automated re-writing or re-rendering to improve interpretability (for reading, watching films, listening to music)	Vocabulary and grammar assistance tools for a patient with some language disorders
CO: Communication	Automated emails, social media posts; improved intelligence in communication; effective spreading memes or ideas	Effective communication tools for people with Asperger's
ES: Emotion and self-control	Systems that predict and inform us of our emotional states and those of others, help us detect fake emotions in others, or trigger our emotional responses	Systems recognizing emotional states of people for patients with autism
NV: Navigation	GPS apps (e.g., Waze), building associations between places, routes, and our cognitive states to help with route-finding, safe-walking, or orientation	Route assistants to safely navigate surroundings for people with dementia
CL: Conceptualization, learning and abstraction	Machine learning apps helping find new categories, concepts or possibilities, e.g., new patterns about daily or public events; new personalized learning strategies	Personalized learning assistants for people with learning disorders or disabilities
QL: Quantitative and logical reasoning	AI systems that process uncertainty (e.g., risk or number of accidents), or quantities (e.g., people in a room) in real time	Diet analyzers and estimators for patients suffering from anorexia

Table 12.1 (continued)

Capability	General examples	Clinical examples
MS: Mind modeling and social interaction	Modeling social networks to help anticipate decisions, actions, and interests of other people	Apps determining double meanings in conversations, for people with Asperger's
MC: Metacognition	Self-tracking and analysis can help identify the potential and limitations of users, making users more aware of their own capacities	Systems monitoring self-esteem and confidence in depression episodes

many daily tasks and are expected to generate a range of applications as soon as AI can enhance or compensate them.

Having limitations for one or more of them can indirectly (i.e., as a side effect) cause many mental conditions, such as stress, depression, or loss of self-esteem, as the subject feels unable to do things that other people do easily. All these capabilities have effects on cognition and development, so they are linked to mental and behavioral disorders in one way or another (i.e., the consequences and side effects can be numerous). In other words, any extender using AI that can alleviate or compensate some cognitive limitations could have, in principle, positive effects on some of these conditions. This is sometimes referred to as the "mental capital—the cognitive and emotional resources that influence how well an individual is able to contribute to society and experience a high quality of life," and increasing this capital could "mitigate the risk of disorders such as depression, substance-use disorders, bipolar disorder and dementia" (Sahakian, p.c.) [38].

In order to determine the way AI extenders can impact on various mental conditions, we analyzed one standard classification of mental and behavioral conditions, the WHO's ICD-10 Classification of Mental and Behavioral Disorders [3]. This classification (as explained in the ICD-10 "blue book") is a comprehensive list including the clinical descriptions of the conditions. One shortcoming is that it excludes some related conditions, such as Parkinson's disease, which are classified as diseases of the nervous system, but whose symptoms might nonetheless be improved by AI extenders addressing the 14 capacities in Table 12.1 (e.g., changes in communication, or sensorimotor function).

From this list of conditions (each with an ICD assigned code starting with "F"), we identified those that are associated with each of the cognitive capabilities shown in Table 12.1. By "associated" we mean that a variation in the cognitive capacity, i.e., either an increase or decrease, will have a direct effect on the mental or behavioral condition. For each capability, we searched through the ICD-10 for a series of tokens.[3] For instance, for sensorimotor interaction (SI), we looked for "sensor*" and "moto*," and checked (manually) whether the reference made sense (e.g., was it actually talking about an association in the form of a cause or a symptom?). Table 12.2 shows the result of this analysis. It highlights which of the 14 cognitive capabilities have a direct effect on mental health conditions.

There are a few things to note about Table 12.2. First, it must be understood as showing "direct" effects of AI extenders rather than any potential side effects they

[3] The full list of items used for each capability can be found in the Appendix.

Table 12.2 Association between mental conditions (rows) and the capabilities that can be extended by AI (columns, as per Table 12.1)

ICD-10 Condition	Cognitive Capability													
	MP	SI	VP	AP	AS	PI	CE	CO	ES	NV	CL	QL	MS	MC
F00-F03 Dementia (Alzheimer's, Vascular, Others, Unspecified)	■	■	■		■				■	■	■	■		
F04 Amnesic syndrome	■										■			
F05 Delirium	■		■	■	■		■			■	■			
F06 Other disorders due to brain issues	2		0	0							7			
F07 Personality and behavioural disorder due to brain issues	2			0		0	0		0		2			2
F1x Mental and behavioural disorders due to brain issues			0,5	0,5					2		6			
F20 Schizophrenia			■	■			■	5	■				■	
F21 Schizotypal disorder				■										
F22 Persistent delusional disorders				■										
F23 Acute and transient psychotic disorders			0	0,3										
F25 Schizoaffective disorders														0
F30 Mania			1	1										1,2
F32 Depressive episode				3	■				■					■
F40 Phobic anxiety disorders													■	1
F43 Reaction to severe stress, and adjustment disorders					0				23	0			■	
F44 Dissociative [conversion] disorders		6	6		3				■	0				
F51 Nonorganic sleep disorders										3				5
F60 Specific personality disorders						4	4		■				■	■
F63 Habit and impulse disorders									■					
F71 Moderate mental retardation							■				■			
F73 Profound mental retardation			■				■	■			■			
F80 Specific developmental disorders of speech and language				0			0,1,3				■		2	
F81 Specific developmental disorders of scholastic skills	0		■	0			0				2	2		0
F82 Specific developmental disorders of motor function		■												
F84 Pervasive developmental disorders (e.g., ASD, Asperger's)					4		0	■	■		■		■	
F90 Hyperkinetic disorders (or "attention deficit disorder")					■									
F91 Conduct disorders													■	■
F94 Disorders of social functioning (children and adolescents)													■	

A black solid cell means that the whole category of mental conditions is affected by the capability, a gray cell means that this is the case only for some of the subcategories (the code of the affected subcategory or subcategories appears in the cell), and empty cells mean no association

might have on mental conditions. For instance, a visual processing system (so providing VP capabilities) that objectively describes what is on the patient's scene might help discard the false perceptions that come from hallucinations (as captured by "F05 Delirium"). This applies not only to visual and auditory inputs but also some other misperceptions (e.g., "that person is looking at me all the time" or "he is following me"). In this way, AI systems can be an alternative source to perceive reality, which can help, in some cases, discard those false perceptions (sensory, emotional, etc.) that are common in many mental conditions. Note that in this example, the AI extender has a direct effect on mental conditions that involve hallucinations. These same systems might also have the side effect of improving conditions that involve one's reasoning or planning, e.g., personality disorders, but we do not include these in the corresponding right-hand column.

Second, for many of these conditions, the ICD-10 explicitly states that the cause is unknown, and the clinical descriptions include lists of symptoms and diagnostic guidelines that are based on an assessment of the presence or absence of certain features or characteristics. Hence, in many cases, the best that we can predict is that AI extenders will have a direct effect on alleviating the symptoms (rather than addressing the causes) of the conditions listed.

Finally, Table 12.2 can be used to recognize the potential of an AI extender featuring a capability (or research in one particular area of AI) for a range of mental conditions. For instance, it is no surprise that MS (mind modeling and social interaction) is associated with many conditions, but it was less expected perhaps that AP (auditory processing) had such a number of repercussions. This is especially interesting as the state of the art of AI in auditory perception has improved significantly during the years, and its integration with hearing aids may be on its way. Table 12.2 can also be read in the other direction. If we want to treat or improve the state of patients having some particular condition, we must look at the matrix and see what cognitive capabilities we need to imbue on a system. For instance, sleepwalking (or "somnambulism") could be treated with some device that, through the use of AI, could follow where the patient is moving and check for obstacles and hazards. The table is the first approximation, but it can be valuable to have a first understanding of the many possibilities of AI (and AI extenders in particular) for mental health.

From all the abilities in Table 12.1, metacognition is perhaps the most critical one to discuss. This is for two reasons: first because of the methodology, we had to employ in searching the ICD-10 bluebook for associated conditions in Table 12.2, and second because of how it is (we believe) associated with so many different disorders. In the first case, we note that the term "metacognition" does not appear at all in the ICD-10 bluebook.[4] Nonetheless we believe that metacognition has wide associations with many of the conditions listed because of how a patient must realize their own limitations related to their particular condition. Many patients, for example, improve simply by being diagnosed ("Now I understand what is happening to me"). Relatedly, treatment and care are much easier when this is known [39]. In the context of cognitive extension, however, it is very important that the person realizes how the added AI extenders change the person's capabilities, so what the person does and what the person *thinks* she or he does—which the diagnosis clarified—are kept aligned with the use of AI extenders. A planned and temporary removal of an AI extender can be very helpful for this alignment, in the same way that hearing-impaired people realize how bad their condition is when they compare hearing with and without their hearing aids. This is also related to a placebo effect that may appear with the use of AI extenders, simply because the person thinks that he or she now has "superpowers" or a subsystem to rely on, which may boost his or her confidence.

Other AI extenders may be more focused toward monitoring and intervention rather than enhancing or replacing some cognitive capabilities. A monitoring system using machine learning to determine when a person is more likely to have an outburst or a crisis can be considered an AI extender as much as it extends our self-awareness, in the sense of an internal perception of indicators in our bodies that we can understand and react to accordingly. If the tool also makes recommendations or interventions, we can still consider it an AI extender, which helps the patient with self-control, awareness of the situation, or simply suggesting the best actions, and

---

[4] We instead had to search for related terms such as awareness, capabilities, limitations, consciousness, self-confidence, etc. (see Appendix for a complete list).

so forth. In other words, monitoring and recommendations can be seen as extenders at the metacognition and decision-making levels.

12.6 Potentialities and Challenges of AI Extenders

There are many potential benefits of the use of AI extenders for mental health—both for helping those who are cognitively impaired (which is our focus in this chapter) and for healthy users, who rely on digital devices as cognitive enhancers. In this section, we will focus on five benefits, followed by five risks.

1. AI extenders inherit all the benefits of using noninvasive treatments, something that is shared with (physical) orthopedics, in terms of flexibility, updates, repair, and removal. The use of machine learning can (a) improve the degree of personalization, as systems can learn and improve their behavior from the information they collect, and (b) make sense of a wider data set about one's lifestyle (i.e., one that looks beyond the biological individual) than a doctor ever could—including information about one's social life, screen time, the environments one spends time in, etc. Collecting and analyzing this wider data set could eventually allow for a better understanding, assessment, and treatment of mental health conditions. These benefits are available for both cognitively impaired and cognitively health users.
2. Under the lens of the EMT, we can consider the use of an AI extender as a genuine cure provided the system is reliable and integrated. Traditionally, a "cure" is some intervention that aims to directly address the cognitive deficit by making underlying mechanistic functions work better or by limiting their negative effects (what we describe as "restorative" strategies above). With an AI extender, however, we instead aim to design a system that is able to compensate for those malfunctioning underlying mechanisms. With the right kind of device integration (or "coupling"), if a person gets used to giving a description of a scene or determining false memories or perceptions from true ones using a device, this could ultimately be incorporated as part of their cognition, and help cancel or replace those malfunctioning biological processes. When this happens, we argue, the new situation can be assimilated to being "cured" or "back to a safe condition."
3. AI extenders may be a good option for those cases where restorative strategies through internal interventions, such as medications (e.g., antipsychotic drugs) for improving conditions like meta-cognition or mind modeling, may not yet be available. There is no known restorative strategy for dementia, for example, but Ienca et al. note that a wide range of intelligent assistive technologies are being developed to provide general cognitive support aimed at "empowering" adults with dementia [40].
4. The tight coupling of AI extenders makes it easier to give "cognitive credit" to the person for their accomplishments. Returning to the case of Lewis, who relies on a device to help him cope with symptoms of ADHD that affect his learning, the EMT allows us to still credit Lewis as learning, while the regular presence of the extender makes it easier (like a pair of glasses). We described Lewis as being allowed to use his assistive device during examination, for example, which also makes sense under the EMT, as the device is really part of the substrate of his

cognition. Under the EMT we would have to give a similar analysis of the cognitive accomplishments of cognitively healthy users of AI extenders as well: they deserve credit for what they achieve with their device.

5. Finally, AI extenders can provide more sophisticated resources than regular extenders. Consider again the case of BPD discussed above. Bray had hypothesized that people with BPD tend to rely on others in order to compensate for their internal deficits of executive functions because this is the only available option to them, and that this explained their characteristic fears of abandonment, and losing their autonomy [29]. But AI extenders could potentially provide the same support for meta-cognitive deficits as other people could, only with increased stability and reliability. Again, this can be a benefit both in clinical settings and for the cognitively healthy, looking to enhance their abilities.

The negative side effects of AI extenders for mental health can be varied. Some of them are also shared by other extenders or cognitive enhancers and are related to the four basic principles of medical ethics—respect for autonomy, justice, beneficence, and nonmaleficence [41, 42], but others are more specific to AI extenders (when used both in clinical settings by the cognitively impaired and for enhancement purposes by the cognitively healthy). The reason is that the use of AI technology and the tight coupling of an extender can make interactions less predictable. We will focus on five areas of concern:

1. The first consideration is *autonomy*. In medical ethics, the principle of autonomy includes respect for both an individual's right to decide and for the freedom of whether to decide [43]. One risk is that, for the sake of having the patient under control, some AI extenders will make use of interventions and nudges that effectively bypass the agent's right to decide. By encouraging actions without appealing to the agent's rationality (e.g., by presenting them with reasons to act), these devices could risk becoming *manipulative*. These scenarios become particularly concerning when we consider the technology to be a genuine part of the person's mind—the innermost space of private information, where one's intentions are formed and decisions are made [44]. As such, any manipulative interventions would clearly deviate from the maxim of nonmaleficence. Indeed, in the worst case, some of these systems could be hacked and used with malicious purposes.
2. In another important sense, autonomy should also protect one's ability to safely act in the ways one decides, ensuring short-term and long-term *reliability*. We can imagine cases of overreliance in which a person is put in risky situations (in terms of mental health), by becoming overly dependent on an AI extender which is liable to unexpectedly fail, as any technology can. This goes beyond the classical problems of cognitive laziness and atrophy that may be caused by the use of AI extenders [45, 46]. A somewhat related concern is a scenario in which patients feel so integrated with the extender that they resist changes to the system, as these would imply a change of personality and cognitive capabilities.
3. The third problem derives from an *unregulated or recreational use* of these AI extenders for mental health, where the appropriate validation and certification of procedures and tools do not follow the standards of medical practice with some other treatments, putting beneficence (good practice) at risk. This is particularly

worrisome when mentally healthy people experiment with AI extenders, leading to some pathological mental situations (e.g., similar to either substance abuse or dependence syndrome), but with some technological and AI components that could be new to the analysis. This is also related to the above points on autonomy, as an overreliance would negatively affect one's autonomy.

4. The fourth problem regards *moral status and privacy*. This goes beyond the risk that an extender may be stolen or accessed by the third party (or by the clinician or family beyond some established parameters)—a risk that applies to essentially any medical device. Under the EMT, an AI extender really is a part of the person's mind, and hence gaining access to the personal information stored in a device would be like reading the brain of a person, especially as these extenders may contain memories, experiences, decisions, and other very sensitive information [47, 48]. This is the classic double-edged sword in AI: while collecting more information about the individual can fuel powerful and highly personalized predictions (a benefit we discuss above), it also threatens personal privacy.
5. Finally, there may be problems with their *allowance in the public space* caused by a misunderstanding (or strong disagreement) of the EMT. This may lead to limitations on when and where these devices are allowed (exams, recruitment, etc.), and for how long they can be removed (airport security, other hospital treatments, etc.). This is of course related to the medical ethical principle of justice, and to the question (discussed above in "benefits") of whether we should consider AI extenders as cures.

There is a broader concern worth being mentioned, which is common to many kinds of enhancement. A widespread use of AI extenders can change our conception of what humans are capable of, and in the particular case of mental health, our notion of "mental normality." As more and more capabilities can be enhanced or modified with these devices, the diversity of behaviors and capabilities may change as people can increasingly choose what cognitive profile they prefer for themselves. The principle of justice also demands that we consider future scenarios that could arise for society—such as a moment when everybody has access to enhancements.

Determining what profiles are safe for the person (typically in the long term) and for society is going to require a deeper understanding of what mental health is, to what degree mental conditions are pathological, and what enhancements people should be allowed to make. Such considerations are well beyond the scope of this chapter, but what is clear is that the notion of a "standard" or "normal person," only comprising what the brain can do, if it ever made sense, will likely have to be completely discarded, especially when associated with a goal of being "cured."

12.7 Recommendations

In the previous sections, we have argued that a widespread use of AI extenders, and their understanding as such, may have important implications in the analysis and practice of mental health. For instance, the attachment of a patient with their AI extender can be so close that any change on the device or its software may require a deeper consideration for which the professionals involved may not be used to yet. It is then these professionals—the designers of AI extenders, coming from different areas of engineering and especially AI, and the clinicians, from physicians to nurses and other careers—who need a re-understanding of what these AI extenders mean for the evolution of the mental health and all the possible side effects on a patient.

The most urgent recommendations should be addressed to AI designers. The regulations and expectations that are put on an app or another kind of "software" or "hardware" extender should be no less stringent than those put on drugs or other kinds of treatment. The reference to take here is similar to the area of orthopedics, where manufacturers must include diverse research and development teams, including clinicians, and perform careful tests. Likewise for any digital monitoring app, development teams must determine an ethically acceptable way of designing these systems so that we can avoid these potentially negative effects [40]. But AI extenders must be more reliable than physical orthopedics. If Helen or Lewis's AI extenders fail, the consequences may be serious and even dangerous; hence, designing for safety and reliability is essential. But, on top of this, from the point of view of cognitive extensions, the manufacturer must understand that the software and the hardware become part of the mind, so no updates, discontinuations, or access to the data can be done without informed consent. Under a strict interpretation of the EM thesis, modifying an AI extender should be compared to modifying the brain.

Clinicians, too, must be aware that new gadgets imbued with obscure AI are going to become a regular part of their repertoire of diagnostic, treatment, and monitoring tools. They need to understand their basics, and how they couple with the human mind in order to create some new behaviors unseen in their careers. A good starting point for training and information for clinicians could be based on the six issues raised by Bauer et al.: (1) decide when to recommend an extender, (2) observe what other extenders the patients use (and consider how different extenders might potentially interact), (3) understand how their monitoring works, (4) explain the effects to the patients, (5) keep themselves informed about the state of the art of AI extenders, and (6) scrutinize and validate them [49]. With the inception of technology, and especially AI, human minds are changing, and mental health must change too, in terms of categories and the consideration of normality. Even if clinicians are not familiar with the philosophical underpinnings of the EMT, they know well what orthopedics is, and understand the feeling of many patients that an artificial arm, say, is a real arm. A similar analogy can be used for AI extenders, but going beyond the idea of mimicking the original functions exactly in the same way that a titanium leg may be more effective and elegant than a more realistic plastic prosthetic.

Finally, there are many future directions for research for a better understanding of AI extenders in the context of mental health, for which this chapter is just a beginning. Table 12.2, and future refinements, can be used to spot gaps and limitations or ways in which some devices can be used for some other conditions. But beyond each particular set of capabilities and conditions, we need more general guidelines, methodologies, and well-designed experiments to help in the development of the future AI extenders used for mental health. The EMT can leverage this research, but we also need better structural incentives to create intelligent assistive health technologies, rather than focusing only on biological causes and cures.

Acknowledgments We thank Richard Heersmink and Jacopo Domenicucci for their valuable comments and suggestions. JHO was supported by the EU (FEDER) and the Spanish MINECO under grant RTI2018-094403-B-C32, the Generalitat Valenciana PROMETEO/2019/098, and the Future of Life Institute under grant RFP2-152. KV was supported by the Leverhulme Centre for the Future of Intelligence, Leverhulme Trust, under Grant RC-2015-067.

Appendix

The following table includes the search tokens that we used to look for the conditions in the ICD-10 blue book for each capability, in order to construct Table 12.2 in the chapter

Cognitive capability	Tokens used for the search in the ICD-10
Memory processes	*memor**
Sensorimotor interaction	*sensor*, moto**
Visual processing	*visu*, percept**
Auditory processing	*audi*, percept**
Attention and search	*atten**
Planning	*plan*, organiz**
Comprehension and expression	*expres*, compre**
Communication	*commun*, lang**
Emotion and self-control	*emot*, control* and affect**
Navigation	*orient*, naviga**
Conceptualization, learning, and abstraction	*learn*, conceptual**
Quantitative and logical reasoning	*calculat*, mathemat**
Mind modeling and social interaction	*social**
Metacognition	*Aware*, capab*, limitations, conscio*, self*, incompetent*

References

1. Clark A, Chalmers D. The extended mind. Analysis. 1998;58:7–19.
2. Hernández-Orallo J, Vold K. AI extenders: the ethical and societal implications of humans cognitively extended by AI. In: AAAI/ACM Conference on Artificial Intelligence, Ethics, and Society (AIES 2018), Honolulu, Hawaii, USA. January 27–28; 2019. p. 507–13.
3. World Health Organization. ICD-10: international statistical classification of diseases and related health problems. 2nd ed. Geneva: World Health Organization; 2004.
4. Hurley SL. Vehicles, contents, conceptual structure and externalism analysis. Analysis. 1998;58:1–6.
5. Wheeler M. Reconstructing the cognitive world. Cambridge, MA: MIT Press; 2005.
6. Rupert RD. Challenges to the hypothesis of extended cognition. J Philos. 2004;101(8):389–428.
7. Sprevak M. Extended cognition and functionalism. J Philos. 2009;106:503–27.
8. Sutton J. Exograms and interdisciplinarity: history, the extended mind, and the civilizing process. In: Menary R, editor. The extended mind. Cambridge, MA: MIT Press; 2010. p. 189–225.
9. Menary R. Cognitive integration and the extended mind. In: Menary R, editor. The extended mind. Cambridge, MA: MIT Press; 2010. p. 267–88. https://doi.org/10.7551/mitpress/9780262014038.003.0010.

10. Rowlands M. The new science of the mind. Cambridge, MA: MIT Press; 2010.
11. Heersmink R. Dimensions of integration in embedded and extended cognitive systems. Phenomenol Cogn Sci. 2015;14:577–98. https://doi.org/10.1007/s11097-014-9355-1.
12. Wegner DM. Transactive memory: a contemporary analysis of group mind. In: Mullen B, Goethals GR, editors. Theories of group behavior. New York: Springer-Verlag; 1987. p. 185208.
13. Wegner D, Raymond P, Erber R. Transactive memory in close relationships. J Pers Soc Psychol. 1991;61(6):923–9.
14. Sutton J, Harris CB, Keil PG, Barnier A. The psychology of memory, extended cognition, and socially distributed remembering. Phenomenol Cognit Sci. 2010;9(4):521–60. https://doi.org/10.1007/s11097-010-9182-y.
15. Chalmers D. Foreword to Andy Clark's supersizing the mind: embodiment, action, and cognitive extension. Oxford: Oxford University Press; 2008. p. ix–xvi.
16. Hutchins E. Cognitive artifacts. In: Wilson RA, Keil FC, editors. The MIT encyclopedia of the cognitive sciences (MITECS). New ed. Cambridge, MA: MIT Press; 1999. p. 126–7.
17. Spohrer J, Banavar G. Cognition as a service: an industry perspective. AI Mag. 2015;36(4):71–86.
18. Cole E. Cognitive prosthetics: an overview to a method of treatment. NeuroRehabilitation. 1999;12:39–51.
19. Kolodner JL. Cognitive prosthetics for fostering learning: a view from the learning sciences. AI Mag. 2015;36(4):34–50.
20. Derian M. Cognitive prosthetics. Oxford: Elsevier; 2019.
21. Mihailidis A, Fernie GR, Barbenel JC. The use of artificial intelligence in the design of an intelligent cognitive orthosis for people with dementia. Assist Technol. 2001;13(1):23–39.
22. Simpson RC, LoPresti EF, Schreckenghost D, Kirsch N, Hayashi S. Solo: a cognitive orthosis. In: AAAI Spring Symposium: persistent assistants: living and working with AI; 2005.
23. Fulmer R. Artificial intelligence and counseling: four levels of implementation. Theory Psychol. 2019;8(4):11800. https://doi.org/10.1177/0959354319853045.
24. Weizenbaum J. ELIZA—a computer program for the study of natural language communication between man and machine. Commun ACM. 1966;9(1):36–45.
25. Luxton DD, editor. Artificial intelligence in behavioral and mental health care. 1st ed. San Diego: Academic Press; 2015.
26. Juan MC, Alcaniz M, Monserrat C, Botella C, Baños RM, Guerrero B. Using augmented reality to treat phobias. IEEE Comput Graph Appl. 2005;25(6):31–7.
27. del Valle ACA, Opalach A. The Persuasive Mirror: computerized persuasion for healthy living. In: Proceedings of the 11th International conference on human-computer interaction; 2005.
28. Levy N. Neuroethics: challenges for the 21st century. New York: Cambridge University Press; 2007.
29. Bray A. The extended mind and borderline personality disorder. Australas Psychiatry. 2008;16:8–12.
30. Vold K. Overcoming deadlock: scientific and ethical reasons to embrace the extended mind thesis. Philos Society. 2018;29(4):489–504.
31. King C. Learning disability and the extended mind. Essays Philos. 2016;17(2):38–68.
32. Heersmink R, Knight S. Distributed learning: educating and assessing extended cognitive systems. Philos Psychol. 2017;31(6):969–90.
33. Drayson Z, Clark A. Cognitive disability and embodied, extended minds. In: Wasserman D, Cureton A, editors. Oxford handbook of philosophy and disability. Oxford: Oxford University Press; 2019.
34. Garner JB, Campbell PH. Technology for persons with severe disabilities: practical and ethical considerations. J Spec Educ. 1987;21(3):122–32.
35. Kirsch NL, Levine SP, Fallon-Krueger M, Jaros LA. Focus on clinical research: the microcomputer as an "orthotic" device for patients with cognitive deficits. J Head Trauma Rehabil. 1987;2(4):77–86.
36. Calvo R, Peters D. Positive computing: technology for wellbeing and human potential. Cambridge, MA: MIT Press; 2014.

37. Krueger J, Maiese M. Mental institutions, habits of mind, and an extended approach to autism. Thaumàzein. 2018;6:10–41.
38. Cooper C, Goswami U, Sahakian BJ. Mental capital and wellbeing. Chichester: Wiley-Blackwell; 2009.
39. Shergill SS, Barker D, Greenberg M. Communication of psychiatric diagnosis. Soc Psychiatry Psychiatr Epidemiol. 1997;33(1):32–8.
40. Ienca M, Wangmo T, Jotterand F, Kressig RW, Elger B. Ethical design of intelligent assistive technologies for dementia. Sci Eng Ethics. 2018;24(4):1035–55. https://doi.org/10.1007/s11948-017-9976-1.
41. Beauchamp TL, Childress JF. Principles of biomedical ethics. 7th ed. New York: Oxford University Press; 2013.
42. Gillon R. Medical ethics: four principles plus attention to scope. Br Med J. 1994;309:184–8. https://doi.org/10.1136/bmj.309.6948.184.
43. Burr C, Morley J. Empowerment or engagement? Digital health technologies for mental healthcare. 2019. https://papers.ssrn.com/sol3/papers.cfm?abstract_id=3393534. Accessed 5 Sep 2019.
44. Reiner P, Nagel S. Technologies of the extended mind: defining the issues. In: Illes J, Hossain S, editors. Neuroethics: anticipating the future. Oxford: Oxford University Press; 2017. p. 108–22.
45. Carr N. The glass cage: where automation is taking us. London: Random House; 2015.
46. Barr N, Pennycook G, Stolz JA, Fugelsang JA. The brain in your pocket: evidence that smartphones are used to supplant thinking. Comput Hum Behav. 2015;48:473–80.
47. Blitz MJ. Freedom of thought for the extended mind: cognitive enhancement and the constitution. Wis Law Rev. 2010;4:1049–119.
48. Søraker J. The moral status of information and information technology: a relational theory of moral status. In: Hongladarom S, Ess C, editors. Information technology ethics: cultural perspectives. Hershey: Idea Group; 2007. p. 1–19.
49. Bauer M, Glenn T, Monteith S, Bauer R, Whybrow PC, Geddes J. Ethical perspectives on recommending digital technology for patients with mental illness. Int J Bipolar Disord. 2017;5:1–14.

Part III

AI in Neuroscience and Neurotechnology: Ethical, Social and Policy Issues

The Importance of Expiry Dates: Evaluating the Societal Impact of AI-Based Neuroimaging

13

Pim Haselager and Giulio Mecacci

13.1 Introduction

The combination of artificial intelligence and neuroimaging (AINI), specifically the combination of machine learning (ML) and functional magnetic response imaging (fMRI), has increased the possibilities for brain reading, i.e., decoding mental content or identifying behavioral dispositions from brain activity. Although brain reading currently is at the beginning stage, consisting mainly of proofs of principle, further progress could lead to societally relevant applications. For instance, in the domains of finance, law, health, and sexuality, the usage of such applications requires a careful analysis of the associated ethical, legal, and societal implications (ELSI). In order to contribute to such an analysis, we will focus, in this chapter, on the following themes. First of all, we will discuss the quandaries inherent to the early assessment of technological impact: how can we meaningfully analyze future implications of a technology that is currently being developed? Second, we will analyze the implications of the combination of artificial intelligence and neuroimaging for brain reading. What does AI add to the standard computational processing involved in acquiring imagery data and statistical analysis? What consequences could this have from the perspective of our evaluative framework? Finally, we will conclude the chapter by providing some recommendations for regulation and further research. In a situation of "moral overload" [1], responsible research and innovation should bring about technology that overcomes the trade-off between incompatible values, e.g., a person's mental privacy and public security. While technology is being developed, we suggest that various forms of expiry dates (for

P. Haselager (✉) · G. Mecacci
Department of Artificial Intelligence, Donders Institute for Brain, Cognition and Behaviour, Radboud University, Nijmegen, The Netherlands
e-mail: w.haselager@donders.ru.nl; g.mecacci@donders.ru.nl

F. Jotterand, M. Ienca (eds.), *Artificial Intelligence in Brain and Mental Health: Philosophical, Ethical & Policy Issues*, Advances in Neuroethics,
https://doi.org/10.1007/978-3-030-74188-4_13

informed consent, data storage, and data analysis) could be, at least temporarily, useful in avoiding some undesirable implications of AINI.

13.2 The Risks of an Early Technology Assessment

When examining potential ethical, legal, or societal implications of a rapidly developing science and its associated technologies, there is a risk of either over or underestimating the potential of a technology within a certain timeframe. Overestimating the possibilities of a developing technology can create hype: exaggerated claims or excessive publicity or discussion surrounding specific research field, technology, or application (e.g., [2–6]). Hype can occur both in relation to the expected positive and negative possibilities or consequences. Positive hype leads to unrealistic expectations, negative hype to unjustified concerns. Both should be avoided as much as possible.

The problem, of course, is that no one knows exactly what can be expected. Oftentimes technological developments occur much slower than expected (leading to, e.g., "AI winters," [7]), although recently the field of AI has also experienced the opposite, in that certain results were achieved faster or in different fields than expected (e.g., moving from beating the world champions in chess to beating them in Go). Given the potentially large economic and political implications of scientific and technological developments, the difficulty of technology extrapolation has sometimes led to accusations of "hyping" (the implication sometimes being that this would be a deliberate or personal interest-based form of distortion). It is not clear, however, whether such statements can be based on more than the observation that the discussed technological developments are currently unlikely to be achieved, or at least seem still far removed from being achievable. Although one may debate to what extent future developments can be accurately assessed, it is not immediately apparent that overestimating the possibilities of a certain technology or research field amounts to hyping. Moreover, ethical debates and societal procedures to regulate the application of technology generally require substantial amounts of time. Hence, the risk of being "too late" in discussing the broader implications of a particular technological innovation is not negligible. This is especially the case when the relative novelty of a certain technology does not give raise to direct and immediate concerns, making responsible stakeholders prone to the so-called "delay fallacy" [8], which results in repeatedly postponing risk-related discussions. For example, the necessity of "horizon scanning" has been mentioned ([9], p. 292; *see also* [10]), as an activity important to anticipate coming developments and potential (legal) implications before a new technology is used in court. Elsewhere [11], one of us discussed the possibility of an "*ELSI gap*" between what currently is possible given the state of science and technology and the kind of ethical, legal, or societal implications that could be considered relevant to discuss before the technology "gets there." After all, the price society may have to pay for being too late, can turn out to be quite high. Technology tends to contribute to the creation of habits and practices that are hard to change once they are established, even if there is agreement about their undesirability. Hence, in what follows, the cases and analyses are to be taken as future-oriented, with all the uncertainties and risks involved in such an enterprise.

The difficulties of estimating the societal impact of a technology, especially under time constrains, are not the only concern. In the context of developing and controlling technological innovation in a societal context, estimations are only the first part of a double-bind problem, as highlighted by the so-called "Collingridge Dilemma" [12]. On the one hand, the impacts of a certain technology, and the kind of potentially affected stakeholders, can hardly be predicted before such a technology is widely distributed and adopted. On the other hand, once the technology has become entrenched, it becomes harder to change or control it. Different methods have been proposed in the philosophy of technology to contain the risks of this dilemma. As emphasized by the advocates of value-sensitive design and responsible innovation [1, 13], articulating moral and value considerations at the earliest stages of technological development, together with an early involvement of the greater possible number of stakeholders, are essential contributors to the successful development of a technology. This is the case despite the uncertainties that such early stage brings about. The risk of mistaken extrapolations and not entirely accurate optimism or concerns regarding future technologies is the natural result of a still partial or insufficient knowledge. We accept our share of responsibility for them while discussing this topic.

13.3 Evaluating Implications of AI in Combination with Neuroimaging (AINI)

A realistic technological risk assessment requires sufficient knowledge of what a given technology can and cannot do. AI and neuroimaging—we will focus particularly on machine learning and fMRI here—are relative well-established fields. However, the combination of AI and neuroimaging is relatively young and multidisciplinary, making it a fast moving—and slightly blurred—target when trying to define its state-of-the-art and current possibilities. This partially prevents a meaningful discussion about its ethical, legal, and societal implications. To make a step in the direction of building up relevant knowledge for a sound assessment, we proposed a conceptual framework composed of five criteria, specifically aimed at investigating the potential implications of brain reading technology for privacy [14, 15]. Those criteria are meant to aid a potential assessor of the technology in isolating a manageable set of values that are relevant for that inquiry. Their abstractness allows to apply them to both current and future technology, as they do not depend on the specific characteristics of any given brain reading method.

Brain reading technology entails a risk for privacy in the sense that it might potentially make public, either against a subject's will or unbeknownst to them, personal information about their thoughts, opinions, dispositions, traits, and so on. We argued that the extent to which this might be made possible for each given technology is highly dependent on five aspects: accuracy, reliability, informativity, concealability, and enforceability. It will not be necessary in the present context to enter into the nitty gritty details, but we think it is useful to point out briefly what those aspects represent. *Accuracy* represents the performance of a given technology: its capacity to correctly (in probabilistic terms) identify a mental state, and its resilience to false positives and false negatives. *Reliability* represents how well a certain technology's accuracy is preserved both between different subjects (in spite of structural and functional variability) and within the same subject over time (in spite of functional plasticity). Whereas accuracy and reliability are quantitative aspects, the other three are qualitative ones. *Informativity* has to do with the quality of the information that can be produced by the means of a certain brain scanning technology. For instance, recent developments in cognitive neuroimaging have achieved a relatively solid accuracy in recognizing certain mental states. However, these mental states consist of categories that are very general, and the level of detail of the information extracted does not consent to obtain an important amount of sensitive information, if any. *Concealability* represents the degree to which a technology can be applied to subjects without them being aware of it. A subject might be partially or fully aware of being scanned but not of the type of information that is being extracted. The context might be such that a subject is misled by an ill-intentioned applier of the technology, or it might be that data extracted in bona fide reveals to be very sensitive after further analysis, or thanks to future analytical instruments and methods. We will consider this aspect more in detail later in this text. Finally, *enforceability* indicates the extent to which a technology can be used against one's will. Current neuroimaging methods, e.g., lie detection or to identify one's sexual orientation can be intentionally disrupted by a noncompliant subject with relative ease. But if such disruptions can be reliably detected, the threat of sanctions upon discovery may lead to enforceable practices.

There are several aspects of AINI that deserve discussion. Below we will limit ourselves to three points. First of all, what is the *promise* of using AI in neuroimaging? We will focus here on the promise of improved analysis of brain data due to AI's capacity to process them in large quantities, and the possibility to find informative new clustering via unsupervised learning. Machine learning plays a vital role here. Second, there are questions regarding the *standards* that AINI has to fulfill in order for its insights to be usable in societally relevant scenarios. How accurate, reliable, and informative need the results of AINI be, e.g., in the context of neuroprediction, before the results can be considered legitimately actionable in real-world applications such as insurance, mortgage, or in a court of law? Third, we suggest that the *enforceability* of AINI-based information is relevant for the possibilities that this technology might grant in terms of mental privacy invasion. The expected rapid developments in the capacities of AI and machine learning methods, together with its increasing use in surveillance, e.g., to promote national security and counterterrorism [16], might lead to potential abuse of brain reading technology. This, in turn, could lead to unbalancing the already precarious trade-off between privacy and security [17]. Hence, we will consider the implications of applying new analysis methods to old data, from the perspective of informed consent and data ownership.

13.4 AINI in Society: Are We Ready Just Yet?

Computational processing has long played a vital role in measuring brain structure and function. It is required for collecting and cleaning the brain data, transforming the data into formats that are more understandable for human brains (e.g., by turning them into pictures), and for performing the complex statistical analysis of data in relation to hypotheses. Increasingly, the role of computers is extended by allowing artificial intelligence to interpret data, e.g., through machine learning. In this section, we will try to sketch some implications of this combination of AI and neuroimaging in the attempt to find relevant patterns in brain data, and by suggesting interpretations of the data that can assist in generating explanations or predictions of cognition and behavior.

AI techniques, especially machine learning, in combination with neuroimaging data can predict psychiatric phenomena and/or treatment response, in increasingly better and novel ways (see also [9], p. 304–305 and p. 307). In a recent review of AINI applications for psychiatry, Nielsen et al. [18] indicate that supervised learning can be applied to predict, e.g., attention-deficit and hyperactivity disorder (ADHD), autism, depression, schizophrenia, and Tourette syndrome. Age-related deviations from standard development may be identified, and this can be used to predict real-world risky behavior [19]. Furthermore, unsupervised learning techniques may result in the identification of novel subgroups of patients, despite differences in neuroimaging data. For instance, subtypes of depression that are responsive to a particular treatment, such as transcranial direct current stimulation (tDCS) could be identified [20]. Just et al. [21] indicate that highly accurate classifications of suicidal youth can be obtained by applying machine learning to neural representations of suicide and emotion concepts.

However, the extent and exact nature of the improvements in psychiatric predictions is debated [22], and the appropriate standards and methodologies for AINI have not crystalized yet. For example, Vilares et al. [23] indicate that AINI might enable a potentially legally relevant differentiation between knowing and reckless subjects, i.e., by establishing on the basis of brain scanning whether suspects knew about the illegality of their behavior or were (partially) unaware of this, but took the risk. Different legal consequences, including greater punishments, are applied to individuals who act in a state of knowledge about the consequences of their actions, compared with a state of recklessness. They indicate that AINI can be of help in making this distinction, but they also emphasize that the differentiation is quite context dependent. Vieira et al. ([24], p. 17) note inconsistent results of applying ML to neuroanatomical data (in relation to first episode psychosis detection), perhaps related to small sample sizes, single-site studies, and notice that various important methodological issues are recently becoming clearer. As they say:

> "Despite the high level of interest in the use of machine learning (ML) and neuroimaging to detect psychosis at the individual level, the reliability of the findings is unclear due to potential methodological issues that may have inflated the existing literature. (...) Findings from this study suggest that, when methodological precautions are adopted to avoid overoptimistic results, detection of individuals in the early stages of psychosis is more challenging than

originally thought. In light of this, we argue that the current evidence for the diagnostic value of ML and structural neuroimaging should be reconsidered toward a more cautious interpretation."

The performance considerations made in the previous paragraph become especially important in cases where legally relevant distinctions or predictions are involved. The Daubert standard is a rule of federal law of the USA regarding the admissibility of expert witness testimony. It explicitly states (Rule 702, clause b & c, as amended Apr. 26, 2011, eff. Dec. 1 [25]) that a testimony of an expert witness has not only to be based on sufficient facts or data, but also that the testimony is the product of reliable principles and methods. Given discussions in the literature as provided above, it stands to reason to suggest that at least some forms of AINI in some types of applications do not yet hold up to requirement (c) of the Daubert standard.

Moreover, Paulus et al. [26] and Jollans et al. [27] discuss the intricacies of AINI applied to predicting individual behavior, in particular to find an optimum in the bias-variance tradeoff, i.e., avoiding both overly complex models that fit the data (too well) but do not generalize, and overly simple models that do not fit the data (*see also* [28]). The most accurate predictions are usually obtained through an intensive training of a single subject on tasks that evoke clear and extensive neuronal activation. However, this comes at the price of a reduced reliability, which is the applicability to a larger pool of different brains, and to a constantly transforming brain at different times. Good predictions should be sufficiently specific to reliably recognize a certain mental task, and sufficiently generic to recognize across the many ways this can be represented in one, or even different, brains. Paulus et al. suggest that while there is clear room for progress,

> "there is no single generically applicable machine learning tool, or one that performs uniformly better than others, to make individual level predictions based on neuroimaging data ([26], p. 660).

In addition, they note that although a specific approach can be optimized, "this often comes at the cost of difficult interpretability of how the various features of the workflow contribute to the predictions." (ibid.). Hence, even in such cases, there still is a trade-off between making practically accurate, reliable, and informative predictions, and a clear understanding how such correct predictions come about.

13.5 AINI and Expiry Dates

Taken together, the increased capacity for brain reading due to rapidly improving AINI, as well as the potential risk of misunderstanding the current possibilities, leads to questions regarding the current practices of informed consent. The European Court on Human Rights (ECHR) has indicated that medical interventions against the subject's will or without the free, informed, and express consent of the subject constitute an interference with his or her private life ([9], p. 299). As a first observation, because of the rapidly growing possibilities granted by technology, informed consent requires a further analysis of its temporal dimension. Simply put, as AI improves and novel types of information may be extracted from once sampled brain data, how long can the consent given at the moment of sampling be considered to remain valid? We suggest that the possibility of *an expiry date for informed consent* about the storage and analysis of personal data should be further explored, at least for the time that the trade-off between accuracy and interpretability is not optimized.

Secondly, our discussion above illustrates the importance of an EU policy regarding AINI. Article 5 of the EU General Data Protection Regulation ([29]; https://gdpr.eu/article-5-how-to-process-personal-data/) states that personal data should only be collected and processed for a legitimate specific *purpose* and that the data should only be retained and stored for as long as necessary to satisfy the specified purpose. The question raised here is whether progress in ML will require a further specification of the concept "purpose." Do improvements in ML amount to qualitatively similar types of processing of brain data or could the differences in analyses be so substantial that it should be evaluated as implying a difference in purpose? How do we establish a move from more of the same to substantially different in the context of AINI? More generally, what *standards* for the interpretation of AINI results should be formulated and what *procedures* should be followed to establish that they have been achieved? More specifically, how are the criteria of accuracy, reliability, and informativity to be operationalized and contextualized? Given the fast developments in the field of AINI, it seems timely to start working on such questions, especially if we consider the further possibility that, under certain sociopolitical circumstances, brain reading technology might be utilized against one's awareness or consent.

Third, the implications of AINI in the light of the recent responses to terroristic threats to national security deserve attention. Surveillance methods are recently turning into large-scale data collections as opposed to targeted espionage and intelligence gathering [30]. The novel possibilities to gather personal information that might be offered by, e.g., consumer neurotechnology [31], are a risk that should be carefully evaluated in this particular sociopolitical scenario. In our framework, this aspect is addressed in relation to the criterion of enforceability. As a starting point, we are in full support of Ienca and Andorno's [32] defense of an individual's right to the protection from coercive collection of brain-based information about a person's mind, and its further storage and use. However, as Ligthart [9] indicates that societal interests might sometimes override such basic rights, and he provides an interesting analysis of coercive neuroimaging in the context of EU criminal law. He distinguishes physical from legal coercion. Regarding the physical form of coercion, Ligthart suggests that it is unlikely to prevent countermeasures, because practically speaking it is very easy to disturb the brain measurements by, e.g., minute movements of head or tongue, voluntarily focusing on random thoughts. Although this in itself is true, we suggest that the issue of enforceability may not just be the (lack of) effectiveness of the physical coercion, but also whether resistance can be reliably discovered. Via minute movements, or by not complying with instructions, the data subject may sabotage the measurement, but if the sabotage is consistently detectable, the mere fact of resistance will be noticed and potentially has consequences. The knowledge that one's resistance can be discovered may in itself form a deterrent against resistance. Therefore, the issue underlying enforceability is not only, or even primarily, the effectivity of physical coercion, but rather the detectability of countermeasures. As it is not obvious that there will be significant limitations on AI-based countermeasure detection techniques, the issue of physical coercion is more important than it may appear on first sight.

Legal coercion implies foreseeable negative consequences for withholding consent, such as punishment, diminished chance for parole, or adverse inferences regarding the innocence of a subject. Article 8(2) of the European Convention of Human Rights [33] holds that legal coercion may be justified when in a democratic society it is in the interest of, e.g., national security, public safety, or the protection of the rights and freedoms of others. Important here is the question whether the legally enforced obtained brain information could be used in the future for other applications or analyses than the ones that originally led to or justified the legal coercion, undergoing what has been called a "function creep" [34]. This phenomenon has been identified in relation to, e.g., big data [35, 36] and DNA information, to refer to "changes in, and especially additions to, the use of a technology" [37]. Ligthart notices that "the results that may be obtained from DNA and fingerprints may be used in the future to demonstrate information about someone's involvement, or future involvement, in a crime" and goes on to propose that "the same applies mutatis mutandis to coercive forensic neuroimaging" (p. 296–297). Although we are in general agreement with the author, we think that the latter claim might downplay the potential of AINI. One of the earliest cases of genetic profiling in combination with structural neuroimaging turned into a now notoriously controversial legal case in Italy [38]. During this legal proceeding, a lighter sentence was granted to the defendant in virtue of a certain genetic makeup that was deemed responsible for seemingly abnormal brain structures [39]. However, DNA analysis provides close to no information in regard to the ontogenesis of certain behavioral dispositions that are acquired due to external influences, e.g., as a consequence of a certain upbringing. Information about specific brain structures is usually insufficient to univocally determine behavior and leaves space for great subjective variability. However, the complexity and type of cognitive information that can be derived via AINI goes far beyond the use of DNA and fingerprints for identification purposes. Machine learning-enabled functional neuroimaging could enable the decoding of occurring mental *states,* such as thoughts, desires, and intentions. These might be detected not only during their occurrence (real time) but also recorded and timestamped in the collected data, potentially becoming accessible once adequate analytical methods become available. This makes AI-enabled neuroimaging qualitatively different from genetic proofs, and its data significantly more complex, detailed, and individualized. It is precisely in this area that the to-be-expected significant advances in AI could have important implications for the analysis of earlier recorded and archived data. Here too, therefore, a temporal window for brain data usage deserves consideration. Hence, we propose that an *expiry date for the storage and analysis of brain data* should be investigated, especially in cases where the newly derived information or classification goes beyond the purposes for which the original data was collected. We need to establish to what extent increased ML power could lead to the disclosure of relevant information about a subject's cognition or behavioral disposition that extends beyond the original informed consent declaration. If such be the case, the introduction of an expiry date on the obtained brain data is recommended. Actions could range from guaranteed (irreversible) anonymization to, in sensitive cases, the obligation to delete data.

The exact relation between our proposal for a brain data analysis expiry date and the neuro-specific right to mental privacy recommended by Ienca and Andorno [32] requires further analysis. Our current thoughts are that whereas the neuro-specific right to mental privacy is a very general and principled one, asserting the basic protection of an individual's right to resist extracranial externalization of brain derived mental information, our suggestion may be seen as a more specific way (one among several) to ensure the exercise of that right. In other words, our suggestion could be taken as a means toward the end specified by Ienca and Andorno [32]. In passing, we would also like to draw attention to the potential parallel that can be drawn between our suggestion for brain data expiry dates and the right to be forgotten. Of course, Articles 5 and 17 of the GDPR (https://gdpr.eu/article-17-right-to-be-forgotten/) are complex and specify various grounds for both its legitimate invocation and its exclusion. We merely mention it here to indicate that if a right to be forgotten exists in relation to ICT data, at the very least expiry dates regarding the potentially even more sensitive brain data should be open for discussion.

13.6 Conclusion

In a nutshell, our main argument is that, because of the expected rapid progress in AI, especially the field of machine learning, it cannot be excluded that AINI could reveal significantly more and/or different types of information than originally thought of or aimed at on the basis of past or currently obtained brain data. We suggest that this could imply a usage of brain reading unbeknownst to the subject (at the time of the scanning), and possibly also against somebody's will (at the time of the scanning). In addition, these later occurring possibilities could also occur without the knowledge or will of the researchers or technicians involved in the brain scanning and data collection. Hence, in relation to this possibility, we have suggested that brain measurements should come with expiry dates for brain data-related consent, storage, and analysis.

In essence, this finishes our contribution. However, just before submitting this chapter, the EU white paper appeared [40]. We notice that the topic of AINI is not yet discussed in it, and we would like to take the opportunity to make a plea to do so in further versions or follow-ups. We welcome the attention given to importance of setting standards:

> "The following requirements relating to the data set used to train AI systems could be envisaged: Requirements aimed at providing reasonable assurances that the subsequent use of the products or services that the AI system enables is safe, in that it meets the standards set in the applicable EU safety rules (existing as well as possible complementary ones). For instance, requirements ensuring that AI systems are trained on data sets that are sufficiently broad and cover all relevant scenarios needed to avoid dangerous situations." ([40], p. 19)

Moreover, the temporal aspects of data control we emphasized are mentioned as well:

> "The records, documentation and, where relevant, data sets would need to be retained during a limited, reasonable time period to ensure effective enforcement of the relevant legislation." ([40], p. 19).

In our view, such statements emphasize the importance of standards and temporal aspects of data expressed in our chapter as well. If anything, AINI deals with even more sensitive data than other types of AI. That, in our view, implies that the EU needs to address the societal implications of AINI with urgency.

References

1. Van den Hoven J. Value sensitive design and responsible innovation. In: Owen R, Bessant J, Heintz M, editors. Responsible innovation. Chichester: Wiley; 2013. p. 75–83. https://doi.org/10.1002/9781118551424.ch4.
2. Abetti PA, Haldar P. One hundred years of superconductivity: science, technology, products, profits and industry structure. Int J Technol Manag. 2009;48(4):423–47. https://doi.org/10.1504/IJTM.2009.026688.
3. Caulfield T. Ethics hype? Hastings Cent Rep. 2016;46(5):13–6. https://doi.org/10.1002/hast.612.
4. Caulfield T. Spinning the genome: why science hype matters. Perspect Biol Med. 2018;61(4):560–71. https://doi.org/10.1353/pbm.2018.0065.
5. Caulfield T, Condit C. Science and the sources of hype. Public Health Genomics. 2012;15(3–4):209–17. https://doi.org/10.1159/000336533.
6. Fox S. Irresponsible research and innovation? Applying findings from neuroscience to analysis of unsustainable hype cycles. Sustainability (Switzerland). 2018;10(10):1–16. https://doi.org/10.3390/su10103472.
7. Hendler J. Avoiding another AI winter. IEEE Intell Syst. 2008;23(2):2–4. https://doi.org/10.1109/MIS.2008.20.
8. Hansson SO. Fallacies of risk. J Risk Res. 2004;7(3):353–60. https://doi.org/10.1080/1366987042000176262.
9. Ligthart SLTJ. Coercive neuroimaging, criminal law, and privacy: a European perspective. J Law Biosci. 2019;6(1):296–316. https://doi.org/10.1093/jlb/lsz015.
10. Nadelhoffer T, Sinnott-Armstrong W. Neurolaw and neuroprediction: potential promises and perils. Philos Compass. 2012;7:631–4.
11. Haselager WFG. Implications of neurotechnology: brain recording and intervention. In: Hage J, Brożek B, Vincent N, editors. Cambridge handbook on law and the cognitive sciences. Cambridge: Cambridge University Press; 2020.
12. Collingridge D. The social control of technology. London: Frances Pinter; 1980.
13. Friedman B, Hendry DG, Borning A. A survey of value sensitive design methods. Found Trends Human Comput Interact. 2017;11(23):63–125. https://doi.org/10.1561/1100000001.

14. Mecacci G, Haselager WFG. Identifying criteria for the evaluation of the implications of brain reading for mental privacy. Sci Eng Ethics. 2017;25(2):443–61. https://doi.org/10.1007/s11948-017-0003-3.
15. Mecacci G, Haselager WFG. Five criteria for assessing the implications of NTA technology. Am J Bioeth Neurosci. 2019;7740(5):20–3. https://www.tandfonline.com/doi/full/10.1080/21507740.2019.1595781. https://doi.org/10.1080/21507740.2019.1595781.
16. Verhelst HM, Stannat AW, Mecacci G. Machine learning against terrorism: how big Data collection and analysis influence the privacy-security dilemma. Sci Eng Ethics. 2020;26:2975–84.
17. Bird SJ. Security and privacy: why privacy matters. Sci Eng Ethics. 2013;19(3):669–71. https://doi.org/10.1007/s11948-013-9458-z.
18. Nielsen AN, Barch DM, Petersen SE, Schlaggar BL, Greene DJ. Machine learning with neuroimaging: evaluating its applications in psychiatry. Biol Psychiatry Cognit Neurosci Neuroimag. 2020;5:791–8. https://doi.org/10.1016/j.bpsc.2019.11.007.
19. Ponseti J, Granert O, Jansen O, Wolff S, Beier K, Neutze J, Deuschl G, Mehdorn H, Rudolph MD, Miranda-Domínguez O, Cohen AO, Breiner K, Steinberg L, Bonnie RJ, et al. At risk of being risky: the relationship between "brain age" under emotional states and risk preference. Dev Cogn Neurosci. 2017;24:93–106.
20. Drysdale AT, Grosenick L, Downar J, Dunlop K, Mansouri F, Meng Y, et al. Resting-state connectivity biomarkers define neurophysiological subtypes of depression. Nat Med. 2017;23:28–38.
21. Just MA, Pan L, Cherkassy VL, McMakin DL, Cha C, Nock MK, Brent D. Machine learning of representations of suicide and emotion concepts identifies suicidal youth. Nat Hum Behav. 2017;1:911–9. https://doi.org/10.1038/s41562-017-0234-y.
22. Reardon S. The painful truth. Nature. 2015;518(7540):474–6. https://doi.org/10.1038/518474a.
23. Vilares I, Wesley MJ, Ahn WY, Bonnie RJ, Hoffman M, Jones OD, Morse SJ, Yaffe G, Lohrenz T, Montague PR. Predicting the knowledge-recklessness distinction in the human brain. Proc Natl Acad Sci U S A. 2017;114(12):3222–7. https://doi.org/10.1073/pnas.1619385114.
24. Vieira S, Gong QY, Pinaya WHL, Scarpazza C, Tognin S, Crespo-Facorro B, et al. Using machine learning and structural neuroimaging to detect first episode psychosis: reconsidering the evidence. Schizophr Bull. 2020;46(1):17–26. https://doi.org/10.1093/schbul/sby189.
25. Daubert Standard, Rule 702. 2011. https://www.law.cornell.edu/rules/fre/rule_702.
26. Paulus MP, Kuplicki R, Yeh HW. Machine learning and brain imaging: opportunities and challenges. Trends Neurosci. 2019;42(10):659–61. https://doi.org/10.1016/j.tins.2019.07.007.
27. Jollans L, Boyle R, Artiges E, Banaschewski T, Desrivières S, Grigis A, Martinot JL, Paus T, Smolka MN, Walter H, Schumann G, Garavan H, Whelan R. Quantifying performance of machine learning methods for neuroimaging data. Neuroimage. 2019;199:351–65. https://doi.org/10.1016/j.neuroimage.2019.05.082.
28. Poldrack RA, Huckins G, Varoquaux G. Establishment of best practices for evidence for prediction: a review. JAMA Psychiat. 2019;77:534–40. https://doi.org/10.1001/jamapsychiatry.2019.3671.
29. EU General Data Protection Regulation. 2018. https://gdpr.eu/.
30. Bigo D, Carrera S, Hernanz N, Jeandesboz J, Parkin J, Ragazzi F, Scherrer A. National programmes for mass surveillance of personal data in Eu member states and their compatibility with Eu law. In: Liberty and Security in Europe Papers, No. 61; 2013. https://papers.ssrn.com/sol3/papers.cfm?abstract_id=2360473.
31. Ienca M, Haselager P, Emanuel EJ. Brain leaks and consumer neurotechnology. Nat Biotechnol. 2018;36(9):805–10. https://doi.org/10.1038/nbt.4240.
32. Ienca M, Andorno R. Towards new human rights in the age of neuroscience and neurotechnology. Life Sci Society Policy. 2017;13(1):1–27. https://doi.org/10.1186/s40504-017-0050-1.
33. European Convention of Human Rights. 2010. https://www.echr.coe.int/Documents/Convention_ENG.pdf.
34. Innes M. Control creep. Sociol Res Online. 2001;6(3):13–8. https://doi.org/10.5153/sro.634.
35. Brayne S. Big data surveillance: the case of policing. Am Sociol Rev. 2017;82(5):977–1008. https://doi.org/10.1177/0003122417725865.

36. Wisman T. Purpose and function creep by design: transforming the face of surveillance through the internet of things. Eur J Law Technol. 2013;4(2):1–19.
37. Dahl JY, Sætnan AR. "It all happened so slowly"—on controlling function creep in forensic DNA databases. Int J Law Crime Justice. 2009;37(3):83–103. https://doi.org/10.1016/j.ijlcj.2009.04.002.
38. Feresin E. Lighter sentence for murderer with "bad genes". Nature. 2009. https://doi.org/10.1038/news.2009.1050.
39. Rigoni D, Pellegrini S, Mariotti V, Cozza A, Mechelli A, Ferrara SD, Sartori G, et al. How neuroscience and behavioral genetics improve psychiatric assessment: report on a violent murder case. Front Behav Neurosci. 2010;4(10):160. https://doi.org/10.3389/fnbeh.2010.00160.
40. EU White paper on Artificial Intelligence. Brussels, 19.2.2020 COM(2020) 65. 2020. https://ec.europa.eu/info/sites/info/files/commission-white-paper-artificial-intelligence-feb2020_en.pdf.

Does Closed-Loop DBS for Treatment of Psychiatric Disorders Raise Salient Authenticity Concerns?

14

Ishan Dasgupta, Andreas Schönau, Timothy Brown, Eran Klein, and Sara Goering

14.1 Introduction

Deep brain stimulation (DBS) is used by over 160,000 people worldwide to treat a range of neurological disorders including Parkinson's disease (PD), essential tremor, and dystonia [1]. By implanting electrodes in specific brain regions, clinicians are able to provide stimulation in order to manage symptoms that are not responsive to traditional pharmacological approaches. Despite these successes, thus far DBS technologies have shown limited efficacy for treating psychiatric disorders such as depression, eating disorders, and addiction [1]. This has sparked immense interest in developing the next generation of DBS technologies to treat psychiatric disorders. Where the current DBS systems are *open-loop*—applying a constant or intermittent electrical current to the brain—the new generation of DBS devices will utilize artificial intelligence technologies, such as machine learning algorithms, to facilitate *closed-loop* implants that are adaptive and continuously modified by neural feedback [2]. Closed-loop DBS devices read neural data, interpret the signals to make a clinical decision, and stimulate the brain dynamically. This process occurs continuously and without active input from users or clinicians, allowing for far fewer adjustments and improving treatment specificity.

I. Dasgupta (✉) · A. Schönau · T. Brown · S. Goering
Department of Philosophy, University of Washington, Seattle, WA, USA

Center for Neurotechnology, University of Washington, Seattle, WA, USA
e-mail: idasg@uw.edu

E. Klein
Department of Philosophy, University of Washington, Seattle, WA, USA

Center for Neurotechnology, University of Washington, Seattle, WA, USA

Department of Neurology, Oregon Health Sciences University, Portland, OR, USA

F. Jotterand, M. Ienca (eds.), *Artificial Intelligence in Brain and Mental Health: Philosophical, Ethical & Policy Issues*, Advances in Neuroethics,
https://doi.org/10.1007/978-3-030-74188-4_14

Since we invite the possibility of automating the modulation of our brain, we must also consider the ways in which such devices may change the end user and the way they interact with their world. These worries are backed up by empirical evidence showing that at least some users of open-loop DBS report changes to aspects of their personal identity, agency, or sense of self [3–6]. These reports have been the catalyst for a robust and still ongoing debate over implanted neurostimulators and the nature of the impact they have on users [7]. For example, is it worth treating an illness if people no longer feel like themselves? What if friends and loved ones no longer recognize the person they knew? Some of these worries are not entirely new—they have been raised around the issue of authenticity and the use of antidepressants [8] or open-loop DBS devices [9]. On the other hand, some have noted that closed-loop DBS technologies raise *salient* authenticity concerns, especially when they are used to treat psychiatric conditions [10, 11].

In this chapter, we explore the ways in which closed-loop DBS systems can introduce changes to the self that are different from open-loop DBS. At each step of the closed-loop process—reading neural signals, interpreting data, and stimulating the brain—new complexities around authenticity are introduced by the increased reliance on automated systems. Threats to privacy, traditional clinical relationships, and agency raise concerns about whether users can still live authentically. These concerns can be mitigated by developing new ethical guidance to address the unique setting of closed-loop DBS.

14.2 Deep Brain Stimulation and Psychiatric Disorders

The treatment of psychiatric disorders has taken many forms over history. Starting with psychoanalytic psychotherapy, the field of psychiatry eventually developed pharmacological and surgical interventions. The emergence of DBS technologies in the early 1990s [12, 13] introduced an alternative to ablative psychosurgeries for patients with a wide range of refractory psychiatric disorders. DBS offered the ability to deliver electric stimulation "into specific targets within the brain and the delivery of constant or intermittent electricity from an implanted battery source" [1]. Over the last 30 years, the use of DBS has been investigated in the treatment of psychiatric conditions such as major depressive disorder (MDD), obsessive compulsive disorder (OCD), bipolar disorder, Tourette syndrome, schizophrenia, addiction, and anorexia [1].

Although the exact mechanism differs according to the disease being treated, the basic concept behind open-loop DBS for psychiatric disorders is similar. Researchers identify a candidate for surgery based on a symptom-based tool like the DSM V, a cognitive measure, such as the Montgomery-Asberg Depression Rating Scale (MADRS), and history of inadequate response to other forms of treatment. In the well-known RECLAIM trial for MDD, participants must have had failed treatment trials with at least four antidepressants [14]. If the patient meets the criteria for a clinical trial, electrodes (one or two, depending on the study) are placed in a specific region of the brain thought to be responsive to stimulation. After the surgery, stimulation is turned on and the voltage set in order to manage the symptoms of the condition. Stimulation is adjusted in trial visits based on assessments of symptoms and side effects. This stimulation is either continuous or follows a predetermined schedule. In certain cases, users may have individual control over their stimulation and may be able to turn the device on and off, as well modulate the voltage setting.

DBS offers large improvements over prior psychosurgeries, such as being able to directly interface with brain circuitry while avoiding permanent lesioning. Due to its efficacy, DBS, as opposed to ablative psychosurgery, is now considered the treatment of choice for some individuals with treatment refractory PD, essential tremor, and epilepsy [15]. Regrettably, there has been limited success in using DBS to treat psychiatric disorders. Initial studies using DBS for OCD and MDD showed promise, but randomized clinical trials failed to show similar efficacy [14, 16]. Researchers argue there are explanations for the discrepancy in results [2, 17, 18]. Our understanding of neural circuitry responsible for mental illness is lacking when compared to motor disorders. Additionally, psychiatric disorders are often the result of multiple dysfunctional circuits as opposed to one network. For instance, what we classify as MDD may actually be a collection of unique conditions. Furthermore, the lack of reliable biomarkers for symptoms makes it difficult to determine whether modulation is responsible for clinical success. Taken together, researchers argue that more targeted stimulation that is modulated according to neural feedback across different neural circuits may be a solution to using DBS to treat psychiatric disorders.

Closed-loop DBS systems have been proposed as an alternative to open-loop DBS and can succeed where their predecessors failed [17]. While open-loop DBS works in a unidirectional fashion by providing a constant level of stimulation to targeted brain areas without integrating any sort of neural feedback, the new generation of DBS devices is "closing the loop" by recording neural data, analyzing it for salient features utilizing machine learning algorithms, and using these analyses to alter stimulation parameters like amplitude and frequency. This loop of reading data, analyzing data, and stimulation addresses some concerns about authenticity raised by open-loop DBS but raises salient issues in the closed-loop context.

14.3 Authenticity and Treatment of Psychiatric Disorders

In a colloquial sense, to be authentic means to be "true to oneself." For many, being true to oneself is not a complicated calculus. We all can intuitively point to something about ourselves that is more foundational than everything else. For example, one can change one's hair color easily, but it is much more difficult to become an outgoing person if one is born as an introvert. On the other hand, we also understand that people change their conceptions of self over time and are not permanently bound to traits they acquired early in life. A once lazy student who becomes more disciplined about coursework by using a new mindfulness technique can reinvent her identity to become a hardworking person. As long as she is able to conform her actions to her new conception of self as a responsible student, it seems uncontroversial that most people in her life would be willing to grant that she has changed a part of her former identity. This example illustrates that choices affecting authenticity run a spectrum from minor to fundamental. It also shows that there are different ways of thinking about authenticity. Since a comprehensive discussion of all types of authenticity are beyond the scope of this chapter, below we highlight three ways that have been prominently discussed in the neuroethical literature on DBS.

14.3.1 Sense of Authenticity

As early as 2006, Schupbach et al. reported that patients receiving deep brain stimulation (DBS) for PD felt alienated after receiving their device even though they experienced a reduction in measured clinical symptoms: "Now I feel like a machine, I've lost my passion. I don't recognize myself anymore" [5]. Recent studies of DBS for psychiatric indications have echoed similar concerns. A study by De Haan et al. discusses one DBS user with OCD who felt like a different person because of the change in their libido following implantation: "I did not like that at all...No, that clearly didn't fit with who I am...it was really too much; that really wasn't me, you know; I really felt as if there was someone [else] standing next to me..." [11]. Goering et al. interviewed end users, some of whom expressed that they were no longer able to act in ways that were consistent with their pre-DBS self: "I can't really tell the difference. There are three things—there's me, as I was, or think I was; and there's the depression, and then there's depression AND the device and, it, it blurs to the point where I'm not sure, frankly, who I am" [10, 11]. This empirical data suggests, at the very least, that DBS can cause changes to users that make them feel as if they are no longer like their prior self.

One way we think about authenticity is in terms of our sense of self *in a particular moment*. Does changing the color of my hair make me feel more like my authentic self or like an imposter? This sense of authenticity captures the way in which an individual experiences their self from the first-person perspective. A sense of authenticity is paramount to ensuring that people feel comfortable living their everyday lives. It is also the type that is most intuitively thought of when discussing DBS technologies. For example, after a patient undergoes implantation, how they feel once stimulation begins is of significant concern. DBS devices differ from pharmacological interventions in that the effects are typically sudden and can be jarring in nature rather than slow and iterative [13]. It is possible that feelings of alienation are due to the lack of time users have to adjust to major changes in mood. Many traditional pharmacological treatments, such as SSRIs, take weeks to months to reach their full effects allowing users to experience changes more gradually [19]. In this context, it may be difficult to determine whether feelings of inauthenticity are caused by the stimulation from the DBS itself or from changes the user experiences to their personality due to amelioration of a disease condition. That is to ask, is the device stimulating areas of the brain in ways that cause these new changes, or are these changes the result of reducing the negative effects of depression? [11]. Regardless of the cause of the changes, there is a very real sense in which end users of open-loop DBS experience feelings of inauthenticity.

14.3.2 Narrative Authenticity

A second way of thinking about authenticity is how the person aligns their actions with their true self [20, 21]. On a practical level, some of the worries about end users' feelings of inauthenticity after receiving DBS are tied to the prevalence of the concept of autonomy in modern bioethics. Generally, the worry is that a person will be autonomous, capable of understanding the consequences of their actions and choosing freely, but not authentic [22]. For example, in a classic example from the literature, a man undergoing DBS for PD becomes manic and develops a new interest in gambling [23]. When the device is turned on, the man claims that this new version of himself, where he is more impulsive and risk taking, is his real self and that his former self was an inauthentic version. Thus, if our gambling man is deemed to have retained his autonomy, under a traditional bioethical inquiry it may be inappropriate to deny him the right to make all medical decisions on his behalf. On the other hand, giving autonomy more weight than authenticity may presume an inappropriate dichotomy between the two. Under this view, it may be impossible for one to act autonomously if they cannot act in a way that is consistent with their true self. If we are to subscribe to this view, however, in what ways would we integrate authenticity into our current legal system for informed consent and medical decision-making? For example, at present, we have an established framework for determining whether a person is capable of autonomous behavior and can be legally responsible for their own medical decision-making [24]. Integrating authenticity would require that in cases like the man who gambles, clinicians would have some power to restrict the man's decision-making if his actions are not consistent with his prior self.

14.3.3 Assessing Authenticity

A third way of thinking about authenticity is how to determine whether someone is truly acting authentically. The way one approaches this question has to do with their views regarding the nature of the true self. Those who hold an essentialist view of the self argue that the self is composed of an unchanging and fixed core that represents one's true self [25]. Any departures from this fixed self would result in an inauthentic self. Those with existentialist views generally believe that the self is ever-changing and is constantly redefined by the individual through self-determination [25]. Under this view, people are free to change their true self throughout their life and can reinvent themselves if they choose. Finally, according to relational conceptions of the self, identity is constructed through repeated recalibration based on feedback from others, rather than in isolation [26]. These views are not mutually exclusive, and most people will likely subscribe to some hybrid version of the self that incorporates aspects from each framework. We do not attempt to settle this long-standing debate here. Rather, we argue that irrespective of which view one holds, in practice there are often disagreements between the user, their family members, and clinicians about whether or not the person is acting authentically after receiving treatment for psychiatric disorders.

Although there is an understandable preference to defer to the end user in these cases, there may be practical value in involving others in clinical decision-making when determining authenticity. Feedback from friends and family may help a clinician determine whether an end user will effectively integrate a device into their life. Given that most patients receiving these devices are likely to continue to have caregivers substantially involved in their daily lives, if DBS creates changes in personality that are too extreme, it may not be advisable to go through with the procedure even if it offers reduction of clinical symptoms. That being said, it is worth asking whether we are only concerned about feelings of inauthenticity when third-party perspectives disagree with one's first-person narrative? If the gambling man had become more thoughtful and loving, would clinicians and family members object? These are some of the questions that arise when considering issues related to assessing authenticity.

As discussed above, one way to evaluate feelings of inauthenticity is to rely on first-person reports from end users. Whether one believes the true self is fixed or changing, the end user will ultimately be in the best position to determine what is her true self and in what ways she wants to change that self and embrace a future identity. For example, if the gambling man is content with his new identity and is not breaking laws or harming himself or others, some may argue that he has a right to reinvent his new identity (even if those around him would prefer that he not). Furthermore, if this change is coupled with an improvement in clinical outcomes, one might argue that the change in the man's personality is a risk worth taking given the benefit he is experiencing from the DBS. On the other hand, what is to be made of the man's values and desires before he received the implant? For example, what are we to do if upon turning the DBS off we find that the man reflects upon and rejects his recent impulsive self? We are then left with two first-person accounts—both of which feel like they are authentic. Along these lines, a difficulty has emerged in the neuroethics literature about how to objectively determine when an end user's self-reported claims of inauthenticity are to be trusted [27]. How are clinicians and family members to resolve the discrepancy between the calm and thoughtful person they knew before implantation and the rash and offensive man they see before them now?

14.3.4 Why Authenticity?

Before starting our discussion of the differences between open and closed-loop DBS technologies, we must briefly consider why authenticity, as opposed to other ethical principles, should be the focus of our inquiry. We focus on authenticity for a few reasons. First, the empirical evidence, as we have seen above, illustrates that when end users talk about changes to their self, many intuitively utilize the language of authenticity. Second, authenticity is often a central concern expressed by people suffering from psychiatric disorders [28]. Psychiatric disorder can often lead to people feeling less like themselves, and treatment with DBS is sought in order to improve authenticity. In this way, when we consider closed-loop systems for treatment, we must examine both how they might reduce symptoms of the disease that implicate authenticity and the ways in which the device itself introduces new concerns about authenticity. Third, authenticity can offer practical utility in guiding clinical decision-making [29]. Since traditional bioethical inquiry heavily favors concepts such as autonomy as being paramount for medical decision-making, authenticity may offer a way to check our typical intuitions about treatment decisions. Finally, using DBS for psychiatric disorders, rather than motor or sensory disorders, raises an additional layer of complexity because treatment will seek to change the users' beliefs, desires, and emotions—key components of their self—directly rather than such changes occurring as an unintended consequence [11]. Therefore, authenticity, rather than autonomy, or other bioethical principles, may help us better understand the journeys end users take in coming to terms with the changes to the self-introduced by DBS.

14.4 Authenticity and Closed-Loop DBS

In what follows, we discuss three different technological improvements made by closed-loop DBS and how they may alleviate or exacerbate the types of authenticity we discussed earlier.

14.4.1 Reading Data

Unlike other treatments used for psychiatric disorders, closed-loop systems will directly measure neural data from the surface of the brain as opposed to measuring symptoms based on clinical observation and anecdotal reports [2]. This introduces a new range of possible data that is considered when treating psychiatric disorders. Traditionally, clinicians are limited to subjective evaluations and indirect behavioral assessments to monitor clinical progress. For example, when a patient with MDD is prescribed a new SSRI, a clinician typically waits for some time before conducting a clinical assessment of the drug's efficacy. Similarly, in open-loop DBS, the clinician is using clinical data to adjust the stimulation parameters in order to get the desired reduction in symptoms. The process, in both pharmacological and open-loop DBS, involves much trial and error and can be burdensome for end user and clinician alike [30].

In contrast, in closed-loop DBS, the device will measure neural activity related to a disease symptom directly from the brain. For instance, neural signals gathered by implantable ECoG electrodes can potentially be used to measure dysfunctional states in neural networks associated with psychiatric disorders [2]. The goal would be to develop a list of common phenotypes that are associated with the disorder and then try to identify these phenotypes autonomously. The ability for closed-loop devices to read neural data in this way can help to improve the treatment of psychiatric disorders and thereby reduce issues related to authenticity that are the result of the disease.

First, since the device is continuously reading out brain data, the gathered information is valuable for providing ongoing health monitoring and screening. This real-time characteristic of the system may be used to offer preventive care. Here, individual brain readings are potentially useful for early detection and diagnosis of other related diseases that may be leading to feelings of inauthenticity in the end user. For instance, if a patient is implanted with a closed-loop DBS to treat depression, the monitoring of neural data may alert the clinician to a comorbid condition that needs treatment. An early intervention can help the patient avoid further feelings of inauthenticity as a result of their disease. The additional neural data can also be used for reconfiguring settings (see below) that make the patient feel less alienated, rather than having to wait for an appointment where one has to describe the ways in which they feel different. Second, on a large scale, aggregation of neural data on a population level can accelerate our understanding of neuroscience by harnessing big-data and machine learning algorithms [18]. This gathered data might be used to understand the underlying biology of psychiatric disorders [18]. In the long run, this cumulative knowledge from numerous patient cohorts might help to generate a better conceptual understanding of the brain functions that implicate feelings of inauthenticity. This could improve treatment by tailoring it towards the end user's reports of how they feel.

However, one current obstacle, especially for psychiatric disorders, is the lack of reliable biomarkers for usefully differentiating between pathological and healthy states [18]. Closed-loop devices will only be as useful if they are able to measure data that is relevant to a user's condition. If these devices are unable to measure accurate neural phenotypes, they may be as (in)effective as our current observation-based and self-reported measures of mental illness.

Another issue raised in the literature when it comes to ongoing recordings is the exploitation of sensitive data. In order to process the large amounts of data collected and for clinicians to be able to make remote adjustments, closed-loop devices may have to be connected wirelessly to external systems that have more computing power or allow clinician access. This may make them vulnerable to hackers, who can potentially access private information and even control the device [31]. In the literature, this risk is discussed under the umbrella term of "neurocrime" that refers to individuals aiming at "illicit access to and manipulation of neural information and computation." [32]. Here, hackers might gain access to information that represents the patient in a way he feels uncomfortable in being represented with. Exploiting this information can change interpersonal dynamics of privacy that may result in authenticity issues. If we take seriously the likelihood that closed-loop devices will introduce new security issues, hackers could not only steal private information but also alter stimulation (see below) in a way that affects motor function, impulse control, and emotions [33]. Here, the ability to externally change stimulation patterns can be exploited to intentionally impact the end user's experience of authenticity. Designing future medical devices with a specific standard of neurosecurity that implements relevant security principles could help to provide neural devices with adequate security mechanism [34].

14.4.2 Analyzing Data

Once the patient's data is read, it will be analyzed through an algorithm that utilizes artificial intelligence or machine learning technologies. The closed-loop device must correctly interpret the user's neural data and associate it with a corresponding disease state. If the appropriate criteria are met, the system will determine that the user is experiencing symptoms of their illness and that stimulation needs to be provided. In the pharmacological and open-loop context, a clinician would utilize the DSM to tabulate a patient's anecdotes in order to make a diagnosis or would rely on observation.

There are two important implications for this stage of the process. First, making disease assessments using various biomarkers and data sources instantaneously allows for the system to make individualized diagnosis based on a user's unique brain activity. Second, population level data will be utilized to train the system so that it recognizes brain activity that is considered dysfunctional as indicated by it falling outside the normal range found in the training set. Both implications mark a considerable step forward compared to the manual tinkering necessary in open-loop devices. Patients report that the trial-and-error phase for adjusting the stimulation to meet their individual needs is one of the most frustrating aspects during clinician visits [3].

Closed-loop devices allow us to directly tackle this issue. Here, the device takes the role of the clinician in changing parameter settings but does so infinitely more times than a person could. In theory, the patient can benefit from less personally demanding visits for recurring parameter adjustments in clinical settings [35]. At the same time, the adaptivity of the system may result in more effective, personalized treatments that prevent potential overstimulation and unintended side effects [18]. For instance, in the case of DBS for PD, stimulation can be turned off when the patient is at rest while it can be turned back on when a new movement is recognized. Here, tailoring the system towards the current needs of the individual through adaptive data analysis is a tremendous improvement from the ongoing stimulation provided by open-loop DBS. Since the system permits the user to forget about the ongoing stimulation, the discontinued need to constantly reflect about their current state might prove helpful to provide a background setting in which the user feels comfortable, presumably resulting in a more authentic state.

On the other hand, there are ways in which we can imagine a system that makes diagnoses based on machine learning algorithms may introduce bias against certain types of people. For example, what data-set will the artificially intelligent systems be trained on? There are ethical concerns about how our existing societal biases might be integrated into machine learning algorithms and become further automated, making them more difficult to recognize or address [36]. These biases are not limited to gender or race and can include negative views regarding people who do not exhibit neurotypical behavior [37]. There are ways in which closed-loop systems can undermine a user's authenticity if they misdiagnose legitimate feelings or emotions as being pathological rather than justified in a particular situation. For example, what if a user is angered as a result of experiencing racial injustice. Would a closed-loop device trained on users who never experienced that type of stress be able to distinguish between anger felt in response to a legitimate trigger from anger that is due to dysfunction in neural networks? These types of misdiagnoses have the potential to make some users feel as if the system is forcing them into behavior that is inauthentic.

14.4.3 Stimulation

Adaptive systems will utilize the results of data analysis to formulate a treatment plan. This will be based on two levels of data: personal data indicating an individual's normal variation in symptoms and population level data indicating the range of functional neural activity. In this way, the device can calculate the stimulation needs of an individual user within the proven range for the therapy. This will allow the closed-loop system to provide a more precise stimulation at multiple brain locations at varying voltages without the need for clinical manipulation. As in prior sections, the promise of closed-loop devices is that they may mitigate concerns about authenticity by offering better treatment of underlying psychiatric disorders.

Executing this treatment plan, however, relies on the human brain to simultaneously adapt to ongoing changes in stimulation. This immediate connection between end user and device can lead to problems of authenticity that may be manifested in feelings of alienation [6] or autonomy concerns [22]. From a phenomenological perspective, some patients report an impact on their perception of themselves and their bodies, resulting in statements that they are feeling like a robot or like an electric doll [5]. Here, the pressing neuroethical issue consists in the difficulty a user has in differentiating between what he is doing and what the device is doing.

In the literature, this concern is addressed as the potential danger of closed-loop DBS to undermine agency in a way that the individual's capacity to regulate mood may not allow the normal range of emotional responsiveness [10, 11]. Imagine a patient with a psychiatric DBS who attends a funeral but is not able to produce the expected emotion of crying. Since the stimulation is delivered automatically, the patient would not be able to control the device or change any settings intentionally. Even worse, the patient might not even be able to recognize whether the current emotional setting is a consequence of the device's adaptive stimulation or actually a part of his "original" self without the device. As a result, the device being always on with its adaptive, self-learning mechanisms may put the user into a state of constant uncertainty. While a user can always stop taking prescribed pills when problems occur, an equivalent action of stopping the stimulation may not be possible while using closed-loop DBS [10, 11]. While a drug cannot change its method of modification mid-treatment to account for unwanted side effects, a closed-loop device could make it so the user never even feels alienated. The AI-assisted device, so it seems, is always in control since it autonomously determines the time and intensity of the stimulation. Even when it is not stimulating, it is doing so for a reason that is unbeknownst to the user. This shows that, especially in psychiatric DBS, the implementation of an AI is not necessarily beneficial for the individual's experience of living with the device. Instead, the addition of another layer of treatment decisions utilizing an integrated AI potentially exacerbates the user's uncertainty about who is responsible for mainting their well-being.

This uncertainty may also motivate concerns about whether the user is in control over their actions. Imagine a closed-loop DBS causing a cramp by overstimulating a patient who, as a consequence, unintentionally turns the steering wheel of his car. Who is responsible for the resulting accident? The patient, clinician, or the programmer of the AI? This general issue of responsibility ascription in stimulating neural devices is coined in the literature as a "responsibility gap" [38]. The integration of AI into this control scheme adds additional complexity to the already existing black box nature of neural devices, which makes recognizing and understanding the inner workings of the closed-loop device and the influence it has on the patient immensely difficult. For our current discussion, we are less concerned about who is morally or legally responsible for an involuntary action caused by a closed-loop DBS, but rather how the user may *feel* less in control.

There are two ways in which the user may feel less control. First, they may feel as if the device is acting instead of them leading to feelings of alienation. Second, if one feels less in control, they may also be less motivated to enact positive behavioral change in their lives thereby implicating issues related to narrative authenticity. For example, it is possible that users, even if they feel that they are in control over their actions, may nonetheless cede responsibility to the device because they believe that an artificially intelligent system will address all their treatment needs. A patient suffering from depression who may have exercised in the past to improve their mood in conjunction with an SSRI may now feel less motivated, since they have a system that provides more precise stimulation and obviates the need for the user to take personal responsibility for improving their mood. This raises concerns about whether someone is still authentic when they no longer have to put in the hard work of improving themselves [8].

In this case, one might note that the patient can always check in with the physician if there are problems with the stimulation parameters. As a response, the clinician might check whether the device is stimulating correctly according to its readings, but will be unable to significantly change the programming, since that will probably require updates or changes made by the manufacturer. Furthermore, since the device is stimulating solely on the basis of neural data that does not take the patient's unique experience of having a psychiatric disorder into account for providing treatment, it is unlikely to be able to handle the task of processing the patient's phenomenological experience when calculating stimulus [17]. Here, clinicians need to be sensitive about keeping a personal relationship with their patients by including their anecdotally related symptoms into the therapeutic process [35].

14.5 Future Directions

We have identified several features of closed-loop DBS that may impact users' feelings of authenticity in morally salient ways. Our discussion elucidated that closed-loop devices, if developed to their full potential, can reduce concerns about authenticity by reducing disease-related symptoms. In this way, users may feel more like themselves, be able to better construct their own narrative identities because of improved mood or control over their behavior, and others may start to recognize the person they were before being impacted by disease. Conversely, this discussion also highlighted the ways in which new technological aspects of closed-loop systems raise salient concerns about authenticity. Reading of neural data can raise issues about whether the correct information is being recorded and opens the door for unauthorized access to neural data. The machine learning systems involved in analyzing the data can introduce a variety of biases into the systems, thereby automating many of the problems we already have in treating and diagnosing mental illness. Finally, adaptive stimulation can exacerbate existing worries about DBS causing changes in the self, as well as introducing new worries about whether a user will be in control over their own actions and future behavior.

To ensure safe and responsible use of this emerging technology, it is essential that we not only predict these future issues, but also flesh out initial steps towards possible solutions. Here, a potential first step would be to create ethical guidelines that structure the ongoing debate in a way that any anticipated negative consequences of closed-loop technology are prevented or mitigated significantly. One way forward could be to recognize that authenticity has an important function in medical decision-making without giving it more power than it currently holds. Since our legal systems are not yet equipped to deal with issues related to authenticity, we can attempt to deal with feelings of inauthenticity through additional care for end users while keeping autonomy as the dispositive concept when it comes to deciding whether a person is able to make their own decisions.

Second, since closed-loop systems make use of algorithms that require training and including users in the training process can either foster authenticity or exacerbate inauthenticity, potential end users should get a thorough explanation of what will (or might) happen to them and what they can do if they feel alienated, isolated, or odd post-surgically. In order to achieve this, clinicians should make sure that their patients are well-informed about potentially occurring changes in authenticity by offering background knowledge about all three stages (reading, analyzing, and stimulation) of the neural device. Additionally, ethical guidance should focus on the ways in which developers of closed-loop systems are cognizant of the different biases that can be introduced into their devices if careful attention is not paid to the data sets they are trained on. Guidelines should be developed in order to guide developers when they are creating these devices so that they include enough training data in order to capture a multitude of diverse neural data.

Third, as touched on above, the relationship between the patient and clinician will need adapting as closed-loop systems become introduced in clinical practice [38]. Here, patients should be given the option of participating in the tuning process if they desire it. This may include deeper training on how their stimulator works, how its algorithms work, what data is collected, how the data is analyzed, and how the stimulation is influencing different parts of the brain. In terms of privacy, this may include an overview on the measures that are in place to protect their recorded brain data and secure their implant from security breaches. In terms of alienation, this could include thoroughly informing the patient about possible changes on a psychological level as well as providing guidance on who end users can turn to if they experience a sudden change in authenticity.

Another way clinicians can help users guard against potential unwanted changes to the self is to utilize legal instruments like advanced directives. For example, Klein has argued that in cases like the gambling man, the user may use a Ulysses contract to prospectively note behaviors they find unacceptable and situations where they would want their device turned off, even if this goes against the will of the post-DBS person: "If I ever become a compulsive gambler, please intervene" [39]. Named after the hero from Homer's Odyssey who famously instructed his crew to leave him tied to the mast and ignore his future self, Ulysses contracts could serve as a DBS analog to advanced directives used in patients with dementia. Utilizing Ulysses contracts in conjunction with preimplantation consultation with family members and loved ones could help the user establish a series of conditions under which clinicians would remove the treatment against the patient's wishes. This could include the ability for people close to the user to raise a red flag if they sense the person's behavior is changing but the user themselves does not notice the change or does not mind the change. This can guard against concerns about assessing authenticity and determine which autonomous self to respect when there is a conflict between a person before and after they receive a closed-loop DBS.

Finally, we saw earlier that authenticity, along with identity more broadly, is a relational concept; people evaluate the authenticity of their actions with the help of others, and what counts as an authentic action is constrained by others as well. Closed-loop DBS users could look to other users to figure out if their actions are authentic to them, or if they are a by-product of how the device operates. Research on and development of closed-loop devices should facilitate communication between users of these devices by coordinating meetups and recording honest testimonials (not just in the form of positive marketing materials).

While these and other considerations still need to be discussed more thoroughly in the ongoing debate, once fleshed out and put into place, they offer valuable support for end users to successfully adapt to living with closed-loop devices.

References

1. Lozano AM, Lipsman N, Bergman H, Brown P, Chabardes S, Chang JW, et al. Deep brain stimulation: current challenges and future directions. Nat Rev Neurol. 2019;15:148–60.
2. Widge AS, Malone DA, Dougherty DD. Closing the loop on deep brain stimulation for treatment-resistant depression. Front Neurosci. 2018;12:175.
3. Klein E, Goering S, Gagne J, Shea CV, Franklin R, Zorowitz S, et al. Brain-computer interface-based control of closed-loop brain stimulation: attitudes and ethical considerations. Brain Comput Interfaces. 2016;3:140–8.
4. de Haan S, Rietveld E, Stokhof M, Denys D. Effects of deep brain stimulation on the lived experience of obsessive-compulsive disorder patients. PLoS One. 2015;10(8):e0135524. [cited 2015 Dec 10]; http://philpapers.org/rec/DEHEOD.

5. Schüpbach M, Gargiulo M, Welter ML, Mallet L, Behar C, Houeto JL, et al. Neurosurgery in Parkinson disease a distressed mind in a repaired body? Neurology. 2006;66:1811–6.
6. Kraemer F. Me, myself and my brain implant: deep brain stimulation raises questions of personal authenticity and alienation. Neuroethics. 2013;6:483–97.
7. Mackenzie C, Walker M. Neurotechnologies, personal identity, and the ethics of authenticity. In: Clausen J, Levy N, editors. Handbook of neuroethics [Internet]. Dordrecht: Springer; 2015. p. 373–92. https://doi.org/10.1007/978-94-007-4707-4_10.
8. Elliott C. Enhancement technologies and the modern self. J Med Philos. 2011;36:364–74.
9. Pugh J, Maslen H, Savulescu J. Deep brain stimulation, authenticity and value. Camb Q Healthc Ethics. 2017;26:640–57.
10. Goering S, Klein E, Dougherty DD, Widge AS. Staying in the loop: relational agency and identity in next-generation DBS for psychiatry. AJOB Neurosci. 2017;8:59–70.
11. de Haan S, Rietveld E, Stokhof M, Denys D. Becoming more oneself? Changes in personality following DBS treatment for psychiatric disorders: experiences of OCD patients and general considerations. PLoS One. 2017;12:e0175748.
12. Gardner J. A history of deep brain stimulation: technological innovation and the role of clinical assessment tools. Soc Stud Sci. 2013;43:707–28.
13. Schechtman M. Philosophical reflections on narrative and deep brain stimulation. J Clin Ethics. 2009;21:133–9.
14. Dougherty DD, Rezai AR, Carpenter LL, Howland RH, Bhati MT, O'Reardon JP, et al. A randomized sham-controlled trial of deep brain stimulation of the ventral capsule/ventral striatum for chronic treatment-resistant depression. Biol Psychiatry. 2015;78:240–8.
15. Meidahl AC, Tinkhauser G, Herz DM, Cagnan H, Debarros J, Brown P. Adaptive deep brain stimulation for movement disorders: the long road to clinical therapy. Mov Disord. 2017;32:810–9.
16. Malone DA, Dougherty DD, Rezai AR, Carpenter LL, Friehs GM, Eskandar EN, et al. Deep brain stimulation of the ventral capsule/ventral striatum for treatment-resistant depression. Biol Psychiatry. 2009;65:267–75.
17. Widge AS, Ellard KK, Paulk AC, Basu I, Yousefi A, Zorowitz S, et al. Treating refractory mental illness with closed-loop brain stimulation: progress towards a patient-specific transdiagnostic approach. Exp Neurol. 2017;287:461–72.
18. Lo M-C, Widge AS. Closed-loop neuromodulation systems: next-generation treatments for psychiatric illness. Int Rev Psychiatry. 2017;29:191–204.
19. Posternak MA, Zimmerman M. Is there a delay in the antidepressant effect? A meta-analysis. J Clin Psychiatry. 2005;66(2):148–58.
20. Brudney D, Lantos J. Agency and authenticity: which value grounds patient choice? Theor Med Bioeth. 2011;32:217–27.
21. Mackenzie R. Authenticity versus autonomy in choosing the new me: beyond IEC and NIEC in DBS. AJOB Neurosci. 2014;5:51–3.
22. Kraemer F. Authenticity or autonomy? When deep brain stimulation causes a dilemma. J Med Ethics. 2013;39:757–60.
23. Sharp D, Wasserman D. Deep brain stimulation, historicism, and moral responsibility. Neuroethics. 2016;9:173–85.
24. Berg JW. Constructing competence: formulating standards of legal competence to make medical decisions. Rutgers Law Rev: 53.
25. Bublitz JC, Merkel R. Autonomy and authenticity of enhanced personality traits. Bioethics. 2009;23:360–74.
26. Baylis F. "I am who I am": on the perceived threats to personal identity from deep brain stimulation. Neuroethics. 2011:1–14.
27. Ahlin J. What justifies judgments of inauthenticity? HEC Forum. 2018;30:361–77.
28. Erler A, Hope T. Mental disorder and the concept of authenticity. Philos Psychiatry Psychol. 2014;21:219–32.
29. Sjöstrand M, Juth N. Authenticity and psychiatric disorder: does autonomy of personal preferences matter? Med Health Care Philos. 2014;17:115–22.

30. Choi KS, Riva-Posse P, Gross RE, Mayberg HS. Mapping the "depression switch" during intraoperative testing of Subcallosal cingulate deep brain stimulation. JAMA Neurol. 2015;72:1252–60.
31. Klein E, Brown T, Sample M, Truitt AR, Goering S. Engineering the brain: ethical issues and the introduction of neural devices. Hastings Cent Rep. 2015;45:26–35.
32. Ienca M, Haselager P. Hacking the brain: brain–computer interfacing technology and the ethics of neurosecurity. Ethics Inf Technol. 2016;18:117–29.
33. Pycroft L, Boccard SG, Owen SLF, Stein JF, Fitzgerald JJ, Green AL, et al. Brainjacking: implant security issues in invasive Neuromodulation. World Neurosurg. 2016;92:454–62.
34. Denning T, Matsuoka Y, Kohno T. Neurosecurity: security and privacy for neural devices. Neurosurg Focus. 2009;27:E7.
35. Glannon W, Ineichen C. Philosophical aspects of closed-loop neuroscience. In: El Hady A, editor. Closed loop neuroscience [Internet]. San Diego: Academic Press; 2016. p. 259–70. https://books.google.com/books?hl=en&lr=&id=0ZTBCQAAQBAJ&oi=fnd&pg=PP1&dq=glannon+philosophical+aspects+of+closed-loop+neuroscience&ots=74dkrS5k9z&sig=bOJfqUkD-6FU0tJYOEdy5bQbyFw. Accessed 17 May 2017.
36. Char DS, Shah NH, Magnus D. Implementing machine learning in health care—addressing ethical challenges. N Engl J Med. 2018;378:981–3.
37. Gilbert F. The burden of normality: from 'chronically ill' to 'symptom free'. New ethical challenges for deep brain stimulation postoperative treatment. J Med Ethics. 2012;38:408–12.
38. O'Brolchain F, Gordijn B. Brain–computer interfaces and user responsibility. In: Grübler G, Hildt E, editors. Brain-computer-interfaces in their ethical, social and cultural contexts [Internet]. Dordrecht: Springer; 2014. p. 163–82. Accessed 24 May 2019. https://doi.org/10.1007/978-94-017-8996-7_14.
39. Klein E. Models of the patient-machine-clinician relationship in closed-loop machine Neuromodulation. In: van Rysewyk SP, Pontier M, editors. Machine medical ethics [Internet]. Cham: Springer International; 2015. p. 273–90. Accessed 4 Nov 2019. https://doi.org/10.1007/978-3-319-08108-3_17.

Matter Over Mind: Liability Considerations Surrounding Artificial Intelligence in Neuroscience

15

Gary Marchant and Lucille Nalbach Tournas

15.1 Introduction

Artificial intelligence (AI) has already begun to transform healthcare. While AI is not currently poised to replace human physicians, it is showing tremendous promise in revolutionizing healthcare [1]. From administrative systems to the diagnosis, prognosis, and even treatment of challenging diseases and disorders, AI systems have gained success and have been integrated into more complex medical situations. One emerging area of promise for AI is in neurological devices. As neurological diseases are notoriously difficult to diagnose and treat, the use of AI for their diagnosis and treatment could be uniquely helpful. However, with these potential rewards could come high risk, both clinically and ethically. Specifically, even small errors when manipulating the human brain can be devastating. Additionally, there will be risks associated with the collection of sensitive neurological health data that redefines concepts centering around harm. As such, patients that may benefit from such neuro-technologies maybe both simultaneously eager to utilize them, while being uniquely vulnerable to serious harm.

Liability is one mechanism for addressing such harms. While the enthusiasm for neuro-technology integrated with AI is warranted, it is crucial to consider the unique liability concerns that will come with the inevitable glitches in these systems. This emergence of new technology creates further liability exposure for providers, device manufacturers, and healthcare institutions [1].

G. Marchant
Sandra Day O'Connor College of Law, Arizona State University, Phoenix, AZ, USA
e-mail: Gary.marchant@asu.edu

L. N. Tournas (✉)
School of Life Sciences, Arizona State University, Tempe, AZ, USA
e-mail: Ltournas@asu.edu

F. Jotterand, M. Ienca (eds.), *Artificial Intelligence in Brain and Mental Health: Philosophical, Ethical & Policy Issues*, Advances in Neuroethics,
https://doi.org/10.1007/978-3-030-74188-4_15

As such, the question of who is liable for a glitch in AI technologies becomes increasingly more difficult to answer. The potentially liable party will include the treating physician overseeing the AI, the manufacturer of the AI system itself, or the healthcare institution the patient is receiving said care [1].

The first section of this chapter will begin with an explanation of traditional tort liability in the context of healthcare. The second section will then discuss what exactly is machine learning and what its role will be. Then, the third section will explore neuro-devices and the unique liability concerns posed within. The fourth section will suggest the importance of data collection in redefining traditional harm. Lastly, pathways towards safe, transparent, and responsible use of AI in neuro-technologies will be set forth.

15.2 Traditional Tort Liability

Liability for harm caused by medical errors or device glitches falls under tort law. Primarily, injured patients may recover from physicians, hospitals, as well as medical device manufacturers. Each of the entities mentioned above has a liability scheme: negligence, vicarious liability, and product liability, respectively. This brief foundation is essential in understanding the role of each liability scheme. While these differing schemes will play a significant role in the management of AI in healthcare, they are not always an ideal fit.

15.2.1 Physicians

Medical injuries deriving from physician error are held to a negligence standard. Under this standard, a physician is liable if their conduct "falls below the standard established by law forth protection of others against unreasonable risk of harm."[1] In these cases, a physician's conduct is compared to those of a reasonable physician, which includes similar knowledge, skills, and expertise, under similar circumstances.[2] As this standard of care is based on the skills of peers, it will continuously evolve, including in response to technology changes. This system allows flexibility for physicians to meet an average standard, rather than the highest standard. Unfortunately, it also means physicians may be uncertain of their liability when utilizing emerging technologies, such as tools based on AI to make the decision, that may shift the standard of care.

15.2.2 HealthCare Organizations

Healthcare organizations can be held directly responsible for their own negligence, or they may be held vicariously liable for the negligent acts of its employees, under

[1] Restatement (Second) of Torts §282 (Am Law Inst 1965).

[2] Restatement (Third) of Torts §12 (Am Law Inst 2010).

the legal doctrine of "respondeat superior."[3] Additionally, they may be liable for not properly training employees, which will become significant as hospitals are often the purchasing agents for new technologies. This doctrine is enticing to plaintiffs, as usually, the hospital has more financial resources than an individual physician.

15.2.3 Medical Device Manufactures

Medical device manufacturers are held to a product liability scheme. Here, manufacturers are strictly liable for harm by a defective product. In the case of a device defect, the manufacturers are responsible simply because the injury happened and if there is a reasonably available alternative design. As companies profit from selling these devices that patients rely on, they are held to a high standard in case of harm.

Additionally, the method by which the FDA regulates medical devices has consequences on the manufacturer's liability for product defects. Class III devices are those that pose a high risk to the patient/user. This class typically represents life-sustaining implanted devices such as pacemakers or implanted cardiac defibrillators. While Class III devices comprise less than 10% of all medical devices, their potential for harm warrants extra consideration. As such, most Class III devices require a Premarket Approval Application (PMA) before entering the market. PMA is a determination that the FDA has sufficient valid clinical evidence of the safety and efficacy of the device. Significantly, according to *Riegel* v. *Medtronic*, Inc., once the device has been approved through a PMA, the manufacturer is protected by preemption, which blocks state tort claims.[4]

Occasionally, manufacturers may argue the learned intermediary rule. This defense argues that the physician is the end consumer of medical devices as they are in the best position to consider the risk and benefits and make a recommendation to the patient.[5] The doctrine exists as patients cannot be expected to understand for themselves the risks of sophisticated medical technologies. As such, the scientific knowledge needed to make the best use of these devices is provided to the physicians. If a physician fails to warn an injured patient adequately, they will face liability. This may play an essential role in the AI-backed neuro-technologies that will be discussed below.

15.3 Machine Learning

Machine learning is a method of data collection and analysis on which the system itself learns from data, identifies patterns, and can make predictions and recommendations with very little to no human involvement [2]. The machine can do this process far faster and more effective than the human brain. There is also a smaller

[3] 27 Am Jur 2d Employment Relationship §356 (Thomson Reuters 2002).

[4] Riegel v. Medtronic, Inc., 552 U.S. 312 (2008).

[5] Restatement (Third) of Torts, § 6(d).

subset of machine learning called reinforcement learning in which the machine learns through trial and error, much like how human knowledge is acquired [3]. This type of reinforcement learning is best explained with DeepMind's AlphaGo Zero. DeepMind first created AlphaGo, which became dominant at the abstract strategy game Go after being programmed with rules and all foreseeable moves [4]. While impressive, DeepMind then launched AlphaGo Zero, which was only given the rules of the game and the gameboard to "learn" from [3]. The system played against itself, learning moves that had not yet been found by man. It quickly dominated AlphaGo and any human Go player [3].

These technologies have vast potential to revolutionize many facets of healthcare. For example, an AI system can be loaded with every medical article, electronic medical record (EMR) entry, and clinical trial to offer the most up to date look at disease and treatments. By way of comparison, the average physician only reads 3–4 h of medical journals each month, and often that information is not integrated into their practice [5]. This has massive potential to improve the diagnosis and treatment of neurological and psychiatric disorders, which can be notoriously tricky to diagnose and treat.

For instance, a recent machine learning-based AI system is capable of identifying a variety of acute neurological disorders from a patient's CT scan within seconds [6]. This tool will help physicians prioritize the urgency, diagnosis, and treatment of patients, which allows faster intervention and minimizes damage [6]. The machines are becoming as good if not better than human physicians in several medical areas, including pathology and radiology [7]. Additionally, the collection of large sets of data can be used for data mining, an ability that eclipses their human counterparts to discover trends and extract medically relevant information [8].

While the technology is exciting, it also will likely not fit seamlessly into our existing liability scheme. As these AI systems move from tools utilized by physicians to the decision-makers, some legal and regulatory adaption will be needed. The next section will explore potential liability issues when a glitch allegedly harms a patient.

15.4 Liability Surrounding Neurological Medical Devices That Utilize AI

15.4.1 Physicians

As physicians rely more heavily on algorithms that they do not fully understand but still oversee, they may be particularly vulnerable. This black box situation may be best illustrated by the Mount Sinai program named "Deep Patient." It was created to absorb the data from over 700,000 patients records in order to extrapolate data for high-level predictive modeling. Without expert opinion, Deep Patient became particularly apt at predicting the onset of psychiatric disorders such as schizophrenia. As schizophrenia is notoriously difficult to predict, this is significant for researchers looking to mitigate symptoms or intervene prior to symptoms [9].

This is not without concern, however, as the program is unable to explain its rationale to physicians [9]. This black box issue likely puts physicians in a precarious position. They have a powerful tool; however, they are basing treatment on biomarkers they do not understand. This is counterintuitive to evidence-based medicine. In the case the algorithm makes a recommendation in error, the physician would be unable to double-check the underlying decision-making and fully understand the recommendation used in treatment. If the AI should make a mistake that causes harm to the patient, the physician may face liability for relying on such a flawed system. On the other hand, if some physicians elect not to utilize an AI system because they do not understand its decision-making process, they may face liability if the AI system would have produced a better outcome in a specific patient. Thus, physicians may be put in a "damned if they do, damned if they do not" liability predicament.

While the General Data Protection Regulation (GDPR) in the E.U. requires explainable AI, the U.S. has no such legislation.[6] Although Explainable AI (XAI) is a developing field within machine learning that aims to address how AI systems are making these black box decisions [10]. XAI is a new field in machine learning that aims to shed light on the black box. One method may be to create explainable algorithms. Black box problems have been a trade-off for powerful neural network algorithms; however, programmers have worked to add traceability into them. The Department of Defense (DoD) through the Defense Advanced Research Projects Agency (DARPA) is investing heavily in XAI, to manage concerns over autonomous weapons [11]. Until a major breakthrough, however, physicians are going to be in the position of making decisions without explainable evidence.

Additionally, manufacturers being sued for product defects may try to shift liability to the physician through the aforementioned learned intermediary defense. In the situation of neuro-devices, the learned intermediary doctrine is likely more important than ever, as it will encourage a thorough and suitable informed consent procedure. Informed consent is a process by which a patient is given knowledge of a procedure's potential risks and benefits. It allows patients to receive not only information but also the opportunity to ask questions in order to agree to a specific medical intervention confidently. Even in these cases in which the devices are shifting from simple physician tools to mechanisms that look like the decision-maker, a human physician is still overseeing the devices and going over risks and benefits. Here, the physician needs to be transparent regarding overall device risks, black box issues, and rules of data collection, including which types of data will be collected and the scope of their use.

15.4.2 Hospitals

Interventions utilizing neuro-devices most frequently take place in a hospital setting, which leaves the hospital vulnerable. Robotic surgical tools bring many benefits to the surgeon, and they once again change the liability structure of traditional

[6] 95/46/EC, Article 22, EU GDPR. "Automated individual decision-making, including profiling", http://www.privacy-regulation.eu/en/22.htm

surgery. For example, the da Vinci surgical system offers the surgeon robotic instruments, guided via console, providing a better range of motion, better 3D views, and the ability to operate through smaller incisions.[7] While both the physician and manufacturer are apparent targets for tort claims if an injury occurs, hospitals may also face claims. In one lawsuit, the facts of the case demonstrated that the surgeon chose to ignore the recommendations of the manufacturer and perform robotic surgery on a patient who was not an appropriate candidate.[8] The operation had multiple complications, which left the patient with a poor quality of life and ultimately premature death. Both the jury and court of appeals found that the company adequately warned the physician of the nature of the robotic system, and they did not need to warn the hospital. Despite this, the Washington State Supreme Court extended the company's duty to warn past the physician, to the hospital themselves, as the purchasing agents.[9]

Cases such as this will be particularly significant as there is much promise in the use of robotic assistance to perform cutting-edge brain surgeries. For example, Brain Navi Biotechnology, a Taiwan-based company, has developed a robotic-assisted surgical system specifically for brain surgery, called the NaoTrac.[10] This machine combines AI, computer vision, and robotics to allow physicians to plan a surgery under AI, has the advantage of computer vision during the procedure, and facilitates the procedure with precise robotic arms. The company ran its first human trial in November 2018. The NaoTrac executed external ventricular drainage (EVD) on a hydrocephalus patient. The machine performed the procedure in 30 seconds, and the postsurgical C.T. scan showed that the placement was in the exact location the surgeon had planned.

Similar to the da Vinci surgical system, hospital systems are the likely purchasers of such cutting-edge surgery tools and as such may be vicariously liable for error, even in the case of product defect.

15.4.3 Manufacturers

Neuro-technologies using AI transform the device from a physician's tool to a decision-maker itself. The algorithms used are continually learning, much like the development of a human physician, synthesizing assimilated data over time [12]. However, unlike a human physician, who is held to a negligence standard of liability, which allows a reasonableness standard to apply, the machine will be held to a product liability standard, which means a strict liability standard applies. Here, the device could be punished for learning, even when it may be performing with more accuracy than its human counterpoints [13].

---

[7] Intuitive, About da Vinci Systems, *accessed at* https://www.davincisurgery.com/da-vinci-systems/about-da-vinci-systems

[8] Taylor v. Intuitive Surgical, Inc., 355 P.3d 309 (Wash. Ct. App. 2015).

[9] Taylor v. Intuitive Surgical, Inc., 389 P.3d 517 (Wash. 2017).

[10] Brain Navi Biotechnology, About, *accessed at* brainnavi.com/about

In particular, these types of lawsuits will be particularly alluring to plaintiff attorneys, as device companies and the hospitals that purchase their products may be viewed as having deep pockets and will be subject to strict liability and potential for punitive damages [13]. The example of robotic surgery from above demonstrates this as well.

In the case of NaoTrac, the precision of a robot can aid in lessening neurological problems through human mistakes during surgery, as even microscopic errors can be dangerous.[11] Additionally, it is particularly helpful during extended neurological operations that are plagued by physician fatigue. An in-depth study demonstrated that neurosurgeons were particularly susceptible to burnout and low satisfaction of work–life balance, which simultaneously encourages early retirement and discourages medical students from pursuing the specialty [14]. However, despite these novel aids, the brain is far more sensitive than other organs. Compared to other robotic surgery, robot-assisted neurosurgery could see greater potential harms and lawsuits, with more substantial rewarded damages.

In the instance that the machine malfunctions, it is a clear case of product liability and the company will be sued directly. The majority of claims against da Vinci concern complaints around malfunctions causing leaks of electrical currents and arcing, which is when sparks from a surgical instrument land in the patient's body, causing burns [15]. While these can occur with traditional surgery, it is far more prevalent in robotic surgeries [16]. In 2014, Intuitive, the manufacturer of the da Vinci surgical system, acknowledged that it carved out $67 million to settle approximately 3000 claims related to its device, some dating back to 2012 [15].

These concerns over malfunctions become magnified when looking to future technologies, such as brain–computer interface technology (BCI). An example is Elon Musk's recent announcement that his company, Neuralink, has developed thin "thread-like" electrodes and a "sewing machine for the brain" designed to implant those electrodes directly into the brain through tiny holes in the skull. He aims to test this device in humans within a year. BCIs are machines that allow brain signals to communicate with an external device. While the timeline may be unrealistic, this technology is being developed by Facebook, as well as by Kernel, helmed by Bryan Johnson, another Silicon Valley entrepreneur [17]. These devices aim to treat mental illness, neuromuscular conditions, as well as work on focus and cognition [17]. With such ambitious technology, the risk and responsibility for glitches are significant.

15.4.4 Regulatory Considerations

15.4.4.1 Software

While traditional medical devices have the opportunity to go through a PMA and gain preemption from state tort claims, the software itself poses an exciting challenge. Specifically, software that relies on machine learning is always changing,

[11] Brain Navi Biotechnology, About, *accessed at* brainnavi.com/about

which does not pair well with traditional static regulatory structures. In April 2019, the FDA released a white paper that laid out the agency's outline of managing software as a medical device (SaMD) that uses AI or machine learning.[12] It acknowledges the complexity of AI or machine learning and is taking a risk-based approach to organizing SaMDs based on the intended use. Here, it identifies the importance of the information provided by the SaMD to the healthcare decision as well as the severity of the healthcare situation. It lays out four principles for regulation. First, good machine learning practices (GMLP), which sets the expectation that manufacturers have an established quality system that corresponds to suitable standards and regulations. Specifically, for SaMDs, the FDA proposes relying on software Pre-Cert, in which manufacturers are approved rather than individual products. Second, it outlines a framework for modifications to SaMDs, including an anticipated modification, "SaMD Pre-Specifications" (SPS) and the "Algorithm Change Protocol" (ACP), that would allow for changes in a controlled manner, serving patient needs. Third, it outlines that companies can document modifications on their own unless they are outside the scope of the SPS and ACP, in which case the modification would go before a "focused review." Lastly, the FDA expects manufacturers to agree to the principles of transparency and performance monitoring.

While there is no mention of liability, it is unlikely preemption would be extended to SaMDs as this necessarily flexible process would be quite different from a PMA. In the *Riegel* decision, the Supreme Court noted that "the FDA requires a device that has received premarket approval to be made with almost no deviations from the specifications in its approval application."[13] As mentioned, SaMDs are definitionally in constant deviation, so it is unlikely preemption would be extended to them.

15.4.4.2 Devices

The neuro-devices highlighted above will likely pose a high risk to the patient or will be implanted in the brain. In the case of NaoTrac, it may be classified as a Class II device, similar to the da Vinci surgical system. This would mean it could go to market without a PMA and the company would be liable for any device glitches. Consequently, companies should be aware of potential lawsuits against them as compared to the cost of preemption. With that in mind, the article will move onto the complexities around the liability scheme for software with machine learning, which is far more complex.

While Neuralink's implantable device is a far more likely candidate for Class III categorization, companies like Facebook and Kernel are looking to develop noninvasive interfaces. This may challenge traditional thinking on risk, as devices that merely slip over our head to interact with our brainwaves may be viewed as less risky, even though they may still be quite powerful. For example, Muse is a

[12] FDA, Proposed Regulatory Framework for Modifications to Artificial Intelligence/Machine Learning (AI/ML)-Based Software as a Medical Device (SaMD)-Discussion Paper and Request for Feedback (2019), *accessed at* https://www.regulations.gov/document?D=FDA-2019-N-1185-0001

[13] Riegel v. Medtronic, Inc., 552 U.S. 312,323 (2008).

direct-to-consumer EEG device that interprets mental activity in order to guide the user to a more calm, mindful, focused state.[14] While there has not been any associated harm with the Muse, these types of direct-to-consumer brain wearables may become more advanced, with more potential to harm. However, under recent FDA guidance, such devices targeted at patient "wellness" rather than diagnosis or treatment of specific diseases are outside the scope of FDA regulation. Such devices will be spared the burden of regulatory approval, but will also lack any liability protection that would be provided by an FDA PMA.

15.5 Data Collection Consideration

While the most obvious harm to patients comes in the form of device glitches, it is important to note that there is also a significant risk to patients and consumers through data collection. These neuro-devices will be collecting large amounts of neurological data, which may be particularly vulnerable, including private thoughts and compromising medical information. While those devices used through a physician would be covered by the Health Insurance Portability and Accountability Act (HIPAA), that legislation will likely be too outdated to cover these new technologies adequately.

15.5.1 Invasive BCIs

In the case of an invasive brain–computer interface, as proposed by Neuralink, that will enter the brain and manage a neurological defect the resulting data would fall under HIPAA. However, patients are free to share their own information. Meaning, a patient could theoretically give his or her information to Neuralink to use if they were to set up an associated app, have patient data linked to other researchers and companies, and have the patient sign away rights to said data. Patients may not be aware of what they are protected from and what they are not. As health data is particularly valuable,[15] it is likely companies will try to find a workaround to collect and use their data, even in the most controlled setting. With machine learning, the availability of large data sets is critical to effective applications, so companies will be eager to gain access to such data sets.

15.5.2 Neurowearables

What is even more fascinating is the potential for data collection in noninvasive, direct-to-consumer, neurowearables. These products will generally not be prescribed by physicians, and so the data generated will be outside of HIPAA

---

[14] Muse, How it works (2019), *accessed at* https://choosemuse.com/how-it-works/

[15] Yao, Mariya, Your Electronic Medical Records Could be Worth $1000 to Hackers (April 2017), *accessed at* https://www.forbes.com/sites/mariyayao/2017/04/14/your-electronic-medical-records-can-be-worth-1000-to-hackers/

protections. These devices are small, affordable, and may offer continual brain monitoring [18]. This ease also increases the number of people utilizing them and the amount of data collected, which is immensely valuable to neuroscientists [19]. As mentioned, the Muse device is an EEG that interprets an individual's mental activity. While it is helping the user focus and meditate, it is also connected to the user's mobile device via Bluetooth, offering results and progress. While Muse states that customers' information is anonymized and confidential, one can imagine they, or others in the space, may wish to sell the data they are collecting.

Similarly, 23andMe collects the data of its users and offers a click button to donate your genetic data to the system for research. They now have the genetic data of over five million customers [20]. This has provided the company the ability to partner with GlaxoSmithKline to solve complex medical issues through data mining.[16] While these large data sets will be essential to evolving precision medicine, there is less thought going into the potential concerns around data collection.

The notion of harm, system glitch, and liability will need to be completely reimagined. Cardiff University launched a Data Justice Lab, which aims to record examples of harm caused by big data, including the following categories.[17] First, data breaches may expose individuals to unwanted exposure. Additionally, individuals may face discrimination, imagine that corporations buy data sets from Muse or the like and choose employees based on mental capacity. There may be data errors that incorrectly exclude or include an individual from medical intervention. Lastly, there will likely be social harms. For example, data collection from social media has led to widespread political manipulation.[18] Neurological data will probably be far more meaningful than what we share on social media, and that should not be taken lightly.

15.6 Pathways Forward

While we need to allow AI systems to "learn," much like human physicians, that must be balanced against patient safety. A patient's first approach to implementing AI neuro-technology is central to building a system that is transparent, responsible, and accountable. To this end, we will discuss AI informed consent updated, increased education for physicians, and FDA updates to working with AI.

[16] GSK, GSK and 23andMe Sign Agreement to Leverage Genetic Insights for the Development of Novel Medicines (2018), *accessed at* https://www.gsk.com/en-gb/media/press-releases/gsk-and-23andme-sign-agreement-to-leverage-genetic-insights-for-the-development-of-novel-medicines/

[17] Data Justice Lab, About, *accessed at* https://datajusticelab.org/about/

[18] The Computational Propaganda Project, Computational Propaganda Worldwide: Executive Summary, *accessed at* http://comprop.oii.ox.ac.uk/publishing/working-papers/computational-propaganda-worldwide-executive-summary/

15.6.1 Ethics of Informed Consent

Neuro-technologies bring new ethical considerations centered on balancing the collective good associated with the future of medicine against individual rights. Theoretically, informed consent gives the patient the knowledge to make an empowered and autonomous decision on their health, however, in practice it can be lackluster. Real-life consent forms are often boiler plate, written in a language not easily understood by patients, and rushed through to complete an EMR [21]. This will likely be intensified with the addition of artificial intelligence.

In order to move forward ethically, informed consent needs to be truly centered around the patient. The technological progression cannot move solely from a place of what is best for the future good of medicine, but also respecting the individual rights of those patients who will be used for early adoption. Physicians need to explain traditional risks and benefits, as well as how AI impacts the patient's neurotreatment. In the case of black box issues, patients should be made aware of what the physician is uncertain of in order to assess risk for themselves. Patients should also be made aware of the role of data collection in the AI device. Namely, what types of neurological information will be collected, how it will be stored, how will it be protected, and what it will be used for. Data collection is central for these technologies to improve but introduces the potential for new existential harms. Here, patients should sign specific, informed consent over all collection and uses of their data, separate from the consent they give for treatment itself.

Likewise, for in-home wearable devices, terms and agreements cannot be a simple click box with pages of background information. Data should be explained, with line-by-line checkboxes. Patients should be made aware of whether their data is protected by HIPAA, if the device company owns the data, how the device company is using the data and if the device company will sell the data to other companies.

15.6.2 Neuro Rights

There has also been a growing movement to establish specific neuro rights as an intersection of neuroscience and human rights. Specifically, individuals should have the right to cognitive liberty, the right to mental privacy, the right to mental integrity, and the right to psychological continuity [22]. Knowing that this revolution of neuro-technologies will challenge existing legal and ethical framework, creating this baseline of rights allows human rights to be built into the rapid development of these nascent technologies. These neuro rights will likely have to develop as new concern develop but will serve as a foundation for protecting individuals.

15.6.3 Physician Education

Informed consent can only be improved if physicians are better educated about AI. Physicians should be responsible for understanding all that can be known about the technologies they are utilizing, especially when dealing with neurological AI. They need to be made aware not only of the technology itself but also the data considerations, as that may not seem like their responsibility at first glance. Additionally, they should be responsible for articulating this knowledge to their teams and make sure they are following all standards and guidelines. Even if physicians cannot fully understand the process and output of an AI system, they need to have a sound basis for knowing when they can trust an AI system and when they cannot.

15.6.4 Hospitals

Similarly, a hospital may be vicariously liable for device harm, and they should be responsible for establishing best practices for the use of AI in their hospitals. This includes ensuring physicians are adequately trained, providing continuing education on the product, as well as establishing and implementing standards for cross-checking both physicians and AI.

15.6.5 FDA Updates

While the FDA has outlined frameworks for managing machine learning, they have not yet moved to the draft guidance, let alone rules. Creating a regulatory framework that is both flexible and fosters innovation, as well as puts patient safety first, is not an easy task. Here the FDA should build in privacy and transparency into the rules. Companies developing XAI should be rewarded with the fastest approvals. Similarly, those that make AI with constructs to manage bias and patient privacy should be given more trust.

This task to get guidelines early, while not truly understanding the technology is an illustration of the Collingridge dilemma. This predicament arises when there is a lack of information until the technology is widely used; however, making changes to technology is very difficult once it is well-established [23]. For example, this is one reason managing Facebook has been so difficult. We are playing catch up and trying to control the data of a system that was built to collect data. If privacy concerns were built into the rules at the beginning, it would have been much more manageable. On the other hand, very few people, including Mark Zuckerberg, likely realized the power of big data associated with digital platforms such as Facebook.

Here, the FDA has the benefit of seeing the use of AI and data collection in other areas. They do not need to understand every possible development from neurological AI, but only to recognize its significance and vulnerability, against the goal of patient safety. Much will be learned as these develop, but the setting for standards for data collection, error reporting, and built-in privacy measures will be crucial.

15.7 Conclusion

The use of AI in aiding neuro-technologies is nothing short of revolutionary. These systems offer hope to those unable to walk, those who live in constant fear of an epileptic seizure, and others with debilitating mental illness. As these particular diseases or injuries are particularly challenging to diagnose, prognose, and treat, breakthroughs at this level would be life-changing to patients, as well as caretakers.

All medical treatment carries inherent risk, and striking a balance in how liability is managed can serve as an important tool in both patient care and fostering innovation. These new tools challenge existing tort theory as machine learning transforms the tools used by physicians to the medical decision-makers themselves. Additionally, the much-needed data collections used by these systems will both accelerate medical knowledge and challenge existing notions of harm and ownership.

While all forms of risk will never be removed, as risk is inherent in most high-risk medical interventions, we can build a liability framework to minimize risk and keep patients informed. In the realm of neuro-technologies, patients' need should be at the forefront. As such, algorithms and processes should be created to improve informed consent practices, as well as protect and explain patient privacy.

References

1. Marchant GE, Tournas LM. AI health care liability: from research trials to court trials. J Health Life Sci Law. 2019;18:25–41.
2. SAS. Machine learning and why it matters. 2019. https://www.google.com/url?sa=t&rct=j&q=&esrc=s&source=web&cd=3&cad=rja&uact=8&ved=2ahUKEwjrvJ7aw4XkAhVLnp4KHSL9CngQFjACegQIEBAG&url=https%3A%2F%2Fwww.sas.com%2Fen_us%2Finsights%2Fanalytics%2Fmachine-learning.html&usg=AOvVaw3webX8vLHy6w3Ws1cbjdtV. Accessed 29 Aug 2019.
3. Silver D, Schrittwieser J, Simonyan K, Antonoglou I, Huang A, Guez A, Hubert T, Baker L, Lai M, Bolton A, Chen Y, Lillicrap T, Hui F, Laurent S. Mastering the game of go without human knowledge. Nature. 2017;550(7676):354.
4. Borowiec S. AlphaGo seals 4-1 victory over go grandmaster lee Sedol. The Guardian. 2016;15.
5. Susskind R, Susskind D. Technology will replace many doctors, lawyers, and other professionals. Harv Bus Rev. 2016. https://hbr.org/2016/10/robots-will-replace-doctors-lawyers-and-other-professionals.
6. Ridler C. Artificial intelligence accelerates detection of neurological illness. Nat Rev Neurol. 2018;14(10):572.
7. Jha S, Topol EJ. Adapting to artificial intelligence: radiologists and pathologists as information specialists. JAMA. 2016;316(22):2353–4.
8. Liang H, Tsui BY, Ni H, Valentim CC, Baxter SL, Liu G, et al. Evaluation and accurate diagnoses of pediatric diseases using artificial intelligence. Nat Med. 2019;25(3):433.
9. Knight W. The Dark Secret at the Heart of AI. MIT Technology Review. 2017. https://www.technologyreview.com/s/604087/the-dark-secret-at-the-heart-of-ai/. Accessed 27 Aug 2019.
10. Schmelzer R. Understanding explainable AI. Forbes. 2019. https://www.forbes.com/sites/cognitiveworld/2019/07/23/understanding-explainable-ai/#66ee5ac77c9e. Accessed 31 Aug 2019.
11. DARPA. Explainable Artificial Intelligence (XAI). https://www.darpa.mil/program/explainable-artificial-intelligence. Accessed 31 Aug 2019.

12. Ziad O, Ezekiel JE. Predicting the future—big data, machine learning, and clinical medicine. N Engl J Med. 2016. http://catalyst.nejm.org/big-data-machine-learning-clinical-medicine/.
13. Marchant GE, Tournas LM. AI health care liability: from research trials to court trials. J Health Life Sci Law. February 2019;18:25–41.
14. Busis NA, Shanafelt TD, Keran CM, Levin KH, Schwarz HB, Molano JR, et al. Burnout, career satisfaction, and well-being among U.S. neurologists in 2016. Neurology. 2017;88(8):797–808.
15. Drugwatch. da Vinci Robotic Surgery Lawsuits. https://www.drugwatch.com/davinci-surgery/lawsuits/. Accessed 31 Aug 2019.
16. Espada M, Munoz R, Noble BN, Magrina JF. Insulation failure in robotic and laparoscopic instrumentation: a prospective evaluation. Am J Obstet Gynecol. 2011;205(2):121–e1.
17. Tournas L, Johnson W. Elon musk wants to hack your brain. Slate. 2019. https://slate.com/technology/2019/08/elon-musk-neuralink-facebook-brain-computer-interface-fda.html.
18. Byrom B, McCarthy M, Schueler P, Muehlhausen W. Brain monitoring devices in neuroscience clinical research: the potential of remote monitoring using sensors, wearables, and mobile devices. Clin Pharmacol Therapeut. 2018;104(1):59–71.
19. Medtech Boston. Not just a meditation tool: the muse brain sensing headband in neuroscience research. 2015. https://medtechboston.medstro.com/blog/2015/10/14/not-just-a-meditation-tool-the-muse-brain-sensing-headband-in-neuroscience-research/.
20. Geggel L. 23andMe is sharing genetic data with drug giant. Scientific American. 2018. https://www.scientificamerican.com/article/23andme-is-sharing-genetic-data-with-drug-giant/.
21. Krumholz H. 'Less is more' deepens focus on informed consent. Medpage Today. http://bit.ly/1Bu2MOD.
22. Ienca M, Andorno R. Towards new human rights in the age of neuroscience and neurotechnology. Life Sci Society Policy. 2017;13(1):5.
23. Collingridge D. The social control of technology. New York: St Martin; 1980.

A Common Ground for Human Rights, AI, and Brain and Mental Health

16

Mónika Sziron

16.1 Introduction

There is no shortage of dreams for implementing artificial intelligence (AI) in brain and mental health. Elon Musk's launch of Neuralink is just one of those recent dreams. The possibilities and potentials of AI in brain and mental health are numerous. However, there is also no shortage of fears and criticisms as many dreams of AI in brain and mental health unfold. As these potentials and fears have boiled, it has been the interdisciplinary job of ethicists, social scientists, computer scientists, legal experts, neurologists, and neurosurgeons, to name a few, to try and understand not only how to proceed but also how to understand each other. The current and future challenges of implementing AI in brain and mental health, which are of concern here, are to untangle not only the complexities of human rights but also some of the complexities of these fields working together. There is already extensive work on the implementation of human rights in AI and healthcare. The goal of this chapter is to unravel some of the complexities of guidelines, regulations, policies, treaties, and implementation of human rights for future development of ethical AI in brain and mental health.

To untangle, trail, and read the hundreds of documents, guidelines, regulations, policies, treaties, and ethics codes of AI, let alone healthcare, is an overwhelming assignment. It is an assignment that researchers in these areas are, quite frankly, unlikely to find time to do. It is an assignment that not many students in these fields are required to interpret. It is clear that there are several common threads between many of these documents, principles that require respecting human rights. The term "human rights" is often used but less often explained in detail. International human

M. Sziron (✉)
Department of Humanities, Center for the Study of Ethics in the Professions, Illinois Tech, Chicago, IL, USA
e-mail: msziron@hawk.iit.edu

F. Jotterand, M. Ienca (eds.), *Artificial Intelligence in Brain and Mental Health: Philosophical, Ethical & Policy Issues*, Advances in Neuroethics,
https://doi.org/10.1007/978-3-030-74188-4_16

rights doctrines burgeon after the wake of World War II. The atrocities and disrespect humanity witnesses proves that this sort of regulation and rights for humanity needs to be manifested internationally. Subsequently, the Universal Declaration of Human Rights (UDHR) is proclaimed to be the "common standard of achievements for all peoples and all nations" on December 10, 1948 [1]. The declaration serves as a precedent for subsequent, national, international, and intergovernmental, covenants, charters, and conventions on human rights [2–8]. There are many treaties affirming human rights, and while many nations have signed these various treaties, not all have ratification, acceptance, or approval. A signature binds the nation to follow the treaty so as to not defeat the objective and purpose of the treaty but does not officially bind that nation to the treaty, like that of ratification or acceptance [9]. The United Nations provides documentation of the status of ratification for each nation and treaty in its UN Treaty Body Database [10]. In some countries, the UDHR, as many have realized, is fulfilled in name only. Nations also include inalienable rights in their constitutions, yet these are not always internationally reliable, as there are differences in adaptations.

Human rights-based approaches (HRBA) are surfacing as ways to frame international development, technology development, healthcare, and policy development, to name a few. Using a HRBA means that one does not internalize their development efforts as charity or philanthropic business, but as their duty-bearer obligation to acknowledge humans, as rights holders, claiming their human rights. In a HRBA, humans are not seen as passive subjects of development but as active partners in the process of development due to their being rights holders [11]. It is important to understand, as Broberg and Sano assert, that there is no "one-size-fits-all" HRBA [11]. These variations will play a vital role in the ethical development of AI in brain and mental health. Even if every corporation, nation, and continent were to adopt a HRBA in the development of AI in brain and mental health, there is a great likelihood that the human rights chosen as foundations for these approaches would not be uniform due to variations in culture, ideology, politics, institutions, and resources. Decades of contention between the United States and China pertaining to actualizing human rights provide several examples of how human rights are not globally uniform or implemented due to political, ideological, and cultural differences [12]. The United States Congressional-Executive Commission on China archives an extensive list of purported human rights issues in China. Some countries lack HRBAs due to deficiencies in institutions, resources, or capital devoted to human rights development. Universal human rights will only be achieved when all countries are able to participate in their development, yet the list of least developed countries (LDC), as of May 2021, lists 46 countries. Fortunately, there are efforts to integrate LDCs in the Human Rights Council. In 2012, the Voluntary Technical Assistance Trust Fund to Support the Participation of Least Developed Countries and Small Island Developing States in the Work of the Human Rights Council was established to promote participation from LDCs. The fund supports "activities designed to enhance the institutional and human capacity of least developed countries and small island developing States, to enable their delegations to participate more fully in the work of the Human Rights Council" [13]. A vital resource for a

HRBA is an authority (often a State) that will serve as a duty-bearer providing representation, courts, and law-enforcement for rights holders. HRBAs are not easily applied in scenarios where these resources are lacking. Another vital challenge with a HRBA is educating target groups of their rights, as rights holders, and encouraging them to claim them from duty-bearers, like the State. It should not be assumed that all rights holders are alike, there is no guarantee that once educated, a rights holder will claim their rights, want to claim their rights, or value the same rights. This yields questions like, what human rights principles should guide the developing policies of AI in brain and mental health internationally? Is it possible to develop an international policy of this sort? Should these human rights principles be the same in the developing policies of general AI? How should they be the same or differ?

AI's influence in healthcare is recognized by many. AI has altered the way physicians make clinical choices and diagnoses, and how patient information is stored and retrieved. AI assists physicians by parsing data quickly and more efficiently. AI must "learn" via large data sets that include patient demographics, medical information, lab results, images, and recordings to name a few. Data privacy and patient confidentiality regulations have addressed ethical concerns that have risen from these applications. AI is arguably influencing every sector of healthcare, from radiology to management and pharmaceuticals [14]. AI in healthcare is primarily used in two forms, machine learning and natural language processing, and is used in cancer, nervous system, cardiovascular disease research [15], in medical diagnosis, surgery, hospital management, and even virtual health assistants. However, as AI and brain–computer technologies are increasingly unified, new ethical concerns arise in the areas of AI in brain and mental health. As AI has surged into medical devices, the Food and Drug Administration in the United States has realized the pressing need to regulate AI in medicine. The Academy of Medical Royal Colleges has also realized the urgency of developing guidelines for ethical and safe AI for healthcare [16].

16.2 Human Rights in AI and Healthcare

The vastness of human rights can be overwhelming. Human rights not only have a long and complex history but have been, and still are, the subject of debate [17]. There is a generous number of human rights and a generous number of ways human rights can be interpreted and actualized. These varieties in interpretations can cause confusion and miscommunication. By 2003, human rights-based approaches to development had become so convoluted that the United Nations agencies were compelled to address their own discrepancies, as "each agency has tended to have its own interpretation of approach and how it should be operationalized" [18]. Thus, breaking down the fundamental background of human rights before interpreting the establishment of human rights in AI and healthcare is necessary.

The fundamental background of human rights, which is often overlooked, is the jurisprudence that serves as the foundation for human rights. It is rare to find the mention of jurisprudence in literature pertaining to human rights in AI and somewhat more common in discussions of human rights and healthcare. Examining jurisprudence, as in the philosophy or theory behind law, is a crucial first step in understanding how human rights are applied and from where debates surrounding human rights stem. This is becoming increasingly important as the number of guidelines, ethics codes, and reports coming from nonlegal actors using the concept of human rights escalates. One could spend years defining various jurisprudence concepts and theories, but a quick jurisprudence tool is determining who is using the concept of human rights and identifying their background. In general, the concept of human rights is used by ethicists and legal scholars alike. Ethicists are often interpreting human rights from a moral natural law theory that see human rights as deriving from moral principles and the objective reality of being human. From this perspective, human rights are granted regardless of the State, political order, or positive law, because we are human we have human rights. Legal scholars, in contrast, do view human rights as positive laws that have been approved by a court or State. To avoid such generalizations, more effort should be put on communicating interpretations of human rights beyond the sentiment that they are "universally binding," as we have seen this is not always the case. Distinguishing who is using the concept of human rights is an important indicator of the underlying theories behind their use of the concept. How the concept is being interpreted has important implications for how the specified ethics code, report, or guideline can be actualized.

There is a prevalent consensus that AI already is and will continue to violate human rights [19–21]. The rights likely to be violated are antidiscrimination, freedom of speech, freedom of expression, right to privacy, right to equality, right to security of person, and the right to self-determination. These rights have been, and continue to be, challenged by our technologies, but a pressing concern is that our previous technologies were not as powerful nor have the reach of AI. Concerns are that the values of AI will not align with human values, this has been designated as the value-alignment problem. It has been predicted that AI will bolster the digital divide leading to significant economic and social inequalities. Yet, the power and reach of AI also delivers new positive potentials that humanity has never seen. Thus, we find ourselves trying to balance the concerns and excitement for AI. Using a HRBA has been proposed by several as the scale for this balancing act. Advocates for HRBAs see human rights as the solution that will stabilize the future of humanity and AI. Advocates propose that a HRBA should not only be a part of the regulation of AI but also a part of AI development [21]. Advocates of HRBA believe human rights should be the universally agreed upon values that guide AI development and regulation around the globe [19, 20, 22]. Developing AI to support and respect human rights values would ostensibly resolve the current and future violations of human rights. It would also address the potential social and economic inequalities created by AI, as certain human rights would provide legal language for defense against States and corporations [23]. The Asilomar AI Principles also assert that ethical AI is designed with human values, human dignity, human rights, human freedoms, and cultural diversity in mind [24]. These are aspirational and progressive solutions that seek to benefit humanity. However, after closer inspection designing AI with these characteristics in mind raises its own ethical concerns and challenges.

Deciding which human values and rights to implement and who gets to decide which human values and rights to implement in AI is not an easy task. Fears of ethical imperialism have been vocalized as some AI companies have more money and power than others. There is also concern that different countries will either not design AI with human values in mind or will pick human values that do not align with theirs [21]. This has become a secondary iteration of the value-alignment problem. Not only does AI need to be designed with human values in mind but humans have to align their own values with each other. This second value-alignment problem has highlighted a major issue in implementing a HRBA for AI, that is, what happens when human rights values conflict, cannot be fully actualized, or cannot support every human right? If a certain AI application violates one human right but supports another, what is the solution? This is a problem Cansu Canca, founder of AI Ethics Lab, says can only be solved with the use of ethical reasoning. This becomes particularly important when considering AI in healthcare. If an artificially intelligent BCI will fulfill the right to equality and health, but violate the right to privacy and self-determination, what is the solution? Not only will the solution lead to a subjective answer, but the UDHR, and other regulatory frameworks, do not provide answers on how to go about choosing one human right over another. These decisions are made by using ethical theories like utilitarianism, deontology, or virtue ethics [25].

In 2005, Leslie London stated, "We live in an increasingly globalized environment, characterized by growing tensions between our technological capacities and the abilities of our social policies to meet basic human health needs." Statements like these are ostensibly timeless. Just a decade later, we are still struggling as academics, lawyers, ethicists, politicians, and scientists to cope with our increasingly technological society while maintaining the dignity and rights of the humans that live in it. Human rights in healthcare, similarly to human rights in AI, are debated and not always universally applied. It has been argued that the only example of protecting human rights in biomedical settings at an international level is the European Convention on Human Rights and Biomedicine of 1997. This is due to the convention's focus on developing human rights principles in the biomedical field that makes it such an ideal example [26]. However, its vague use of human dignity as a normative pillar of health law does not sit well universally. This is due, in part, to perspectives that see the normative pillar of health law as healthcare problems, focusing on "rules of civil, criminal and administrative law" [26]. There are also variations in perspectives on health law that stem from varying philosophical perspectives. For example, healthcare in the United States is primarily concerned with principles of self-determination and autonomy, whereas European healthcare is concerned with principles of human dignity and solidarity. There have always been philosophical differences, but there is also another major growing historical difference in outlooks towards healthcare. That is, definitions of health and those that are defining health are changing and in some cases have already changed [26]. Health, at one point in time, was defined exclusively by medical professionals and physicians. The primary responsibility of physicians was to "cur[e] illnesses rather than satisf[y] individuals" [26]. Today, healthcare can be portrayed as a contracted service in which the patient decides, based on transparent information from their physician, how they would like their care to be actualized. It is then the ethical and legal responsibility of the physician to respect the patient's autonomy and follow through with their patient's decision.

At first glance, it may seem that the responsibilities and professional conduct of a computer scientist versus a healthcare professional are quite dissimilar. However, over time these two professions have started to resemble one another more and more, especially from a regulatory perspective. Medical professionals were once the authoritative voice for healthcare, just as computer scientists have been for computer science. The United States in the 1930s began thinking of patients as customers due to rising costs, then the 1960s entertained the patients' rights movement, and today medical professionals are subject to contractual obligations that if not fulfilled may lead to civil lawsuits [27]. While the relationship between medical professionals and patients is different than computer scientists and end users, both can be considered service providers. Medical professionals are obligated to execute good care that is skilled and competent and respects patients' rights. Some of those patients' rights include control over one's treatment, control over their information, the right to nondiscriminatory care, and the ability to cease care [27]. These patient rights are not wildly dissimilar from what citizens are asking of from AI regulation. Perhaps 1 day, rather than directing and defining computer science for themselves, it will be the ethical and legal responsibility of computer scientists to satisfy individual citizens and/or maintain the digital "health" of populations. If AI is to become just as critical to human society as medical care, there may be much more that can be learned about potential regulation of AI by comparing the professions of computer scientists and medical professionals currently, historically, and globally.

Regulation of AI in healthcare does have a growing global network. The International Medical Device Regulators Forum (IMDRF) is a voluntary forum consisting of representatives from Australia, Brazil, Canada, China, Europe, Japan, Russia, Singapore, South Korea, and the United States. The forum works to establish medical device regulatory harmony and convergence. Many instances of AI in healthcare can be defined under the term "software as a medical device" (SaMD). The term was officially defined by IMDRF, as "software intended to be used for one or more medical purposes that perform these purposes without being part of a hardware medical device," and several examples of what can and cannot be considered SaMD are provided [28]. While the forum is a promising step in the direction of global regulation, it only represents one sector of healthcare, medical devices. It is important to note that not all AI that is utilized in healthcare falls under SaMD classifications and regulations. AI that may handle workflow, clinical communication, and patient registration and visits and AI that searches and queries a database for records are not SaMD. Thus rendering the question, who or what is regulating these "other" AI in healthcare settings? It is likely that these AI systems are not regulated with the same global perspective in mind. Since IMDRF's work in SaMD in 2014, individual nations have established and drafted their own regulatory frameworks for SaMD.

Discussions of regulations of SaMD have only recently come to the policy-making attention of the Federal Drug Administration in the United States [29]. However, these discussions are only commencing, as the discussion paper explicitly states, "This document is not intended to communicate FDA's proposed (or final) regulatory expectations but is instead meant to seek early input from groups and individuals outside the Agency prior to development of a draft guidance" [29]. The paper currently does not use a HRBA, or mention rights, instead focuses on manufacturers and risk management. In 2017, the EU passed The EU Medical Device Regulation, which also regulates SaMD, requiring that all manufacturers in the EU single market comply with the regulations by May 2020. The regulation is similar to the FDA regulation in prioritizing safety and risk management for the lifecycle of the device [30]. However, this regulation does mention subjects' rights in development and clinical investigations.

Some corporations and associations have also been active in regulating themselves and producing documentation of these regulations. IEEE, Institute of Electrical and Electronics Engineers, has published the second version of *Ethically Aligned Design* that is devised to "establish ethical and social implementations for intelligent and autonomous systems and technologies, aligning them to defined values and ethical principles that prioritize human well-being in a given cultural context" [31]. Citing several human rights treaties, the first principle of the document is for the consideration of human rights in design. Microsoft has also initiated a human rights impact assessment (HRIA) on their products, detailed in their *Human Rights Annual Report* [32]. The 2018 report mentions that a specific section of the HRIA will be dedicated to AI for the foreseeable future. It was found that, based on a "broad range of AI applications," human rights risks included nondiscrimination and equality; right to life and personal security; privacy, including protecting against unlawful governmental surveillance; freedom of thought, conscience, and religious belief and practice; freedom of expression and to hold opinions without interference; freedom of association and the right to peaceful assembly; right to decent work; and right to an adequate standard of living [32].

Human rights in AI and healthcare are broad topics that unfortunately cannot be fully detailed here. However, from this discussion, there are several key takeaways:

1. It is becoming increasingly important for authors to explicitly mention their interpretation of human rights, as the number of guidelines, ethics codes, and reports coming from nonlegal actors using the concept of human rights escalates.
2. Human rights and HRBA are being contemplated in the regulation of AI on corporate, national, and international scales. However, the success of using only a HRBA to regulate AI is unlikely to solve value-alignment problems.
3. Human rights have played an integral role in healthcare; however, the changing dynamics of the profession overtime have changed the obligations of medical professionals.

4. Regulations of AI in healthcare depend on the purpose of the AI. Some AI technologies used in healthcare are not considered medical devices, which promote changes in health, but rather tools that enhance medical knowledge. Software used for administrative purposes or to store, retrieve, transfer patient data are not categorized as medical devices and are thus regulated differently.

16.3 Future of Human Rights in Brain and Mental Health AI

It is an exciting time for AI in brain and mental health. AI can take on many forms in brain and mental health including artificial neural networks (ANN), machine learning (ML), natural language processing (NLP), machine perception, affective computing, virtual and augmented reality, robotics, implants, brain–computer interfaces (BCI), and supercomputing. While it is true that it is still inconclusive whether AI has more positive than negative outcomes in brain and mental health [33], the restorative capabilities of AI technologies for patients are beyond astounding. While AI and neuroscience have a shared history, we are in the advent of implementing AI in brain and mental health. This of course means that there are still many unanswered questions, ethical concerns, and unknowns. Thus far, there have been several documented benefits of implementing AI in mental healthcare. AI is simply better at some things, like not fatiguing or forgetting. AI has improved self-care and access to mental healthcare. AI has allowed for a greater customization of behavioral and mental healthcare. Finally, AI has numerous economic benefits, reducing labor costs and cost of healthcare in some cases [34].

With novelty comes advantages and disadvantages for regulation of AI in brain and mental health. An advantage is that several regulations that have been tried and tested in AI and healthcare are already developed and can be applied towards brain and mental health AI. A disadvantage is that current regulations of AI may not be suitable for the specificities of the field and reworking will be necessary. While there is a growing list of AI ethics guidelines globally [35], Rafael Yuste and Sara Goering assert that current ethics guidelines for AI are insufficient concerning developments in brain and mental health neurotechnologies. Specifically, new ethical concerns arise in the areas of privacy and consent, agency and identity, augmentation, and bias [36], and, as of late, these are not top concerns in general AI ethics guidelines, which are transparency, justice and fairness, and non-maleficence [35]. As neurotechnologies for brain and mental health continue to evolve, it is likely new ethical concerns beyond these will arise as well. Considering the current regulation status of AI and healthcare and emphases on human rights, it is likely that analyses of human rights will reveal itself for brain and mental health AI regulation. Brain and mental health AI also has the advantage of learning from what AI and healthcare have not done well, namely not including interdisciplinary and consumer/patient perspectives in the process of developing regulations. More work on lived patient experiences will greatly benefit the field. There is still only a small percentage of the human population that have experience living with brain and mental health AI on a daily basis. While we must anticipate the needs of future patients in brain and mental health AI, we must beware that generalizing future human rights and ethics from this currently small percentage will most likely need revisiting if we wish to develop policies that suit the needs of an inevitably more dynamic group of patients in the future. Gilbert et al. have recorded some of the perceptions of lived experience with AI-enabled BCIs thus far. Their results pose interesting questions for applying human rights in brain and mental health AI.

Collecting insights from patients with AI-enabled BCIs found that the technology was able to satisfy the right to self-determination for some patients and violate rights to self-determination for others [33]. The subjective reality is that while some patients may feel their human rights are satisfied, others may feel that those same rights are violated. While lived experiences are subjective in nature, the varied results shed light on the potential violations of human rights and the need for patient perspectives. Many already established regulations could have benefited from the inclusion of a more perspectives. The following is a perspective on including human rights in brain and mental health regulation based on regulations of AI and healthcare.

Human rights are essential for the safety and care of patients. As such, human rights should play an integral role in the regulation of brain and mental health AI. However, human rights should not be the only values that are taken into consideration as, like we have seen, they are sometimes only supported in name, not practice. Thus, ethics still need to be a part of regulation. Ethics will prove to be very important for the regulation of those working in brain and mental health AI. An ethics code should be established that is specifically suitable for the field which acknowledges the variability of human rights globally and which acts as a safety net where human rights may not be developed, implemented, or supported. Human rights and ethics in this field should work together. A generalized definition of what the field is, including the variety of disciplines and studies involved, could spearhead the development of an ethics code. This definition would also determine the initial scope of the field, aptly identifying what is and is not a part of the field. Whoever is using and developing brain and mental health AI should regularly conduct HRIAs on their technologies and adjust their practices according to the human rights risks found. Brain and mental health professionals will need to continue working together nationally and internationally. Despite idealistic goals at the outset, AI's ability to influence the delivery of brain and mental healthcare will ultimately depend on the visions and resources from leaders and governments [37]. Thus, it is important to understand the goals of the field from within and be able to share the possibilities outward.

16.4 Conclusion

Human rights are dynamic and will continuously change. Throughout history, the rights of women, children, minorities, people of color, humans with disabilities, LGBTQIA, etc., have evolved and will continue to do so. The human right to health could very well be altered by developments in brain and mental health AI. Humanity could reach a point when the right to health "highest standard of physical and mental health" means using AI. However, it is less clear to this day how access to AI is a right and in what circumstances. It is inevitable that the regulations of AI in brain and mental health will change and adjust according to technological and societal adaptations. There are not many shared global understandings for many terms and concepts discussed in this chapter. The terms health, human rights, AI, intelligence, and healthcare are all subject of international debate, and it is doubtful that there will ever be a true global consensus. It is not necessarily idealistic, philosophical, or political variations on a global scale that may hamper the ethical development of AI in brain and mental health, these have always been present. Ethical developments may be hampered by not acknowledging these variations, not learning from other perspectives, and failing to identify contrasting values, as there is no one-size-fits-all solution to regulation. These variations, rather than sinkholes, can be the common grounds that guide discussions and promote innovative policies and regulations.

References

1. United Nations. Universal declaration of human rights. 1948. https://www.un.org/en/universal-declaration-human-rights/. Accessed 14 May 2019.
2. African Union. African charter on human and peoples' rights. 1981. https://au.int/en/treaties/african-charter-human-and-peoples-rights. Accessed 14 May 2019.
3. AICHR. ASEAN human rights declaration. 2012. https://aichr.org/wp-content/uploads/2018/10/ASEAN-Human-Rights-Declaration.pdf. Accessed 14 May 2019.
4. ECHR, Council of Europe. European convention on human rights. 1950. https://www.echr.coe.int/Pages/home.aspx?p=basictexts&c. Accessed 14 May 2019.
5. European Commission. EU charter of fundamental rights. 2012. https://ec.europa.eu/info/aid-development-cooperation-fundamental-rights/your-rights-eu/eu-charter-fundamental-rights_en. Accessed 24 May 2019.
6. OHCHR. International covenant on economic, social and cultural rights. 1966. https://www.ohchr.org/en/professionalinterest/pages/cescr.aspx. Accessed 14 May 2019.
7. Organization of American States. American convention on human rights. 1969. http://www.oas.org/dil/treaties_B-32_American_Convention_on_Human_Rights.htm. Accessed 14 May 2019.
8. UNICEF. Convention on the rights of the child. 1989. https://www.unicef.org/child-rights-convention/convention-text. Accessed 14 May 2019.
9. United Nations. Vienna Convention on the Law of Treaties. 1969. https://treaties.un.org/Pages/ViewDetailsIII.aspx?src=TREATY&mtdsg_no=XXIII-1&chapter=23&Temp=mtdsg3&clang=_en. Accessed 14 May 2019.
10. United Nations. Status of ratification interactive dashboard. 2019. http://indicators.ohchr.org. Accessed 15 May 2019.
11. Broberg M, Sano HO. Strengths and weaknesses in a human rights-based approach to international development—an analysis of a rights-based approach to development assistance based on practical experiences. Int J Hum Rights. 2018;22(5):664–80.
12. Qi Z. Conflicts over human rights between China and the US. Human Rights Quarterly. 2005;27(1):105–124.
13. United Nations. Human Rights Council Nineteenth session. 2012. https://documents-dds-ny.un.org/doc/RESOLUTION/GEN/G12/130/62/PDF/G1213062.pdf?OpenElement. Accessed 8 Mar 2020.

14. Tsang L, Kracov DA, Mulryne J, et al. The impact of artificial intelligence on medical innovation in the European Union and the United States. Intellect Prop Technol Law J. 2017;29(8):3–11.
15. Jiang F, Jiang Y, Zhi H, et al. Artificial intelligence in healthcare: past, present and future. Stroke Vasc Neurol. 2017;4:230–43.
16. Academy of Medical Royal Colleges. Artificial intelligence in healthcare. 2019. https://www.aomrc.org.uk/reports-guidance/artificial-intelligence-in-healthcare/. Accessed 25 Jul 2019.
17. Alston P. The populist challenge to human rights. J Hum Rights Pract. 2017;9:1–15.
18. United Nations. The human rights based approach to development cooperation towards a common understanding among UN agencies. 2003. https://undg.org/document/the-human-rights-based-approach-to-development-cooperation-towards-a-common-understanding-among-un-agencies/. Accessed 15 May 2019.
19. Access Now. Human rights in the age of artificial intelligence. 2018. https://www.accessnow.org/cms/assets/uploads/2018/11/AI-and-Human-Rights.pdf. Accessed 24 May 2019.
20. Latonero M. Governing artificial intelligence: upholding human rights & dignity. 2018. https://datasociety.net/output/governing-artificial-intelligence/. Accessed 25 Jul 2019.
21. Risse M. Human rights and artificial intelligence: an urgently needed agenda. 2018. https://www.hks.harvard.edu/publications/human-rights-and-artificial-intelligence-urgently-needed-agenda. Accessed 24 May 2019.
22. Pielemeier J. AI & global governance: the advantages of applying the international human rights framework to artificial intelligence. In: Digital technology and global order. New York: United Nations University Centre for Policy Research; 2019. https://cpr.unu.edu/ai-global-governance-the-advantages-of-applying-the-international-human-rights-framework-to-artificial-intelligence.html. Accessed 24 May 2019.
23. Van Veen C. Artificial intelligence: what's human rights got to do with it? In: Data & Society: points. Medium. 2018. https://points.datasociety.net/artificial-intelligence-whats-human-rights-got-to-do-with-it-4622ec1566d5. Accessed 26 Jun 2019.
24. Future of Life Institute. Asilomar AI principles. 2017. https://futureoflife.org/ai-principles/. Accessed 24 May 2019.
25. Canca C. AI & global governance: human rights and AI ethics—why ethics cannot be replaced by the UDHR. In: Digital technology and global order. New York: United Nations University Centre for Policy Research; 2019. https://cpr.unu.edu/ai-global-governance-human-rights-and-ai-ethics-why-ethics-cannot-be-replaced-by-the-udhr.html. Accessed 25 July 2019.
26. Juškevičius J, Balsienė J. Human rights in healthcare: some remarks on the limits of the right to healthcare. Jurisprudencija: mokslo darbai. 2010;122(4):95–110.
27. Peled-Raz M. Human rights in patient care and public health—a common ground. Public Health Rev. 2017;38:1–10.
28. IMDRF. "Software as a medical device": possible framework for risk categorization and corresponding considerations. 2014. http://www.imdrf.org/workitems/wi-samd.asp. Accessed 20 May 2019.
29. FDA. Proposed regulatory framework for modifications to artificial intelligence/machine learning (AI/ML)-based software as a medical device (SaMD). 2019. https://www.fda.gov/medical-devices/software-medical-device-samd/artificial-intelligence-and-machine-learning-software-medical-device. Accessed 20 May 2019.
30. Council of the European Union. Regulation (EU) 2017/745 of the European Parliament and of the Council of 5 April 2017 on medical devices. 2017. https://eurlex.europa.eu/eli/reg/2017/745/2017-05-05. Accessed 20 Jul 2019.
31. IEEE. Ethically aligned design: a vision for prioritizing human well-being with autonomous and intelligent systems, version 2. 2017. https://standards.ieee.org/content/dam/ieeestandards/standards/web/documents/other/ead_ v2.pdf. Accessed 25 Jul 2019.
32. Microsoft. Human rights annual report. 2018. https://www.microsoft.com/en-us/corporate-responsibility/human-rights. Accessed 26 Jul 2019.
33. Gilbert F, Cook M, O'Brien T, et al. Embodiment and estrangement: results from a first-in-human "intelligent BCI" trial. Sci Eng Ethics. 2019;25:83–96.

34. Luxton D. An introduction to artificial intelligence in behavioral and mental health care. In: Luxton D, editor. Artificial intelligence in behavioral and mental health care. 1st ed. San Diego: Academic Press; 2016. p. 1–26.
35. Jobin A, Ienca M, Vayena E. The global landscape of AI ethics guidelines. Nat Mach Intell. 2019;1:389–99. https://doi.org/10.1038/s42256-019-0088-2.
36. Yuste R, Goering S. Four ethical priorities for neurotechnologies and AI. Nature. 2019;551(7679):159–63. https://www.nature.com/news/four-ethical-priorities-for-neurotechnologies-and-ai-1.22960. Accessed 8 Mar 2020.
37. Patel VL, Shortliffe EH, Stefanelli M, et al. The coming of age of artificial intelligence in medicine. Artif Intell Med. 2009;46(1):5–17. https://doi.org/10.1016/j.artmed.2008.07.017.

Part IV

Epilogue

17 Brain and Mental Health in the Era of Artificial Intelligence

Marcello Ienca

In the decade just beginning, artificial intelligence (AI) is and will increasingly be a fundamental catalyst for medical innovation. Due to its technological novelty, ability to process large volumes of data, capacity for autonomous action, and general-purposive nature, AI holds potential for transforming medicine and healthcare at greater pace and in greater magnitude compared to any other technology. The transformative potential of AI has been deemed "revolutionary" by experts [1, 2], with authors referring to the introduction of AI techniques in healthcare as a socio-technical revolution capable of reshaping entire areas of medicine. In recent years, most attention has been devoted to the use of AI systems in medical domains such as pathology [3] and radiology [4]. In particular, a rapidly growing body of research is showing how approaches to AI such as machine learning (ML) can improve the delivery of healthcare services by improving prognostics, diagnostics, treatment, clinical workflow, and expanding the availability of clinical expertise [5].

Addressing the ethical and social challenges of such socio-technical advances is a complex task, which requires a meticulous scrutiny of both the technology itself and the socio-cultural context in which the technology is embedded. Many of these ethical-social challenges are inherent in the very application of automatic and self-learning systems to medical data, regardless of the physical implementation of those systems (e.g., embodied vs disembodied), the type of data being processed, the institutional setting, medical specialty, patient population, and clinical purpose in which or for which such systems are deployed. For example, aligning AI systems with data privacy requirements, minimizing the effects of algorithmic bias, and achieving transparency have been notably recognized as cross-domain normative requirements which extend to the whole medical domain. However, several ethical and social implications of medical AI are qualitatively dependent on the technological medium, clinical setting, patient group, socio-cultural context as well as on the

M. Ienca (✉)
Department of Health Sciences and Technology, ETH Zurich, Zürich, Switzerland
e-mail: marcello.ienca@hest.ethz.ch

F. Jotterand, M. Ienca (eds.), *Artificial Intelligence in Brain and Mental Health: Philosophical, Ethical & Policy Issues*, Advances in Neuroethics,
https://doi.org/10.1007/978-3-030-74188-4_17

ontological properties, values, power structures, and discourses which characterize those technological mediums, settings, or groups.

Compared to other areas of medicine, the use of AI to improve brain and mental health has not yet received sufficient attention and systematic assessment. This gap in the scientific literature raises both oddity and concern in the light of the intimate historical nexus between AI and the sciences of the mind and brain. In fact, the history of AI is inextricably intertwined with the history of neuroscience and psychology. Since the first conceptualizations of AI, scientists and philosophers turned to the human brain as a source of guidance for the development of intelligent machines [6]. Still today, AI borrows most of its lexicon from neuropsychological categories (e.g., machine *learning*, computer *vision*, *natural language* processing) while many areas of AI research such as artificial neural networks are based on and inspired by neurobiological structures and processes.

Most importantly, brain and mental health constitute a domain of fundamental ethical significance. This is primarily because neural processes and mental phenomena are the closest correlates of fundamental ethical categories such as moral agency, personal identity, and free will. Furthermore, faculties of the brain such as memory, consciousness, and language represent the core set of properties that make us human and through which we self-identify as persons. Therefore, the use of AI in brain and mental health elicits a complex interactive dynamics between artificial and human cognition, whose effects may have profound implications for both individuals and humanity at large. Anticipating and proactively assessing the ethical and social implications of this interactive dynamic between brains, minds, and cognitive technology is of paramount importance to responsibly navigate the AI revolution [7]. A context-specific ethical assessment of AI for brain and mental health is all the more important as people with chronic mental conditions, people with neurocognitive or physical disabilities, elderly adults, and people with dementia all belong to vulnerable groups, and hence experience higher risk of harm and consequently require special protective and some degree of priority consideration, even in the face of severe resource constraints. As AI advances fast, we have a moral obligation to ensure the responsible development and deployment of artificial intelligence for the benefit of millions of neurological and psychiatric patients worldwide.

This book attempted to fill this gap in the scientific and ethical literature by providing a comprehensive overview of the key applications of AI for brain and mental health and a systematic assessment of their ethical and social implications. The various chapters of this volume explored a wide spectrum of AI systems for brain and mental health such as social robots, chatbots, automated text analysis programs, predictive analytics software, brain-computer interfaces, neurostimulation tools, neurorehabilitation aids, smartphone-based mental health apps, neuromonitoring and neurofeedback tools. This comprehensive overview adds to previous work on the ethics of AI-driven technological trends such as intelligent assistive technologies for dementia [8], digital mental health [9], clinical neuroimaging [10], neural motor prostheses [11], and other neural devices [12].

Much editorial attention, in this volume, was devoted to ensuring that such technological innovations were not presented as value-free artifacts but as socio-technical systems embedded in a socio-cultural context, designed for or accessible to specific patient groups, influenced by pre-existing values, and inscribed in a rich grid of ethical-legal norms. The fifteen chapters here contained depict a rich set of AI-enabled opportunities to improve the health and mental wellbeing of both patients and healthy citizens. At the same time, they identify complex areas of ethical problematicity which require careful considerations.

Featuring contributions from world-leading experts from the areas of computer science, robotics, neurology, psychiatry, clinical psychology, bioethics, neuroethics, and the law, this book marks an important milestone in the public understanding of the ethics of AI in brain and mental health. Furthermore, it provides a useful resource for any future investigation in this crucial and rapidly evolving area of AI application.

References

1. Appenzeller T. The AI revolution in science. Science. 2017;357:16–7.
2. Sensakovic WF, Mahesh M. Role of the medical physicist in the health care artificial intelligence revolution. J Am Coll Radiol. 2019;16(3):393–4.
3. Salto-Tellez M, Maxwell P, Hamilton P. Artificial intelligence—the third revolution in pathology. Histopathology. 2019;74(3):372–6.
4. Jha S, Topol EJ. Adapting to artificial intelligence: radiologists and pathologists as information specialists. JAMA. 2016;316(22):2353–4.
5. Rajkomar A, Dean J, Kohane I. Machine learning in medicine. N Engl J Med. 2019;380(14):1347–58.
6. Ullman S. Using neuroscience to develop artificial intelligence. Science. 2019;363(6428):692.
7. Ienca M, Ignatiadis K. Artificial intelligence in clinical neuroscience: methodological and ethical challenges. AJOB Neurosci. 2020;11(2):77–87.
8. Ienca M, Wangmo T, Jotterand F, Kressig RW, Elger B. Ethical design of intelligent assistive technologies for dementia: a descriptive review. Sci Eng Ethics. 2018;24(4):1035–55.
9. Torous J, Andersson G, Bertagnoli A, Christensen H, Cuijpers P, Firth J, et al. Towards a consensus around standards for smartphone apps and digital mental health. World Psychiatry. 2019;18(1):97.
10. Mecacci G, Haselager P. Identifying criteria for the evaluation of the implications of brain reading for mental privacy. Sci Eng Ethics. 2019;25:443–61.
11. Clausen J. Moving minds: ethical aspects of neural motor prostheses. Biotechnol J Healthc Nutr Technol. 2008;3(12):1493–501.
12. Klein E, Brown T, Sample M, Truitt AR, Goering S. Engineering the brain: ethical issues and the introduction of neural devices. Hastings Cent Rep. 2015;45(6):26–35.

Index

F. Jotterand, M. Ienca (eds.), *Artificial Intelligence in Brain and Mental Health: Philosophical, Ethical & Policy Issues*, Advances in Neuroethics,
https://doi.org/10.1007/978-3-030-74188-4

GPSR Compliance

The European Union's (EU) General Product Safety Regulation (GPSR) is a set of rules that requires consumer products to be safe and our obligations to ensure this.

If you have any concerns about our products, you can contact us on ProductSafety@springernature.com

In case Publisher is established outside the EU, the EU authorized representative is:

Springer Nature Customer Service Center GmbH
Europaplatz 3
69115 Heidelberg, Germany

Batch number: 10726060

Printed by Printforce, the Netherlands